Business Models to Promote Technology, Culture, and Leadership in Post-COVID-19 Organizations

Joseph Robert Heinzman
Western Governors University, USA

Debbie S. Judge
Western Governors University, USA

Vincent Franklin
Western Governors University, USA

Jim A. McCleskey
Western Governors University, USA
& Houston Community College System, USA

IGI Global
PUBLISHER of TIMELY KNOWLEDGE

A volume in the Advances in Logistics, Operations, and Management Science (ALOMS) Book Series

Published in the United States of America by
IGI Global
Business Science Reference (an imprint of IGI Global)
701 E. Chocolate Avenue
Hershey PA, USA 17033
Tel: 717-533-8845
Fax: 717-533-8661
E-mail: cust@igi-global.com
Web site: http://www.igi-global.com

Library of Congress Cataloging-in-Publication Data

Names: Heinzman, Joseph R., DATE- editor. | Judge, Deborah, DATE- editor. | Franklin, Vincent, DATE- editor. | McCleskey, Jim, DATE- editor.
Title: Business models to promote technology, culture, and leadership in post-COVID-19 organizations / Joseph Heinzman, Deborah Judge, Vincent Franklin, and Jim McCleskey, editors.
Description: Hershey, PA : Business Science Reference, [2023] | Includes bibliographical references and index. | Summary: "This publication delves into how virtual technology has evolved to create remote offices and remote teaming in Health, Education, Engineering, and other business solutions, exploring culture in businesses and how individuals may interface, communicate, and collaborate in past, current, and future business models"-- Provided by publisher.
Identifiers: LCCN 2022015447 (print) | LCCN 2022015448 (ebook) | ISBN 9781668443583 (hardcover) | ISBN 9781668443590 (paperback) | ISBN 9781668443606 (ebook)
Subjects: LCSH: Business--Technological innovations. | Telecommuting. | Corporate culture.
Classification: LCC HD45 .B873 2023 (print) | LCC HD45 (ebook) | DDC 658.4/06--dc23/eng/20220408
LC record available at https://lccn.loc.gov/2022015447
LC ebook record available at https://lccn.loc.gov/2022015448

This book is published in the IGI Global book series Advances in Logistics, Operations, and Management Science (ALOMS) (ISSN: 2327-350X; eISSN: 2327-3518)

British Cataloguing in Publication Data
A Cataloguing in Publication record for this book is available from the British Library.

For electronic access to this publication, please contact: eresources@igi-global.com.

Advances in Logistics, Operations, and Management Science (ALOMS) Book Series

ISSN:2327-350X
EISSN:2327-3518

Editor-in-Chief: John Wang, Montclair State University, USA

MISSION

Operations research and management science continue to influence business processes, administration, and management information systems, particularly in covering the application methods for decision-making processes. New case studies and applications on management science, operations management, social sciences, and other behavioral sciences have been incorporated into business and organizations real-world objectives.

The **Advances in Logistics, Operations, and Management Science** (ALOMS) Book Series provides a collection of reference publications on the current trends, applications, theories, and practices in the management science field. Providing relevant and current research, this series and its individual publications would be useful for academics, researchers, scholars, and practitioners interested in improving decision making models and business functions.

COVERAGE

- Political Science
- Production Management
- Marketing engineering
- Information Management
- Finance
- Networks
- Risk Management
- Operations Management
- Decision analysis and decision support
- Organizational Behavior

IGI Global is currently accepting manuscripts for publication within this series. To submit a proposal for a volume in this series, please contact our Acquisition Editors at Acquisitions@igi-global.com or visit: http://www.igi-global.com/publish/.

The Advances in Logistics, Operations, and Management Science (ALOMS) Book Series (ISSN 2327-350X) is published by IGI Global, 701 E. Chocolate Avenue, Hershey, PA 17033-1240, USA, www.igi-global.com. This series is composed of titles available for purchase individually; each title is edited to be contextually exclusive from any other title within the series. For pricing and ordering information please visit http://www.igi-global.com/book-series/advances-logistics-operations-management-science/37170. Postmaster: Send all address changes to above address.

Titles in this Series

For a list of additional titles in this series, please visit:
http://www.igi-global.com/book-series/advances-logistics-operations-management-science/37170

Considerations and Characteristics of the Millennial Leader in Today's Global Society
Tymon M. Graham (TMGraham Consulting, LLC, USA)
Business Science Reference • © 2023 • 300pp • H/C (ISBN: 9781668447116) • US $240.00

Moving From R&D Development to Production and Sales in Start-Up Companies Emerging Research and Opportunities
Amiram Porath (APC, Israel)
Business Science Reference • © 2022 • 300pp • H/C (ISBN: 9781799856856) • US $215.00

Advancing Interpersonal Emotion Regulation and Social Regulation
Donta S. Harper (Argosy University, USA)
Information Science Reference • © 2022 • 384pp • H/C (ISBN: 9781668424780) • US $215.00

Handbook of Research on Activating Middle Executives' Agency to Lead and Manage During Times of Crisis
Ann-Marie Wilmot (The University of the West Indies, Mona, Jamaica) and Canute Sylvester Thompson (The University of the West Indies, Mona, Jamaica)
Business Science Reference • © 2022 • 415pp • H/C (ISBN: 9781668443316) • US $325.00

Global Risk and Contingency Management Research in Times of Crisis
Narasimha Rao Vajjhala (University of New York Tirana, Tirana, Albania) and Kenneth David Strang (W3-Research, USA & RMIT University, Australia)
Business Science Reference • © 2022 • 356pp • H/C (ISBN: 9781668452790) • US $270.00

Applied Guide for Event Study Research in Supply Chain Management
Lincoln C. Wood (University of Otago, New Zealand & Curtin University, Australia)
Information Science Reference • © 2022 • 261pp • H/C (ISBN: 9781799889694) • US $215.00

For an entire list of titles in this series, please visit:
http://www.igi-global.com/book-series/advances-logistics-operations-management-science/37170

IGI Global
PUBLISHER of TIMELY KNOWLEDGE

701 East Chocolate Avenue, Hershey, PA 17033, USA
Tel: 717-533-8845 x100 • Fax: 717-533-8661
E-Mail: cust@igi-global.com • www.igi-global.com

Table of Contents

Detailed Table of Contents

Sociotechnical systems theory is important because it brings together technology, behavioral culture, and leadership as it applies in the business environment. To understand where we may be going, it's important to review how past sociotechnical systems have been approached and how those systems were affected by work through scientific management, quality improvement, technological advancements, behavioral focus, and economics. Trust and collaboration are hallmarks of leadership, most agree managers manage things (task) and leaders lead (guide and support) people. COVID-19 has caused challenges to managing and leading, a disruption in our supply chain, manufacturing to adjust, and management to be more focused on innovation and sociotechnical systems in the United States. Manufacturing is a key indicator of employment and jobs health, and the current job loss in industry has been accelerated. Going forward, different skills, education, and industry focus will test innovation and the balance between management of things and leadership of people.

Technology and organizational leadership are a part of most major businesses, and this is no different for businesses in the water sector. Delivery of clean drinking water and the treatment of wastewater are priorities for utilities in the United States as they seek to serve their communities. Existing infrastructure continues to age and faces a challenging future ahead as delivery demands grow. Leaders need tools to adequately prioritize, support, and sustain water assets and services for their communities. Continuous improvement mechanisms that could help achieve desired outcomes point to ubiquitous utility themes, such as best practices, artificial intelligence (AI) applications, and regulatory considerations for pre-COVID-19 and post-COVID-19 periods. This chapter will consider the impact of the pandemic on the already strained water sector to determine what has been working for whom and when; this chapter will discuss interdisciplinary lessons learned and culminate in recommendations for effective utility management and optimization.

This chapter discusses ethical hiring practices in the digital age and argues for the necessity of ethical hiring processes. An examination of hiring before and during the COVID-19 pandemic describes the increased use of digital technologies to aid the ethical hiring process. A particular focus is on the increased use of artificial intelligence (AI) tools to aid in screening and selection. Unethical hiring practices are examined showing the potential adverse outcomes which may result. The chapter concludes with model and a set of guidelines to illustrate the components that enable ethical hiring processes and incorporate the use of AI.

This chapter covers internal controls and segregation of duties needed to protect company assets when a decentralized (remote) workforce is utilized. Company assets focused on in this chapter include cash, customer billings, inventory, computers, data, expense report reimbursements, payroll checks, and time engagement. Also

Preface

OVERVIEW OF BUSINESS MODELS TO PROMOTE TECHNOLOGY, CULTURE, AND LEADERSHIP

A model is a way to view the world, a simplified representation of how a more complex system works. The model train is not a train, but it provides an insight into the class of objects called trains and enhances our understanding. In that same sense, this book presents a set of chapters designed to provide insight into how organizations integrate technology, culture, and leadership in a post-COVID-19 environment.

Technology has always been important to organizations and crucial to their continued development. Digital communications, the internet, social media, and artificial intelligence (AI) represent potentially disruptive technologies, and their value and impact accelerated during the pandemic. Organizations that fully integrated the latest technology and provided effective leadership had a smoother transition during the COVID-19 pandemic, while turmoil and crisis conditions occurred in organizations that lacked effective leadership. Leadership that effectively utilizes technology is vital in the post-pandemic environment. Further study on this type of leadership and practice helps ensure that businesses remain prepared and capable despite an unsure future in the new normal of the post-COVID-19 reality.

The authors presented here delve into how virtual technology has evolved to create remote offices and remote teaming in health, education, engineering, and other settings. The book also presents culture in business, and the authors discuss how individuals may interface, communicate, and collaborate in past, current, and future business models. Other authors address issues related to the environment, ethics, leadership decision-making, resilience, and emotional intelligence. Most of the authors included here present thoughtful discussions of how the pandemic affects the work of organizations and individuals.

THE NEW NORMAL

The COVID-19 pandemic transformed the daily life of millions of people worldwide by dramatically limiting physical interactions during the shutdowns resulting in remote work arrangements, an emphasis on digital communications, and virtual delivery of a wide variety of services. After most of the shutdowns ended, the world did not immediately return to the former methods of conducting daily life. Companies found that remote work saves money and is just as effective, or in some cases more effective, for the organization. Employees found that eliminating commutes and working from the comfort of home increased job satisfaction and organizational commitment. While these new realities forced organizations and people to adapt, surprising best practices often emerged. What may have been lost in social interaction is being replaced by virtual connectedness. What may have been lost in collaboration is being replaced by efficiency and time management alongside work-life balance and job satisfaction. These are the characteristics of the new normal as organizations and individuals continue to move through a period of adjustment. A staggering death toll, uneven economic performance, high inflation, the great resignation, shifting demographics, political disagreements, and other challenges loom large on the current landscape. Despite these issues, people profess a willingness and desire to move forward into this new normal and continue to adjust as needed. More than that, many have once again begun to thrive.

WHO SHOULD READ THIS BOOK

The topics under discussion here include management, public health, education, and culture, and this reference work is ideal for business owners, managers, human resource professionals, supervisors, scholars, researchers, academicians, practitioners, instructors, and students. Indeed, anyone interested in the future of technology, work, and organizations will find ample food for thought within these pages. The work presented here casts a wide net of subject matter, viewpoint, and author background. The result is eclectic, yet themes emerge around systems, models, leadership, ethics, culture, and technology. A brief overview of the chapters below should be sufficient to tempt the potential reader.

THE CHAPTERS CONTAINED HERE

The authors included in this work represent an exciting sample of professionals, educators, business people, and organizational experts. Each chapter contributed is briefly summarized below.

Chapter 1: Sociotechnical Systems and Work Approaches – Technology, Culture, and Leadership

In Chapter 1, J. Robert Heinzman presents a discussion of sociotechnical systems theory, a review of previous applications of the theory, a discussion of scientific management, quality improvement, technological advancement, and behavioristic approaches, and addresses a wide range of topics that impact today's workplace including the move toward reshoring of manufacturing jobs.

Chapter 2: Impact of COVID-19, Technology, and Organizational Leadership Business Considerations in the Water Sector

In Chapter 2, Regina Cousar and Alma Beciragic discuss our water supply and the technology and leadership required to continue providing adequate safe drinking water to our communities in a post-COVID-19 environment. The authors recommend continuous improvement approaches and the use of AI applications and discuss regulatory considerations related to the water supply.

Chapter 3: Ethical Staffing in the COVID-19 Digital Age – Are a New Set of Ethical Practices Needed to Guide Selection?

In Chapter 3, Kevin Jones and Nazish Saba discuss ethical staffing in the post-COVID-19 environment. The authors address the need for a new set of ethical staffing practices given the increased use of digital technologies in the staffing and selection of employees. The discussion includes the potential adverse effects of using AI for screening and selection.

Chapter 4: Leadership Using Internal Controls for a Decentralized (Remote) Workforce

In Chapter 4, Lisa Bushur-Harris discusses utilizing internal controls to protect company assets given a remote decentralized workforce. Company assets include cash, customer billings, inventory, computers, data, expense report reimbursements, payroll checks, and employee time engagement. The author also addresses company reputation, service levels, social media presence, and CSR in a post-COVID-19 environment.

Chapter 5: Leadership and Strategic Decision Making Under Exogenous Shocks Such as COVID-19

In Chapter 5, Jim McCleskey addresses three topics from leadership and strategy, including the effect of individual and group decision processes on leadership decision-making, the newest directions and developments in the strategic leadership literature, and the effect of exogenous shocks such as COVID-19 on leadership decision-making. The author also describes the impact of exogenous shocks on small and medium-sized enterprises, crisis management, stress on decision-makers, effective decision-making, and leader behaviors that impact crisis decision-making.

Chapter 6: Tailored Leadership as a Post-COVID-19 Opportunity for Enhanced Performance

In Chapter 6, Rebecca Melton and Kimberly Brooks discuss leadership styles and provide a framework for leaders to tailor their approach to a task, issue, or employee challenge. The authors posit that utilizing this framework may enhance the organization's performance through a leader's consideration of the context of the scenario and its interaction with an employee or team. The authors present their framework through the lens of servant leadership.

Chapter 7: Learn, Unlearn, and Relearn – A Step Towards Bringing Resilience in Business Organizations in the Post COVID-19 Regime

In Chapter 7, Nabanita Ghosh, Anubha Srivastava, Sunita Kumar, and Daniel Gnanaraj discuss the impact of COVID-19 in India and describe the changes as a series of paradigm shifts. These include the shift in leadership style in organizations, the shift in the cultural climate in organizations, the shift towards greater use of technology and away from more traditional work, and the ability to sustain the new behaviors in a post-COVID-19 environment.

Chapter 8: In Times of Uncertainty, Organizational Cultural Shifts Appear Seamless With Emotionally Intelligent Transformational Leaders at the Helm

In Chapter 8, Lezlie Banks asserts that difficult times call for emotionally intelligent leaders. The author discusses understanding the origins of emotions, recognizing the benefits of emotional intelligence and transformational leadership styles, drawing attention to how leader behaviors create learning organizations and proposing best practices for these turbulent times.

Chapter 9: Impacts on Helping Professionals' Preparation During the COVID-19 Lockdown Through the Perspective of Two Professional Program Educators

In Chapter 9, Crystal Walcott and Deborah Judge share insights into the impacts of COVID-19 on teacher and nurse education. The authors present a historical perspective on the state of nurse and teacher education from pre-COVID through post-COVID environments. The authors emphasize the development of empathy and caring in teacher and nurse candidates in the post-COVID-19 environment.

Chapter 10: Physician Perspectives Regarding Use of Electronic Health Records in Public Health Disease Reporting – COVID-19 Reflections

In Chapter 10, Michelle Stewart reviews public health disease reporting practices and advocates for integrating technology and the use of electronic health records for infectious disease reporting. The author presents a case study conducted with physicians and offers a policy recommendation regarding Meaningful Use legislation and health information technology integration.

Chapter 11: Shifting Moral Values

In Chapter 11, Wiley Hughes discusses the impact of the pandemic and subsequent shutdowns on morality in the United States. Moral norms often shift during a crisis. The author emphasizes how changes in church attendance, work, and environmental interactions have contributed to the formation of a new, less moral culture. Ultimately, the author posits that if the culture can move away from morality, it is also possible and preferable to move it back toward a more traditional moral character.

CONTRIBUTIONS AND A LOOK AHEAD

A work like this book probably presents more questions than answers. The unyielding march of technology, changing paradigms in leadership, organizational culture, education, and morality, and the characteristics of a new normal that is still coming to fruition all combine to heighten anxiety and leave readers wondering where they fit into all of this. Perhaps then, asking the right questions is an ideal place to start, and this book may guide its readers toward the questions they must answer for themselves and their organizations in the future. What are the appropriate business models to promote technology, culture, and leadership in post-COVID-19 organizations? What leadership styles, practices, and approaches will best serve the individuals and organizations in the near- and long-term in a post-COVID-19 environment? What is the new normal, and what adjustments must be made to work and live in it effectively? We leave the readers to ponder these questions using the chapters ahead as a small but well-illuminated set of guideposts. Enjoy!

Joseph Robert Heinzman
Western Governors University, USA

Deborah S. Judge
Western Governors University, USA

Vincent S. Franklin
Western Governors University, USA

Jim A. McCleskey
Western Governors University, USA & Houston Community College System, USA

Acknowledgment

I humbly thank my fellow editors Debbie Judge, Vincent Franklin, and Robert Heinzman. The professionalism, humor, and diligence of each have made this a wonderful project. I extend a heartful thanks to the staff at IGI for their support.

I dedicate this project to my sons Logan, Shane, and Reilly. You and your children will live in interesting times, facing a new normal of organizations and people forever altered by COVID-19 and the resulting pandemic and shut down. May you prosper in it. As always, many thanks to my wife and partner of the past 27 years, Yvonne. Your support makes everything possible.

Jim A. McClesky

A special thank you and dedication to all the healthcare workers and teachers on the frontlines during the pandemic who created innovative ways to provide the best care to patients and students. Their dedication and commitment to putting the needs of others first, learning new technical skills, and maintaining a positive attitude throughout unexpected times are commendable.

Deborah S. Judge

This work is dedicated to the 'front line' health care personnel who worked so diligently and performed so magnificently to save so many and to keep the rest of us safe and healthy during this pandemic. The heroism and selflessness displayed by these individuals are rarely witnessed outside of incidents of armed conflict.... GOD Bless each and every one of you! Also, and as always, a wave to my Mother and Father who remain my life's anchor.

Vincent S. Franklin

Seldom do we find ourselves in a time that dramatically changes the face of cultural impact as COVID-19. I want to thank all the contributors to this effort for staying on course and delivering such thought-provoking and important information during a time when we have all been impacted whether through shutdowns, shifts to remote, lockdowns, masks, and loss. To my fellow editors Jim McCleskey, Vincent Franklin, and Debbie Judge thank you for hanging in and for your continued commitment to completing this important project.

I dedicate this project to my children Joseph, Jonathan, and Jessica. Your future will find continued cultural shifts as each generation comes to terms with the lasting effects of COVID-19. Thank you to my best friend, partner, and love, my wife, Michelle. Without your support, this would not be possible.

Joseph Robert Heinzman

Introduction

COVID-19 has forced the world's population to endure one of the "Interesting Times" highlighted by the *Chinese curse* or *English proverb* (depending upon where one believes the saying originated). The pandemic's impact and consequences are recent and (arguably) not yet fully evident. However, this epidemic, in ways only witnessed during other worldwide catastrophes and events, has forced human beings (over and above experienced pain, suffering, trauma, and loss) to adapt socially, professionally, and personally in ways not noted in a generation. Concurrently, COVID-19 altered (throughout the world) organizational cultures, technology usage, financial priorities (personal and business), and the interactions between leaders and subordinates within organizations.

The way that individuals 'interact' with each other is 'different'. This book works to identify and examine some of those differences. The contributors share observations of the manner in which work has changed; infrastructure use has and is continuing to evolve; staffing and the use of computer systems have further developed; decentralized decision making gaining additional prominence, and the observation that anxiety in the workforce remains high in environments still challenged by the virus.

The impact of the pandemic has been broad and will force continued organizational evolution. That said, the impact must be examined efficiently and expediently, which the hard-working and dedicated authors and contributors to this work have endeavored to accomplish.

Vincent S. Franklin
Western Governors University, USA

Chapter 1
Sociotechnical Systems and Work Approaches:
Technology, Culture, and Leadership

Joseph Robert Heinzman
https://orcid.org/0000-0003-2295-1731
Western Governors University, USA

ABSTRACT

Sociotechnical systems theory is important because it brings together technology, behavioral culture, and leadership as it applies in the business environment. To understand where we may be going, it's important to review how past sociotechnical systems have been approached and how those systems were affected by work through scientific management, quality improvement, technological advancements, behavioral focus, and economics. Trust and collaboration are hallmarks of leadership, most agree managers manage things (task) and leaders lead (guide and support) people. COVID-19 has caused challenges to managing and leading, a disruption in our supply chain, manufacturing to adjust, and management to be more focused on innovation and sociotechnical systems in the United States. Manufacturing is a key indicator of employment and jobs health, and the current job loss in industry has been accelerated. Going forward, different skills, education, and industry focus will test innovation and the balance between management of things and leadership of people.

DOI: 10.4018/978-1-6684-4358-3.ch001

SOCIOTECHNICAL SYSTEMS AND WORK APPROACHES

Sociotechnical systems theory is targeted at two factors, the things (tools) needed to do work and the people (skills) needed to do the work (Rogers, 2003). The focus is driven by two distinct foci based on management style and perception of how to achieve the objectives within an organization. Traditional theory in the early 20th century points to management focus on a task and later a focus on people (Robbins & Judge, 2010). Sociotechnical systems theory is important because it brings together technology, behavioral culture, and leadership as it applies in the business environment (Scott & Davis, 2007). To understand where we may be going, it's important to review how past sociotechnical systems have been approached and how those systems were affected by work through scientific management, quality improvement, technological advancements, behavioral focus, and economics (Hellriegel & Slocum, 2007).

It's only been the last decade since around 2010 that a clear definition of leadership has begun to help define the difference between what is a leader and what is a manager. Trust and collaboration are hallmarks of leadership, most agree managers manage things (task) and leaders lead (guide & support) people. COVID 19 has caused challenges to managing and leading, a disruption in our supply chain, caused manufacturing to adjust, and management to be more focused on innovation and socio-technical systems in the United States. Going forward different structures and industry focus will test innovation and the balance between management of things and leadership of people.

INNOVATION AND INDUSTRY

To understand the interplay of technology, levels of technology, and the workforce we need to understand where the industry has been and how the interplay of technology and the workforce have changed. Disruptions in the way work are done is not a new phenomenon. The industrial revolution caused technological progress to occur which centered a debate on how technological progress affected workers and how these effects may differ in the short and long run (Davidson & Chismar, 2007).

The Handicraft Era of industrial progress was characterized by high manual skills (Gantt, 1919). The tradesmen or craftsmen were told what was wanted, and at this point, their ingenuity and skill were coupled to produce the required item. This handicraft era can also be earmarked by the lack of standardization of process, product, and price. In this era, the basic variables of how to produce, and what to produce were left to the individual producer and thus it was impossible to provide standards of performance and quality.

As time passed, the industrial situation evolved that was characterized by large groups of congregated labor working for wages in cottages or factories (Kimball, 1939). This form of the industry should not be confused with our present-day manufacturing methods, since history depicts this situation as a grouping of tradesmen or handy-crafters each producing a finished product, rather than the subdivision of tasks into a-highly repetitive chores, aided and abetted by machinery and controlled methods. The distinguishing features of this "cottage production" were congregated labor as well as the fact that product specifications existed, but without detailed instructions in method or standardized conditions.

In the 18th century the "Industrial Revolution" occurred changing the interface between producer and consumer. The producer was now at arm's length from the consumer. At the same time, the task of the individual was reduced from completing a product to that of producing a component, or in many instances to a single operation on a component. This created the simultaneous need for both the designer of a product from an overall standpoint, as well as that of a coordinator of manufacturing from the same overall standpoint.

Inherent in this evolution of industrial progress was the specialization of skills. The first-line foreman was chosen for his position based on his wide experience and skill in producing, while at the same time the product designer progressed through drafting and detailing work. This quite naturally left a void in the organization for those who might appreciate equally the problems of both design and production. The industrial revolution era was focused on the management of things and tasks, there was little regard for how to get the most effective and efficient work out of the workforce. During this period either you will perform, or you will not be here anymore was the common mantra. The perception of how much work should be performed was based on the average of the group.

One of the first individuals to recognize and record a solution to this problem was Frederick Winslow Taylor while working for the Midvale Steel Co. of Philadelphia in 1882 (Taylor, 1911). Taylor suggested what he called "Functional Foremanship" as a means of bridging the gap between design and production. Taylor suggested that this impossible single task of the first-line foreman be divided into two broad areas, first planning, and second, operations.

This method, scientific management, drove two ideas through the planning and operation to increase efficiency and the effectiveness of workers, which would provide an increase in profits. The idea according to Gantt (1919) was to increase efficiency through process methods and provide a better sharing of profit (rewards) with the workforce. This seems to be the first instance of a sociotechnical approach, albeit in its infancy. The focus was mainly on task completion and how to increase the completion of tasks consistently, on time, to a standardized process. At this point in time employees were still considered a tool to complete tasks (Gantt, 1919).

Scientific management helped to improve the baseline of task requirements although it did little to improve motivation factors. These early studies provided a baseline that helped guide the management of things and the leadership of people.

The early 19th and 20th centuries saw continued debate of labor-saving and labor augmentation through technology. Many economists argued that the model of technological progress would marginalize the working-class earnings, others felt that the savings and efficiencies created by the technology would help to increase laborer wages. Knut Wicksell (1901) stated that technological progress would either lower or raise the marginal product of labor, and thus wages, depending on whether the technology replaced or added to the labor. J.B. Clark (1907) made the distinction that the well-being of the workers requires technological advancement, although that same technology would cause temporary displacement of the working class as a necessity. Technological advancement and the focus on tasks continued to be the main focus behind driving capitalism.

During the Great Depression, a period of high unemployment, the attraction of the technological unemployment hypothesis could not be fully resisted. Ewan Clague (1935), a labor economist who was to serve as Commissioner of Labor Statistics for the Department of Labor from 1946 to 1965, stated that "the present outlook is for the rate of displacement of labor to exceed the rate of reabsorption so that technological employment will continue to be large." He concluded by noting that "in time . . . the surplus of older workers will be eliminated by age and death." Part of this worry was based not on rapid innovations in manufacturing, but on labor-saving changes in agriculture such as the tractor, which were one-factor driving massive flows of people from rural areas to the cities. Others such as Edna Lonigan (a US Department of Labor economist), while cognizant of the debates over the effects of labor-saving technology, rejected this argument. Lonigan (1939) stated flatly that "our present unemployment has little to do with machines" and argued that there is no necessary connection between innovation and unemployment. Instead "it is in the failure of the price system . . . to permit the creation of new employment, that the source of the worker's growing insecurity is to be found." Of course, grasping for a secular supply-side explanation of high unemployment is not unique to the depression of the 1930s. Many people have been quick to jump to such an explanation in the Great Recession of 2007–2009 and its aftermath, even when a deficiency in aggregate demand may offer a more plausible explanation.

The focus on technology and the working class as a tool to perform the task continued until roughly around 1927 when how an employee is best motivated to perform was questioned. The Hawthorne studies, in the U.S., which ran from 1927 to 1932, showed that groups have established norms and may be motivated through performance norms (Robbins & Judge, 2010). The studies indicated that workers were reacting to and performing at a high rate due to being observed and listened to. Workers were reacting to the increased attention they were receiving, and their personal performance expectations and norms increased due to the active engagement. A follow-up study researched how workers might react to economic rewards as a motivator. The important output of the study was that workers did not individually maximize output. The worker's output became controlled by the group norm that defined a proper day's work. The workers were concerned that increasing production management would decrease the rate of pay and thus the worker would work harder to obtain the same pay. Also, the workforce was worried that increasing output would cause layoffs and affect job security. The significance of the Hawthorne Studies showed that observation and taking an interest in the employee had a stronger impact on output. This is a significant shift in theory from forcing an individual to perform through coercion to focusing on the well-being of an employee may provide increased output to accomplish tasks.

In 1945 the Ohio State Studies brought together Scientific Management and the Human relations movement by exploring two-factor concepts of leadership behavior (Bratton, Grint & Nelson, 2005). The Ohio State Studies showed that those followers who were high in consideration by their leaders were more satisfied and motivated to perform and had a higher respect for the leadership. The studies were largely ignored until 1992 when a review of 160 studies showed that initiating structure and consideration were associated with effective leadership. According to DuBrin (2007), the Ohio State Studies showed that focusing on structure (task orientation) and People (consideration) is where management should focus. Focus on output is preferred to focus on input, especially in a diverse workforce. Both studies, the Hawthorne, and the Ohio State studies showed that the behavior of individuals was tied to the anxiety and stress they felt in their work environment.

Although Scientific Management is not the first theory of management it is an evolutionary step that may have provided the foundation for the behavioral theory, transitional, and modern theory of management (Wren, 2005). Taylor's work provided the groundwork to set methods and standardized practice in place. Taylor's work was built onto by Mary Parker Follett with her research showing that only through integrated teams, which worked toward finding a solution that was mutually agreed upon by both parties, could a successful outcome occur. Follett's work opened the door for Maslow, Herzberg, and Alderfer in building models of employee needs in the workplace. The transition management era is most recognized by Peter Drucker's work on Management by Objective (MBO), which is a structured tool that sets goals for an organization. Each goal is identified as a strategic or operational goal by management and employees and agreed upon. These goals are then measured and adjusted as needed to meet successful outcomes and achieve the goals (Bratton, Grint, & Nelson, 2005).

Sociotechnical systems are made up of four elements, innovation, communication channels, time, and the social system (Rogers, 2003). Innovation is the ability to adopt an idea, practice, or project and use it in a manner not utilized before. Rogers (2003) discusses that innovation's consequences may create uncertainty and "changes that occur in an individual or a social system as a result of the adoption or rejection of an innovation" (Rogers, 2003).

Communication is the channel through which information is transferred via individuals or organizations. Methods of communication may be oral, written, or non-verbal and are hindered by the filtering and method of transferring that communication. Communication requires strategies and methods in order to be effective. Depending on the level of skill and ability of the communicator and the understanding of the receiver of the communication will drive whether the outcomes meet the goal of the communicator. In order to be an effective communicator, we require knowledge of when to use direct, indirect, or pivoting strategies. We need to understand which method is best to convey the message. Throughout COVID people stopped communicating face to face and began utilizing social media and in many cases biased media from Twitter, Facebook, Instagram, and other media. The pervasiveness of disinformation creates a society of followers and decreases individual thought.

According to Rogers (2003) time is ignored by behavioral theory although the process of entering a market and being successful is driven by the entry into the market. The innovator, early adopter, and early majority have an edge in market success and those later adopters and laggards are less aware of knowledge innovation.

The social system is a set of interrelated units or cross-functional teams, that work jointly to solve a problem. The nature of the social system affects individuals' ability to innovate and work together to solve problems. The social system is tied to culture and will be affected by generational differences, ethnicity, gender, religious belief, lifestyles, customs, traditions, heritage, habits, and values.

The sociotechnical system continues to be a challenge for industry. Throughout the 18th, 19th, and 20th centuries, advances were made in technology that displaced laborers, while behavior studies did help to improve leadership theories and create better teaming opportunities the ultimate result was higher output with less labor. The continued expansion of profits and revenues and increasing the profits while keeping the price at a level that consumers could purchase caused other inefficiencies within the sociotechnical system. Fair wages are not directly a consideration of the sociotechnical system, it does have an effect on labor output.

MANUFACTURING SECTORS

When a system is in balance, everybody benefits. When a system is out of balance someone will lose. The balance in the socio-technical system is a management and leadership challenge. The impacts on labor due to scientific management, quality programs, management, technology improvements, and regulation are best observed by reviewing the manufacturing sector. All sectors of the economy are important and have changed in significant ways over the past few decades. Manufacturing is an important sector to focus on due to it historically providing 20% of all jobs in 1980. U.S. Data on manufacturing employment by the BLS indicates manufacturing labor has declined by more than seven million since 1970. Manufacturing has been traditionally affected by trade and environmental policies causing changes in the economics and driving changes to meet the trade and regulatory requirements. These shocks to manufacturing have labor market effects that spill over into local, regional, and national labor markets. A review of the Bureau of Labor Statistics (BLS), various blogs, the Society of Manufacturing Engineering (SME) library, Boston Consulting, and discussions with local Economic Development Council members indicates that for every job in the manufacturing sector a multiplier anywhere from 4x to 16x occurs. This indicates that a spillover of at least three jobs in other support sectors is created for every one job in manufacturing. Traditionally manufacturing positively adds to the local community by contributing to economic inflow. Manufacturing sectors bring money into the community from outside sources who purchase from the manufacturer. The local economy prospers by providing services which the manufacturing and spill-over workforce patrons.

This author spent a decade supporting the C17 Globemaster III Maintenance Simulation program. Our task was to upgrade the maintenance simulators to match upgrades to the C17 before the next generation upgrade became operational. For each upgrade, the average team required project management, engineers in mechanical, electrical, software, and manufacturing, designers, configuration, quality control, procurement, technicians, manufacturing vendors, and component suppliers. Our internal team usually consisted of 15 individuals, the manufacturing vendor(s) and suppliers varied depending on the upgrade. A two-year project with 10,000 components designed, manufactured, and procured often required at least a dozen manufacturing partners, and a multitude of suppliers. Each manufacturing project generated approximately one hundred manufacturing jobs of various skill levels. Utilizing the multiple of 4x, the real outcome of each project generated four hundred jobs, one hundred manufacturing, and three hundred spillover jobs. According to Jerry Jasinowski (2013), the more advanced manufacturing becomes the higher the multiplier effect. For example, the manufacturing of high-tech products computers, cell phones, and electric vehicles require advanced automation or "As factories get "smarter" and more advanced," the multiplier moves to the 16x level.

According to Charles, Hurst, and Schwartz (2018), the decline in the manufacturing sector may have had an increasingly negative effect due to "employment rates for men aged 21-55 fell 4.6 percentage points and hours worked fell by over 180 hours per year." They go on to explain that uneducated men and women in their prime earning years declined at a higher rate throughout the period between 2000 and 2017. Traditionally the manufacturing sector provided jobs to individuals regardless of education. For example, it was common for an individual, with or without a high school degree, to find work in manufacturing that paid a living wage. A living wage is a level of income that provides a decent living, a decent living may vary by perception but should provide the means for a family of four the ability to afford an auto, home, food, healthcare, and some disposable income. Charles et al (2018) estimate that one-third to one-half of the decline in employment for prime-age wage earners through the 2000s is attributed to the decline in manufacturing jobs. Utilizing the multiplier effect supports that manufacturing as a prime factor would have at least an indirect effect on spillover jobs. The loss of two million manufacturing jobs from 2000 to 2017 would indicate six million spillover jobs for a total of eight million jobs eliminated. While some would leave the workforce some would pursue further training and education, when the means were available or provided, most were forced to take lower-paying jobs.

Figure 1. Monthly U.S. Manufacturing Employment

Note: Figure shows the total employment in the manufacturing industry within the U.S. between 1970 through 2022. The Data comes from the Bureau of Labor Statistics (BLS) Current Employment Statistics and was downloaded directly from the St. Louis Federal Reserve's economic data website.

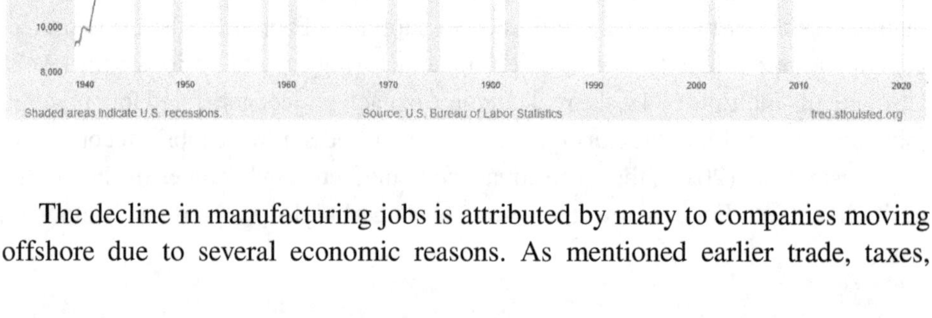

The decline in manufacturing jobs is attributed by many to companies moving offshore due to several economic reasons. As mentioned earlier trade, taxes, and environmental regulations have been blamed for why several manufacturing facilities moved offshore. Those who have lived through the period of downsizing and shifting to offshore manufacturing and suppliers know that some of that is true. What is important to note is that scientific management and total quality circles helped to improve production by 300% before the addition of automation. As technology improved in the factories a shift in the workforce, or a move toward the knowledge worker, caused a different need in U.S. manufacturing. For example, as the computer age and chip manufacturing began to heat up in the seventies due to improvements in miniaturization which allowed a shift from analog to digital control the type of technologist needed in a factory changed. A technician who could work with mechanical devices (analog) would need to be retrained to understand machine code to run and repair digital controlled equipment. These shifts led to other changes, the traditional drilling, milling, and machining gave way to computer numerical controlled (CNC) machines. The first "CNC" machines were programmed by hand to perform single operations, as improvements in control occurred drilling, milling, and turning (multi-Axis) machining reduced the footprint needed for equipment and the need for multiple operators. The next improvement brought on by computer improvements was computer-aided design (CAD), the design could be saved on a disk and inserted into the machine control receptacle. The next step in manufacturing was computer-aided manufacturing (CAM), which removed the need for a machinist with program knowledge, now a design engineer can run the program in software to proof out the operation and then send it directly to the machine to drive the process. The progress in computers, communication, and software has led to computer-integrated manufacturing (CIM). In 2010 Tesla purchased a previous General Motors plant in Freemont California and completely retooled the operation to utilize automated manufacturing, without CNC, CAD, CAM, and of course the digital revolution, computer-integrated manufacturing in an automated or artificial intelligence (AI) factory would not be possible. This is an important step in technology that has caused a major shift in the demographics of the labor force. Technology has allowed manufacturing output to rise while the manufacturing base in the U.S. has declined.

Figure 2 shows the output has been continuous increases in productivity per worker, but something seems to have broken with the Great Recession. First a major drop in productivity, then some progress getting back to trend, and then no progress since about 2010.

Figure 2. U.S. Manufacturing Labor Productivity per Worker
Note: Figure shows the Labor Productivity per Worker in the manufacturing industry within the U.S. between 1987 through 2022. The Data comes from the Bureau of Labor Statistics (BLS) Current Employment Statistics and was downloaded directly from the St. Louis Federal Reserve's economic data website.

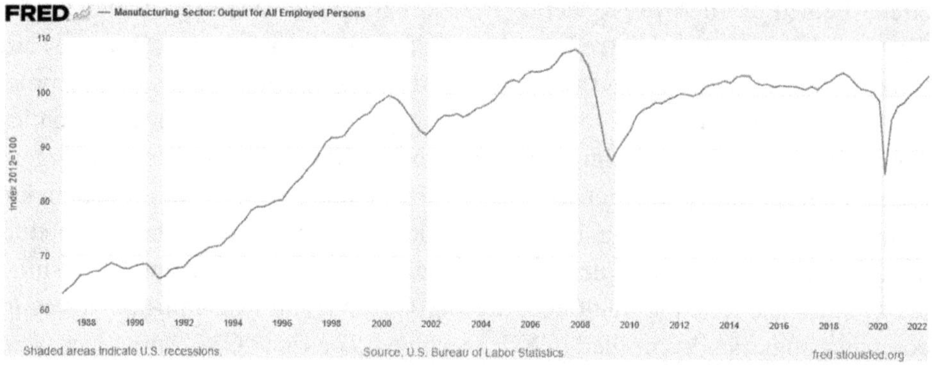

Figure 2 indicates since the Great Recession; manufacturing jobs have offered fewer hours of work or more part-time work. Maybe productivity per hour worked is growing, but figure 3, which shows productivity per hour instead of per person, shows no difference. The cause of this productivity standstill is an indication of a change in the composition of the manufacturing workforce toward lower-productivity work or increased technological progress.

Figure 3. U.S. Manufacturing Labor Productivity per Hour Worked
Note: Figure shows the Labor Productivity per Hour Worked in the manufacturing industry within the U.S. between 1987 through 2022. The Data comes from the Bureau of Labor Statistics (BLS) Current Employment Statistics and was downloaded directly from the St. Louis Federal Reserve's economic data website.

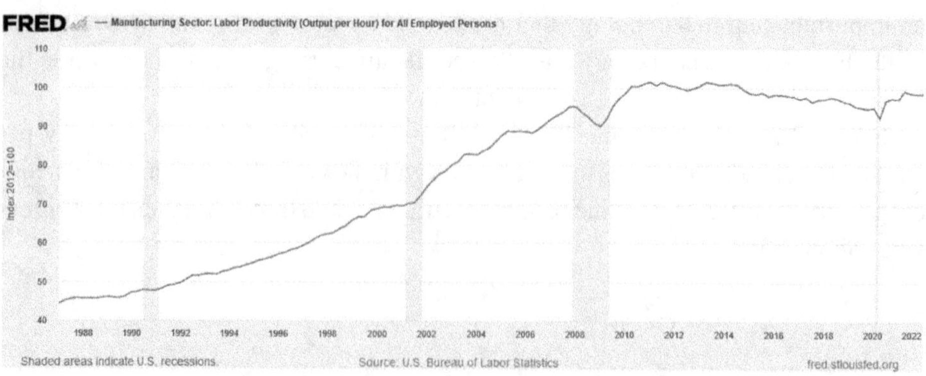

The manufacturing labor force has seen a decrease in participation, although one important measure of labor market conditions labor force participation rate (LFPR), indicates the available workforce has decreased since 2000. The LFPR is equal to the employed plus the unemployed, divided by a measure of the U.S. population. Think of it as those who want to work (i.e., have a job or want one) relative to those who could work (the entire population over age 16 that isn't incarcerated or on active military duty). The incarcerated are not counted because they are not an added value to Gross National Product (GNP). The military, which does provide value through defense, is not counted because it is paid for by the U.S. GNP.

Figure 4 is a graph of the monthly U.S. LFPR starting in 1948. One striking feature is its hump shape, which is related to demographic factors. LFPR fluctuates around 59% until the late 1960s when it starts rising. This rise is attributed to the Baby Boomer generation joining the labor force, as well as to the widespread entry of women into the labor force. The LFPR plateaus at around 67% in the late 1990s and starts to decline in the 2000s.

This decline wasn't just a result of the Financial Crisis of 2007-08. The year 2008 is also when the oldest Baby Boomers started turning 62, the earliest age they could claim Social Security benefits. The Pew Research Center indicates the second largest generation is the Millennial generation at around 73 million, passing the boomer generation as the largest adult population (Bialik & Fry, 2021). The millennial generation (1981 – 1996) was well into the workforce by 2015 with the youngest millennials being 19 years of age. Generation Z, those born between 1996 and 2012 totaling approximately 88 million including growth due to immigration, began to be part of the LFPR data in 2012. Gen Z will surpass millennials as the largest adult population by 2030. 2012 is when the oldest of Generation Z became sixteen years old and began to be factored into the workforce. Generation Alpha (2010 -2026*), which currently numbers around 48 million, will begin to enter the workforce in 2026.

The recent COVID-19 crisis led to the largest drop in LFPR on record: from 63.4% in February 2020 to 60.2% in April 2020. the LFPR has partly recovered, but it is still below its pre-COVID levels. However, the trend line in the graph shows that the post-pandemic LFPR is only 1.5% from its pre-COVID number. *The end date for generation alpha is an estimate.

Figure 4. U.S. Labor Force Participation Rate
Note: Figure shows the Labor Force Participation Rate within the U.S. between 1970 through 2022. The Data comes from the Bureau of Labor Statistics (BLS) Current Employment Statistics and was downloaded directly from the St. Louis Federal Reserve's economic data website.

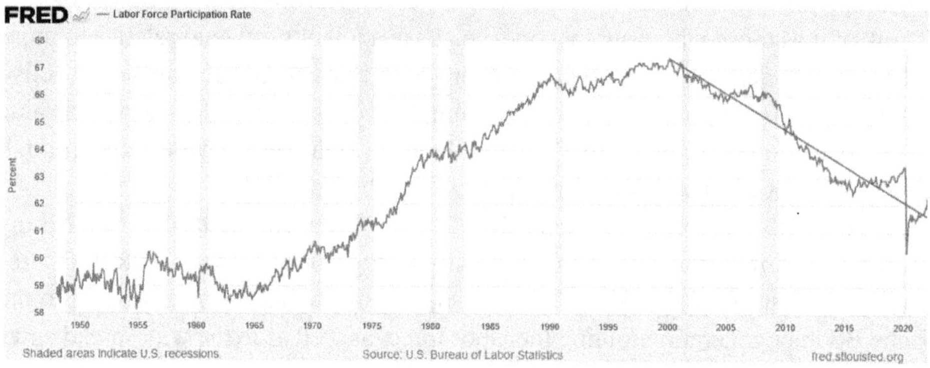

Two of the main forces of the Sociotechnical system to be impacting the current labor force will be innovation and generations' social culture. As boomers (1946 – 64) and gen X (1965 – 1980) leave the workforce the subsequent generations will begin to have a greater effect on the

workplace. Leadership is already struggling with changes in the workforce culture and juggling multi-generational differences. How leadership reacts to the effect of innovation and social culture changes is going to be an important focus going forward. Wages alone, depending on automation and job creation, will not be the main driver for attracting and maintaining a stable workforce.

LEADERSHIP AND CULTURE

Every organization has its own unique culture, some choose to ignore it, and others try to defy it, wise leaders will attempt to understand it. Organizational culture has an impact on the organization's ability to implement a strategy, introduce new technology or increase productivity. By paying attention, in a genuine manner, we will get more positive results and interaction between teams and management. Appreciative inquiry is one method toward increasing relatedness, and positive emotions, when paired with appreciative intelligence the ability to appreciate the positive and reframe thinking creates a better focus toward the future.

Appreciative Inquiry is a four-step method for gathering information: Discover, Dream, Design, and Destiny (Royal, 2006). During the Discover phase, the leadership and employees will focus on the current situation exploring positively focused questions to identify what is working well and what attracts individuals and entities into relationships with the organization.

Following the discussions about the current situation, the dream phase will focus on what could be. Coupled with the knowledge from the Discover and Dream phase the Design phase will turn the focus toward how the team can get from where it is, to where it wants to be. During the Design phase, the leadership, and eventually the team, will create the possibility statements designed to expand the existing circumstances and stretch the boundaries of possibilities. The primary question for this stage would follow, "how can we go about bringing those dreams to fruition?" The final part of the conversation is the Destiny phase where the organization's leaders will decide what will be, not just what they hope will be. It is during the destiny phase that the team will develop processes and procedures to incorporate the possibility statements into the day-to-day operations of the organization.

A leader's ability to reframe and see positive opportunities is an important part of the appreciative intelligence process. The leader creates an environment where an appreciation for the positive and seeing the possible future is applied. Leadership would focus on creating possibilities (opportunities) in the situation and entertain input from employees with an open mind. Leadership should create a genuine atmosphere that invites input with a focus on positive, open-minded (Reframing) and future-oriented discussion. It is important we do not shut followers (employees) ideas down or create barriers to ideas, followers will become dissatisfied and lose commitment to the team and the organization (Arbinger Institute, 2002). As leaders, we must drive a belief in the followers that their input matters, this will build trust and the free exchange of ideas which will help frame a positive environment.

Reframing can be envisioned through the steps toward innovation and through the known and unknown (Mohrman, 2006). This constitutes thought toward accepting the current situation and asking what the future can be. Adversaries are positive focused; they are a voice steeped in passion focused on being part of the solution. Leadership can help followers see where they have genuine input and are part of an important process. Encouraging input and engaging openly in a positive manner encourages followers' belief that sharing ideas is not a risk. The appreciative intelligence approach could help decrease barriers that cause stagnation and hinders growth in organizations.

"Appreciative Intelligence is an individual ability, but it has a significant impact on larger systems. The most successful leaders and innovators extend their Appreciative Intelligence beyond their personal lives and into the very fabric of their organizations. They build an environment, infrastructure, corporate culture, or system that generates success and helps appreciation perpetuate" (Thatchenkery & Metzker, 2006, p. 368).

This author's research shows a higher order of thinking that encompasses Appreciative Intelligence (AI), Social Intelligence (SI), Emotional Intelligence (EI), and Cultural Intelligence (CQ) leads to an Appreciative Environment (AE). AE is about our past, our present, and possible futures. AE is an exploration of understanding that we should put aside the past, we should learn our present strengths, and build the best possible positive future of us. Focus, or more correctly, past focus, present focus, and future focus (PPF) are about reframing our perceptions. As Plato alludes to in Book 1 of Laws, conflict begins with the individual. We must first overcome our own insecurity, balance of ourselves must be obtained, learning who we are. An organization is made up of departments, those departments are made up of teams, and teams are made up of individuals. Everyone is unique and provides value in their unique abilities.

Figure 5. PPF Performance Focus Relation
Note: The figure is from the book Divide You Fall (Heinzman,2016)

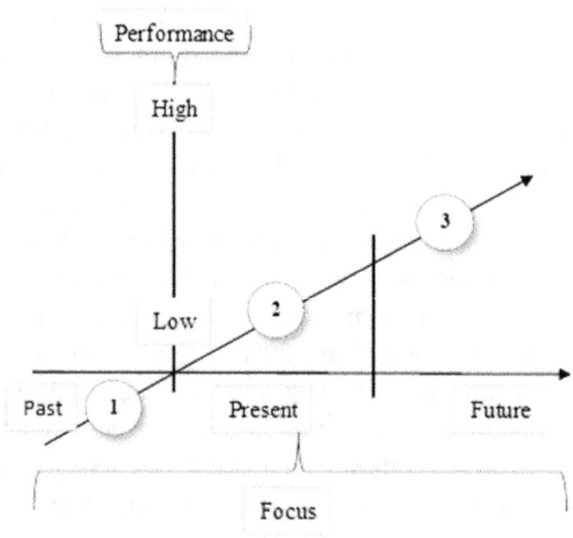

Legend

1. **Problems are Past**
2. **Future is Possibilities**
3. **Built on Present Strengths**

Social Intelligence

AI alone may repress negative occurrences which are important in the change process. AI focuses on positive stories during the discovery phase to decrease the resistance to change and diminish barriers created through the defense of prior activities. Social constructionists point to the need for a social relationship as being an important overlying concept that allows discussion of the future without bringing up the past. Leadership is responsible to create an atmosphere of belonging through social interaction which increases the ability of the employee to engage and adapt (Thorndike, 1920). Engagement and belonging allow stories and reason to be shared openly without hostility. Ability is subjective and depending on individual A's perception in relation to individual B's perception the outcome of individual C will be different (Thorndike, 1913). This encompasses the communication part of Sociotechnical systems.

Thorndike (1913) further explains that ability and outcome are not consistent with intelligence. One may score higher than another on a battery of tests by 30 points. One individual may receive a 70 while the other a 100, does this mean the person can produce at a much higher output? Technical knowledge and book knowledge are not the same. Often, I have students who are competent intelligent working individuals who are learning a new topic they have not engaged in prior. Some students, need extra time to learn the concepts. IQ is generally a measure of cognitive ability; it is not a good measure of overall intelligence.

Subjective measures drive leadership to create social situations where some employees are looked upon more favorably than others. The theory of leader-member exchange, also called LMX or vertical dyad linkage theory, provides insight into how leaders maintain their position through a series of tacit exchange agreements with their members (employees) (Dansereau, Graen, & Haga, 1975). The exchange between the leaders and followers affects the relationship between the employee perceptions of fairness and supervisor-rated performance and organizational commitment. Leadership should look for favorable attributes in followers, this includes the least preferred co-worker in an attempt to create positive outcomes and increase inclusive behavior. If that individual is not able to intrinsically feel they belong they will leave.

Social belonging increases when organizations use teaming methods to explore possible outcomes of events. Organizations learn and understand team members' and individuals' possible actions and why those actions may be taken. When actual events take place the path of addressing issues may evolve differently although outcomes will have higher positive endings due to insight into how each member may perform. Without preparation for alternative futures, organizations may not be prepared to capitalize on opportunities or respond to threats from external environments (Lawler & Worley, 2011). Marlowe & Bedell (1982) determined that social intelligence yielded five dimensions: interest and concern for others, social skills, empathy, emotional expressiveness, sensitivity to others' emotional expressions (sense-making ability), and social anxiety. Depending on a leader's measure, high or low, these factors have an impact on their social intelligence ability. When social intelligence is high the leader should be more apt at creating a sense of fairness and equity among followers.

Activities focused on social structure changes that included increasing interdependency among clinical departments, multidisciplinary cooperation across clinical disciplines, and standardizing clinical decision-making showed an increase in communication by focusing on social cohesion (Lowe, Levitt, & Wilson, 2008). The focus of the study of a new computerized physician ordering system showed role alignment and communication, with a focus on social belonging increased positive interaction across departments and generational diversity.

Emotional Intelligence

The ability to solve practical and social problems has different meanings depending on the context. Creativity is part of a rational ability to decide which knowledge approach may be the best direction based on the circumstances. Behavioral complexities such as emotions are paramount in the intricacies of management and leadership (Dixon, 2010). Emotional intelligence has become a standard in business settings while its predictive powers relative to personnel performance forecasting abilities seems to be negligible. An ability to interact with a diverse employee population, a complex environment, and multifaceted decision requirements would require leaders and followers to have a fully developed emotional presence. Emotional presence is necessary for workers at all levels to be able to accommodate unpredictability, demonstrate adaptability, and perform flexibly within the workplace (Goleman, 1995). Personal intelligence is intuitive and common sense. Edward Thorndike's original research on social intelligence in the early 20th century showed SI to be part of people's practical abilities. Thorndike, Goleman, and Howard Gardner showed that people's intelligence was more than just IQ, which revolved around linguistics and math skills, and that IQ does not provide insight into one's ability to be successful in business. Emotional intelligence is made up of one's self-awareness, the ability to handle emotions, intrinsic motivation, empathy, and the art of relationships. Emotion recognition has been reliably validated within the construct of emotional intelligence as a complicated skill that has societal and workplace consequences (Dixon, 2010). Greg & Camilla (2010) found that emotional intelligence had a significant effect on leader-member exchange and increased the relationship between leadership and followers in a positive manner. The interaction of emotion regulation, well-being, and social relationships has implications for leadership efficacy.

Job Satisfaction

Job satisfaction is the effect of the organization, job design, and culture, which affects how much an individual likes their job. The extent of job involvement and relationship, or leader-member exchange, influences the employees' overall commitment to the organization. A study on the relationship among leadership, organizational culture, and the operation of learning opportunities among employees' job satisfaction levels showed a positive correlation (Su-Chao & Ming-Shing, 2007).

The study, focused on Taiwan industries, provides insight into leadership's effect on followers' job satisfaction in a developing country. This shows an interesting point of the knowledge worker perception of leadership changes, as they move from dependency to understanding and ability to contribute to the organization as knowledge workers. A knowledge worker is defined as intellectual property that is in a form of human capital (Nonaka & Nishiguchi, 2001). Drucker (1999) posited that human capital contributes to the company's intellectual property through the knowledge the employees possess. The human elements of companies have gone through a change in the Twentieth century from laborer to knowledge worker and the 2007 research by Su-Chao & Ming-Sing (2007) shows the same thing happens no matter the country.

Followers, as knowledge workers, look to leaders to model behaviors that encourage innovation, and learning activities and create interest in growth opportunities and monetary incentives. Leadership is responsible to encourage employees to want to learn; this process is necessary for the continuance of the organizational culture that supports organizational learning and increases the ability to transfer knowledge. Further, it is contingent on leadership to enhance the trust between leadership and employees, trust is critical to whether the organization can be successful.

Two ends of the full-range leadership model are transactional and transformational leadership theories. Transactional, or absent, leadership has been shown to affect the organization by causing high absenteeism and stagnation. Transactional leadership follows strategy and process and provides rewards, although is quick to punish employees when they deviate or make mistakes. The outcome is a tendency for followers to avert risk, stagnate innovation, and increase absenteeism. Transformational leadership follows strategy and process, provides rewards, encourages innovation, and empowers employees to make decisions. The method provides a learning opportunity and encourages risk-taking and the growth of followers and the learning organization. Learning organizations can continuously adapt and change through systematic processes. For learning organizations to be successful, leaders of change will come from new sources, not just upper management. The need for leaders from different positions and non-traditional areas with unique methods are needed.

Leadership models have been challenged in remote work environments since COVID-19 caused much of the traditional office workforce to move into a remote setting. Typical questions from management are about how we know if employees are putting in the hours they are being paid for. If the work is getting done, on time, then that should be what good leadership and management are concerned with, not how many hours a person is working. Leadership needs to be concerned with how they will continue to motivate and sustain a workforce accustomed to working from home.

WORKPLACE CHALLENGES

Conservative estimates indicate that 60 percent of those working from home during COVID do not want to return to the office. Can a remote workforce be a reality in manufacturing, logistics, and supply chains, and is the workforce plentiful and skilled to meet the new workplace? The current workforce is limited in new entrants, education, and skills. Conversations in reshoring manufacturing identified as critical to the U.S. economy will continue to stress the current and future workforce. The effect of technology on the workplace provides changes to how employees may be informed, carry out production, skills needed, interact, and work satisfaction. Culture affects leadership styles that may not transcend generational gaps. As technology allows individuals to be more remote and globally located from the central culture of the organization how will culture affect the satisfaction of the employee?

John Hayward (CEO of DiSTI a virtual training systems company) shared some of the challenges his company has experienced due to the pandemic. He shared that their highest priority is employees' health and safety, establishing policies for masking, social distancing, and figuring out our remote work policies. DiSTI experienced similar challenges many organizations have shared of inadequate infrastructure to handle a nearly 100% remote workforce. To meet the challenge the company moved from locally hosted MS Office to Office 365. This included unifying services around Outlook 365, SharePoint 365, and Microsoft Teams. They also selected MS Teams as their primary team collaboration software. DiSTI has been a long-time Jira user to track sprints and progress on our software development efforts, but they found they were inconsistent about what metrics to track, and how they were tracked, presented, and reported.

John stated "While something we would have gotten around to eventually, the Pandemic made it more critical to get consistent project status. Those changes were made and have simplified project reporting." On individual performance John had the following to say:

"Individual Performance tracking, especially for new employees, is critical to our business to ensure we can provide help, training, assistance to those that are struggling and give them opportunities to succeed. We were not doing this before the pandemic – managers largely managed through "Management by Walking Around" or MBWA. This is still a work in progress, but we have improved and understand the amount of work folks are getting accomplished and working with individuals that don't meet expectations. This has included pulling them back into the office and providing face to face training and assistance."

Returning to the office is becoming the second biggest challenge for DiSTI, and from this author's review of current events, it is similar across many businesses who saw their workforce go remote during the pandemic. John shared that there is a desire by DiSTI employees to be nearly 100% remote workers and by some of our management to have them back in the office at least 50% of the time. DiSTI "is continuing to grow and will be impacted if their valuable trained employees decide to depart over our remote work policy, so we have to find a good solution. This is a work in progress" (John Hayward, Personal Interview).

The advent of telecommuting due to technology has increased flexibility, although, for employees with high social needs, telecommuting can increase feelings of isolation and reduce job satisfaction (Robbins & Judge, 2010). One example of the global capability and telecommuting is education. A plethora of online schools and extended online opportunities of brick-and-mortar schools provides an opportunity to explore how educators and students respond to the remote capability provided through technology. The online education system is a good way to explore possible remote leadership challenges in a post-COVID workforce. Technology advances have made it easier for fewer instructors to interface with a larger number of students than in traditional models. With the stress of higher numbers and lack of separation from the office, at-home workers tend to overwork, causing low organizational satisfaction. The reason employees tend to stay in the short term is the reduced cost and convenience. Although employees stay in the long term when they feel a connection with the work environment, believe in the mission and feel leadership supports their career.

Similar to the social research by Thorndike and studies like the Hawthorne and Iowa State showed, employees still need to feel they are recognized, and that management takes an interest in their wellbeing and career path. An understanding of cultural dimensions and work culture provides some indications of how diverse populations are motivated and provides an opportunity to improve job satisfaction.

Geert Hofstede's Cultural Dimensions (1980, 2001) identified five major cultural dimensions: power distance, uncertainty avoidance, individualism-collectivism, masculinity-femininity, and long-term/short-term orientation. Hofstede's studies demonstrated that cultural perception affects work and cross-cultural mingling may cause dysfunctional work teams. In the U.S. the overall culture of the business is a high-power distance, individualism, masculinity, and risk avoidance with a short-term focus. Not all businesses have a common culture, work culture is driven by a business's hierarchical structure, decision-making groups, perception of time, meeting structures, and dress code. As the job environment shifts businesses may need to look outside their local communities, source from other countries in the form of offshore and immigration and hire individuals with transferable skills from different industries to fill positions.

The Global Leadership and Organizational Behavior Effectiveness (GLOBE) studies identify nine cultural dimensions: uncertainty avoidance, power distance, institutional collectivism, in-group collectivism, gender egalitarianism, assertiveness, future orientation, performance orientation, and humane orientation (House, Hanges, Javidan, Dorfman, & Gupta, 2004). The general purpose of GLOBE was to determine whether cultural differences may be related to different leadership views (Yukl, 2006).

The GLOBE study found that different cultures had different leadership profiles, however, the study also found that certain leadership characteristics were valued across cultures, and some leadership attributes were found to be globally undesirable. The desirable characteristics of an outstanding leader, according to the GLOBE studies, are trustworthiness, fairness, foresight, planning ahead, being encouraging, optimism, dynamism, building confidence, motivational, dependability, intelligence, decisive, a good negotiator, win-win problem solver, good communication skills, informed, administratively skilled, coordinating, team builder, and being excellence oriented (House et al., 2004). The attributes found to be globally viewed as obstacles were being a loner, being anti-social, noncooperative, irritability, not explicit, egocentric, ruthlessness, and dictatorial.

Global leaders need to develop cross-cultural competencies and understand the business, political and cultural environments worldwide. Leaders need to understand the perspectives, tastes, trends, and technologies of many other cultures. Leaders need to learn how to deal with diversity and be able to adapt to living and communicating with other cultures to build social culture.

Socialization continues the process by indoctrinating individuals into the organization's culture. According to Robbins & Judge (2010), socialization is the most critical stage of molding an outsider to the culture of the organization. Socialization starts at prearrival when the member learns the skills required for the position. The encounter occurs during the hiring phase where the organization and the individual determine how well a match each is. The final stage is metamorphosis where the company provides either formal or informal training to the member that is hired.

Under the best conditions leadership is not easy, it requires flexibility and intentional focus on the work environment and the available workforce. As the workforce changes due to generational change, population and labor participation rates, education (formal, informal, skills) settings change management and leadership may find it more difficult to find qualified employees.

POST COVID-19

Looking to the future is difficult under the best conditions, as we look forward in the U.S. the labor force participation effect will continue to be challenging at best. Depending on the skills and education as well as the overall population of those considered part of the participating workforce may cause a market where labor has more bargaining, and companies find stiff competition to attract employees.

Current education systems and ongoing professional development focus on traditional degree pursuit, open skills training, apprenticeship programs, and experiential learning opportunities that provide certifications and credentialing. As mentioned earlier, manufacturing is a key labor indicator affecting the economy. The BLS data shows that in 2021 of the 152,581,000 plus employed over 19 million were employed in some area of the production, transportation, and material moving occupations. The current employment multiplier shows that for every one job in durable goods manufacturing approximately 7.5 indirect jobs are created (Report • By Josh Bivens • January 23). This indicates smart manufacturing is progressing, and as smart manufacturing production increases, and moves toward a 16X multiplier should increase.

During COVID-19 the great resignation of employees in office and factory environments found that the pandemic accelerated flexible working opportunities. The U.S. BLS reported close to 13 million people left their jobs in a two-month period between July and September of 2021. During the same time, 10 million job openings were created per day. The U.S. Chamber of Commerce data shows more than 47 million quit their jobs during the pandemic. According to a Washington Post poll conducted around that time, 33% of all workers under 40 considered changing careers during the pandemic. A Pew Research Center study during this period showed that 94% of parents expect their children, 17 or younger, to attend college.

The latest data shows that we have 11.3 million job openings in the U.S.—but only 6.3 million unemployed workers (Ferguson, 2022). We have a lot of jobs, but not enough workers to fill them. If every unemployed person in the country found a job, we would still have 5 million open jobs. At the height of the pandemic, more than 120,000 businesses temporarily closed, and more than 30 million U.S. workers were unemployed. Since then, job openings have steadily increased since January 2020, while unemployment has slowly declined. Overall, in 2021, employers ended up adding an unprecedented 3.8 million jobs. But at the same time, 2.2 million Americans have left the labor force since January 2020. Since November of 2020, the hiring rate has outpaced the rate at which labor has quit. The great resignation may not be an accurate term, it seems more plausible to call it the great reshuffle, hiring is 1.3% above the quit rate on average.

Those individuals who have a college education, advanced training, degrees, credentialing, and other advanced certifications have a higher labor participation rate. This participation helps to provide us some clarity as to why the reshuffle is a better term, financial activities, durable goods manufacturing, and professional and business service had the lowest quit rates at 1.7% and 2.1% and 3.7 respectively, wholesale and retail had a quit rate of 3.8% while leisure and hospitality has a 6.4% quit rate. Of the 1 million to quit leisure and hospitality 1.2 million have been hired since November of 2020. The hardest hit has been durable good manufacturing, it has continued to grow, and new job creation has continued. The challenge is that many who left took early or met retirement age shifted careers, or did not re-enter the labor force. These industries have more openings than skilled workers to fill the jobs.

The current BLS employment projections through 2030 indicate that an increase in jobs from 153.5 million to 165.4 million should be expected. The job creation projection is based on population and participation based on age, as boomers age, their participation will be less causing a continued decline in the LFPR. The population increase according to the BLS data indicates, that even with lower LFPR, the labor force participation population is expected to grow to 169.6 million in 2030.

The higher number of the participating workforce than jobs available should drive more job creation. Also, efforts in reshoring manufacturing may provide further job creation than the population, especially skilled, available. According to Harry Moser (Reshorenow.org) "Since the late 1970s the U.S. has lost manufacturing to a rising tide of imports. The goods trade deficit reached $1 trillion in 2021. More than 50% of the goods we consume are now imported. Approximately 25% of what we import would be sourced domestically if companies and government based their sourcing decisions on total cost (TCO) instead of price. Just doing the math correctly would motivate the reshoring of 3 or 4 million mfg. jobs. As manufacturing recovered, our youth and their parents would see that skilled manufacturing is once again a top career choice." Reshoring has already contributed to 2020 for 160,000 new jobs announcements, with 2021 close to 220,000.

From 2010 to 2019 reshoring efforts and foreign direct investment grew manufacturing jobs by close to 900,000 (Figure 6).

Figure 6. Cumulative manufacturing job announcements
Note: Data show the cumulative reshoring and foreign domestic investment manufacturing job announcements. Used with permission from Harry C. Moser Founder & President of Reshorenow.org.

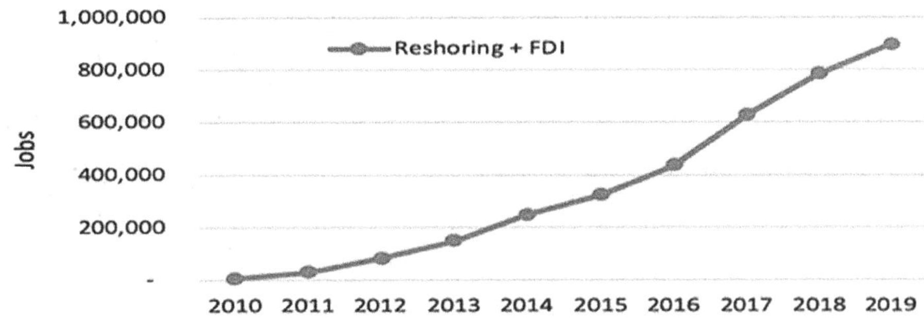

With recent supply chain shortages in critical areas, the reshoring of manufacturing in critical areas makes sense. For example, during COVID the textile research facility at Duke University ramped up to produce cloth used in masks. The textile facility is not set up for continuous production although they pulled together a project team to manufacture critical Personal Protection Equipment (PPE) cloth. Had the textile facility not ramped up, the U.S. would have been left without masks. This is only one instance that shows why reshoring is critical, chips, electric vehicle batteries, electronics, pharmaceuticals, and other high-tech and critical components make sense when viewing the total cost opportunity.

The total cost opportunity model considers risk with offshore, lead times, shipping, quality, loss, and other important factors. Traditional purchasing looks only at price, the cost in other areas is not considered. For example: while working on the 787 Dreamliner one of the offshore partners continued to ship short, ship short means that critical sections of the plan were shipped without all the components installed. This caused a bow wave effect on the downline manufacturing which contributed to overrunning costs and missing key production deadlines.

Figure 7 shows that the TCO calculator (Reshorenow.org) indicates the cost is within $10 when taking all factors into account. It is critical to understand not all reshoring makes sense although it is important to note that if we truly are focused on Green and being environmentally friendly, it is better to have control of those areas where poor EPA (Environmental Protection Agency) focus will be a detriment to the global community. The U.S. has higher standards than China, Mexico, Asia, India, and others. The human factor in these countries shows the workforce is poorly treated, poorly paid, and often includes children as part of the workforce.

Figure 7. Cumulative Cost by Category (TCO)
Note: Data show the cumulative reshoring and foreign domestic investment manufacturing job announcements. Used with permission from Harry C. Moser Founder & President of Reshorenow.org.

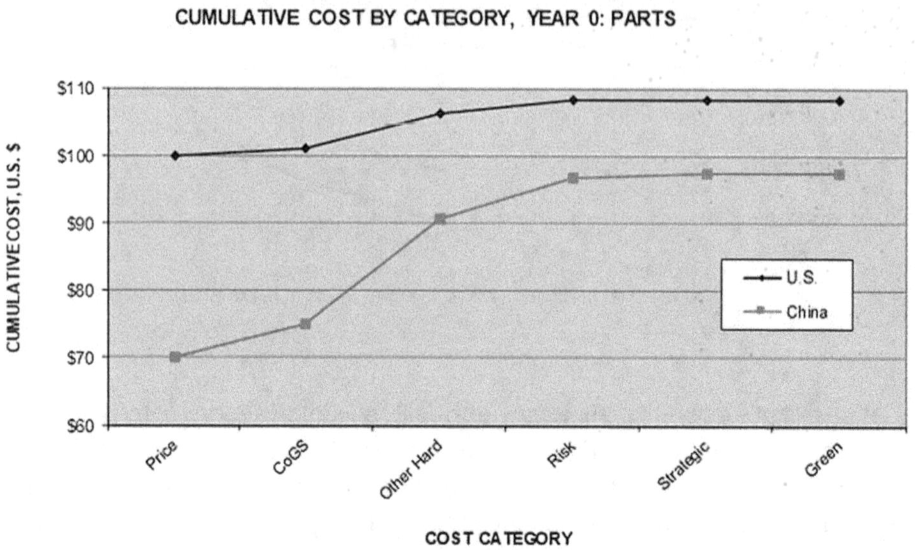

If the reshoring efforts are correct the future job creation may outpace the population of LFPR. If only 3 million new jobs are created in the manufacturing sector, and they tend to fall into smart manufacturing, the result is an added 22.5 million jobs. This will continue to drive a deficit of labor to jobs available per the U.S. Chamber estimates of 5 million jobs unfilled presently. The need for skilled individuals will be greater than in the last twenty years. In the U.S. we are already short on skilled technicians and workers in HVAC, Electrical, Mechanical, Power Transmission, Maintenance, and Manufacturing Industrial areas.

In this author's opinion, the K through 12 system has failed our industrial backbone in this country. The reduction of courses that focus on machining, auto repair, design, electronics, and other skills caused disparity in the technical trades. There has recently been a resurgence of technical schools focusing on trade skills. The National Governors Association is supporting avenues that set up direct-to-work programs through open skills networks to apprenticeship programs.

In a personal interview with Dr. Joe Heinzman, Director of Apprenticeship Programs for Pasco Hernando State College, he had the following to say. The demand for apprenticeships spans a broad range of industries, skills, and trades. An apprenticeship program has two components i.e., formal classroom instruction and On the Job Training (OJT). In Florida, an apprenticeship recognized by the State must be a minimum of one year in length and a maximum of 5 years in length.

Classroom training must have a minimum of 144 hours per year and OJT must have 2000 hours per year. The classroom training provides background information on the skill or trade and provides hands-on labs.
The following apprenticeships illustrate the breadth of the demand.

1. Apartment Maintenance Technician (1 year)
2. Construction Skills, Design, and Project Management (2 years)
3. Electrical Apprentice (resulting in journeyman status) (4 years)
4. Electric Utility Lineman (5 years resulting in crew lead status)
5. Industrial Manufacturing Technician (12 months)

The shortage of skilled craftsmen and women is partly due to the COVID-19 pandemic starting in 2019 and continuing in 2022. Workers found a strong desire to work remotely once exposed to the modality. Workers in the low-paid hospitality and retail industries expanded hobby and part-time businesses into their main mode of making a living. Many students graduating from high school are not working or going to college suggesting that the internet, living at home, and part-time income provide fulfillment. The mass exodus of the baby boomer generation left many skilled positions vacant, and a shortage of a pool of lessons learned in the workforce.

The current desire to re-shore will expand the need for trades and skills culminating in the perfect storm of labor shortages. Apprenticeship programs will lead the way to fill the gaps, but it will take a major source of funding and momentum from federal, state, and local government, industry associations, unions, and other benefactors. The two biggest challenges in the apprenticeship movement today are the ability to attract large numbers of qualified apprenticeship candidates, and the ability to find adequate job openings for placement of the apprenticeships. The desire of employers to accept apprentices as acceptable candidates to fill openings will determine the success of apprenticeship programs.

CONCLUSION

Sociotechnical systems understanding is a critical component of understanding the workplace and the workforce. It is clear that a shift in the workplace is occurring and will continue with more people looking for remote opportunities. As innovation and possible reshoring occur, the expectations in increased technical jobs will require more focus on trades and will drive the education sector to provide better paths to skills. Students and parents need to be informed of the opportunities in the workplace and the paths to get there. Leadership in organizations will need to evolve to trust diverse populations, diverse degrees, and credentials and provide flexible work environments for those positions that may be remote. The use of tools like the TCO should be used to encourage buyers to look at local buying over offshoring when possible. There are two possible outcomes for the workforce, if the BLS is correct the future of the workforce will see a continuous population of unemployed which could drag down the economy. If the reshoring efforts continue and focus on in-country increases, there will continue to be more jobs than the workforce. This can be a positive as the wage will be competitive and help unemployment stay low. Focusing on a Sociotechnical work system which is comprised of individuals interacting with a technical subsystem which is the process, equipment, tools needed to perform the work, and the employee subsystem of people needed to perform the work is required for an organization to be successful. The 6 percent loss of workers (20 million) in the U.S. job market, is yet to be realized.

RECOGNITION

The author would like to recognize the valued input provided by industry professionals especially those who provided direct comments in developing this chapter.

Joe Heinzman DBA - Director of Apprenticeship Programs, Pasco Hernando State Colleges, Fl.

John Hayward – CEO of DiSTI, an Orlando Fl company that creates virtual training systems for the DoD and Commercial customers.

Harry Moser – Founder and President of the Reshoring Initiative, founded in early 2010, helping manufacturers realize that local production, in some cases, reduces their total cost of ownership of purchased parts and tooling.

REFERENCES

Bialik, K., & Fry, R. (2021, May 28). *How millennials compare with prior generations*. Pew Research Center's Social & Demographic Trends Project. Retrieved March 13, 2022, from https://www.pewresearch.org/social-trends/2019/02/14/millennial-life-how-young-adulthood-today-compares-with-prior-generations-2/

Bivens, J. (n.d.). *Updated employment multipliers for the U.S. economy*. Economic Policy Institute. Retrieved March 10, 2022, from https://www.epi.org/publication/updated-employment-multipliers-for-the-u-s-economy/?msclkid=c8d85de2a60f11eca5c7f7fbbbeeab79

Brady, D. (2005, Mar. 28). The Immelt Revolution: He's Turning GE's Culture Upside Down, Demanding Far More Risk and Innovation. *Business Week.*

Clark, J. D., Jones, M. C., & Armstrong, C. P. (2007). The Dynamic Structure of Management Support Systems: Theory Development, Research Focus, and Direction. *Management Information Systems Quarterly, 31*(3), 579–615. doi:10.2307/25148808

Davidson, E. J., & Chismar, W. G. (2007). The Interaction of Institutionally Triggered and Technology-Triggered Social Structure Change: An Investigation of Computerized Physician Order Entry. *Management Information Systems Quarterly, 31*(4), 739–758. doi:10.2307/25148818

DuBrin, A. J. (2007). *Fundamentals of organizational behavior (4th ed.).* Thomson South-Western.

Engineers, . (n.d.). *Reshoring poised to surge 38% to record high.* Reshoring Initiative® IH2021 Data Report. Retrieved March 20, 2022, from https://reshorenow.org/blog/reshoring-initiative-ih2021-data-report/

Ferguson, S. (2022, March 10). *Understanding America's labor shortage.* U.S. Chamber of Commerce. Retrieved March 15, 2022, from https://www.uschamber.com/workforce/understanding-americas-labor-shortage

Heinzman, J. R. (2016). *Divided You Fall, Creating an Appreciative Environment.* Kendall Publishing.

Hellriegel, D., & Slocum, J. W. (2007). *Organizational behavior* (11th ed.). Thomson South-Western.

Jasinowski, J. (2013, November 27). *The Magic job multiplier of manufacturing.* HuffPost. Retrieved March 13, 2022, from https://www.huffpost.com/entry/jobs-multiplier_b_4002113#:~:text=In%20some%20advanced%20manufacturing%20sectors%2C%20such%20as%20electronic,not%20sweetness%20and%20light%20on%20the%20manufacturing%20front

Robbins, S. P., & Judge, T. A. (2010). *Essentials of organizational behavior* (10th ed.). Pearson, Inc.

Scott, W. R., & Davis, G. F. (2007). *Organizations and organizing rational, natural, and open systems perspectives.* Prentice Hall.

U.S. Bureau of Labor Statistics. (n.d.). *Employment projections home page.* U.S. Bureau of Labor Statistics. Retrieved March 10, 2022, from https://www.bls.gov/emp/

Chapter 2
Impact of COVID-19, Technology, and Organizational Leadership Business Considerations in the Water Sector

R. Cousar
Western Governors University, USA

Alma Beciragic
Arcadis, USA

ABSTRACT

Technology and organizational leadership are a part of most major businesses, and this is no different for businesses in the water sector. Delivery of clean drinking water and the treatment of wastewater are priorities for utilities in the United States as they seek to serve their communities. Existing infrastructure continues to age and faces a challenging future ahead as delivery demands grow. Leaders need tools to adequately prioritize, support, and sustain water assets and services for their communities. Continuous improvement mechanisms that could help achieve desired outcomes point to ubiquitous utility themes, such as best practices, artificial intelligence (AI) applications, and regulatory considerations for pre-COVID-19 and post-COVID-19 periods. This chapter will consider the impact of the pandemic on the already strained water sector to determine what has been working for whom and when; this chapter will discuss interdisciplinary lessons learned and culminate in recommendations for effective utility management and optimization.

DOI: 10.4018/978-1-6684-4358-3.ch002

WATER SECTOR CONCERNS BEFORE COVID

What do you really know about your drinking water? What happens to the water that is flushed away? What questions would you ask a water professional if you were given the opportunity? Water production, whether it is drinking water or treated wastewater, is a lifeline resource essential to a well-functioning community and, because it has seemingly always had high reliability in the United States, it can be taken for granted (National Infrastructure Advisory Council, 2016). However, one must remember that water is provided as a business service largely by publicly owned utilities in small areas with limited resources, which can be strained by daily demand and challenged by crises. This vital life service is an essential component of the water cycle in urban and rural areas where water is withdrawn, used, and returned for the good of the environment (EPA, 2004).

Figure 1. Urban Water Cycle.
Reprinted from Primer for Municipal Wastewater Treatment Systems by the United States Environmental Protection Agency. (2004, September). https://www3.epa.gov/npdes/pubs/primer.pdf

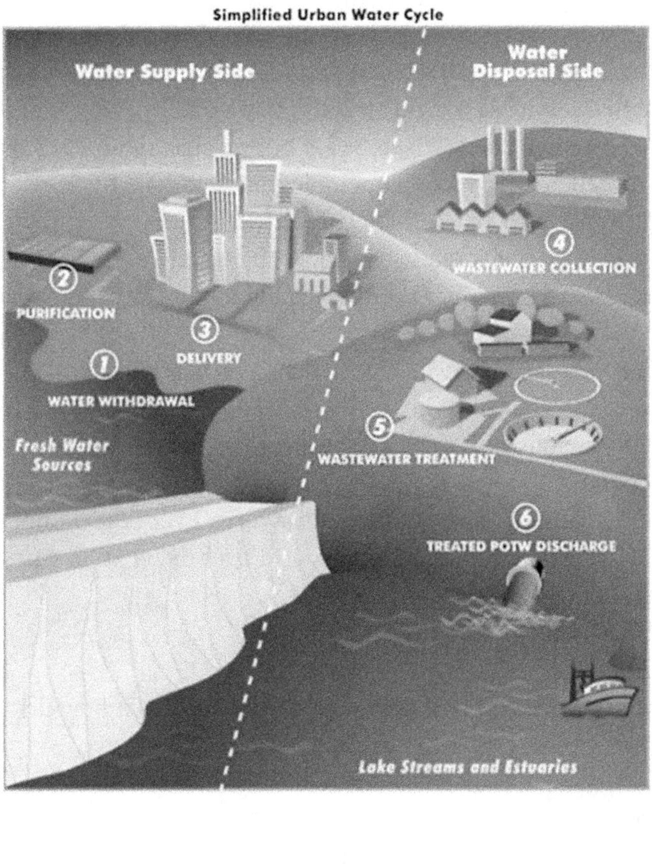

As a business service, the water sector faces similar challenges that other business sectors are exposed to, including being impacted by crisis conditions such as epidemics and security threats while leveraging technology and organizational leadership strategies to survive and thrive. COVID-19, recognized as an infectious disease in 2019, declared a pandemic in 2020, and continuing as a pandemic as of the writing of this work in 2022, is a global health crisis with unprecedented impacts on society, the economy, and the environment. The pandemic has led to more than 2.3 million deaths globally and has sent billions into lockdowns (Usman et al. 2021). By September 2021, COVID-19 was noted as the deadliest disease in American history, surpassing the death toll of the devastating 1918 flu pandemic (McKeever, 2021). While the effects of the COVID-19 pandemic on water and wastewater utilities will continue to unfold, many challenges have already become apparent.

Responses aimed at containing and controlling the impact of the pandemic have varied. As with many other industries, social distancing measures in response to the pandemic have pushed many water utility employees into a virtual world – forcing rapid workstyle and organizational adaptations. Utilities have been figuring out how to simultaneously tackle several challenges: how to operate remotely and securely, how to better protect on-site essential workers, how to deal with unpredictable supply chain challenges, handle labor shortages, and address significant lost revenue due to customers' inability to pay – all while trying to maintain regulatory compliance and safe uninterrupted services (Poch et al. 2020, Sowby 2020). On top of this balancing act, there is an estimated annualized $13.9 billion shortfall for US drinking water utilities due to delayed payments, unemployment, reduced non-residential sales, and slower customer growth (Raftelis, 2020).

Let's consider five quick questions about the water sector in the United States and reflect on if there is an obvious relationship to the COVID-19 pandemic:

1. You drink water. You flush. It gets treated. Where does it go?
    a. Back into your drinking water
    b. Into someone else's drinking water
    c. No clue
2. In the US, drinking water systems that serve over 300 million people need repairs. How much will these repairs cost over the next 20 years?
    a. 100 billion
    b. 300 billion
    c. 400 billion
3. Do periods of drought or extreme wet weather impact how drinking water is treated?
    a. Yes
    b. No

   c. It depends
4. Can COVID-19 be in my drinking water or my wastewater?
   a. Yes
   b. No
   c. It depends
5. Flint, Michigan had water issues. Are other water systems performing such that this cannot happen again?
   a. Yes, it was a one-time event
   b. No idea
   c. No, there are concerns in other systems

Since everyone needs water, those five brief questions may be of concern to you and your family. Water is used to make that wonderful cup of coffee, to freeze that popsicle, and to clean up any spills and upsets. In this chapter, the authors will consider recent practices in the water sector to get to the answers you may need and/or want about water services (and those five questions are addressed by the end of this chapter reflection!). The authors begin with the home water meter and then move back through the entire water system to the source so that everyone can better understand this business sector which is often hidden from view with buried pipes and unseen utility work, but which is absolutely crucial to communities and public health. The first chapter section offers an examination of pre-COVID water sector technologies and organizational leadership methods that are currently effective and remain important to business models. Reviews of available literature have been completed and this pre-COVID examination offers insights about practices up to 2019.

Metering and Smart Water Technologies

Once water is effectively treated by a water utility, it is delivered to homes and businesses. The water will move from the water main to the service line across a meter and then through the customer-owned pipes into the home. The meter, which will monitor water consumption, must be read so that the water utility can request payment for the business service and continue to provide this lifeline service to the community.

Figure 2. Example of a Home Water Meter Location.
Reprinted from Addressing Lead in Drinking Water with the Drinking Water State Revolving Fund by the United States Environmental Protection Agency. (2019, March). https://nepis.epa.gov/Exe/ZyPDF.cgi/P100WD7W.PDF?Dockey=P100WD7W.PDF

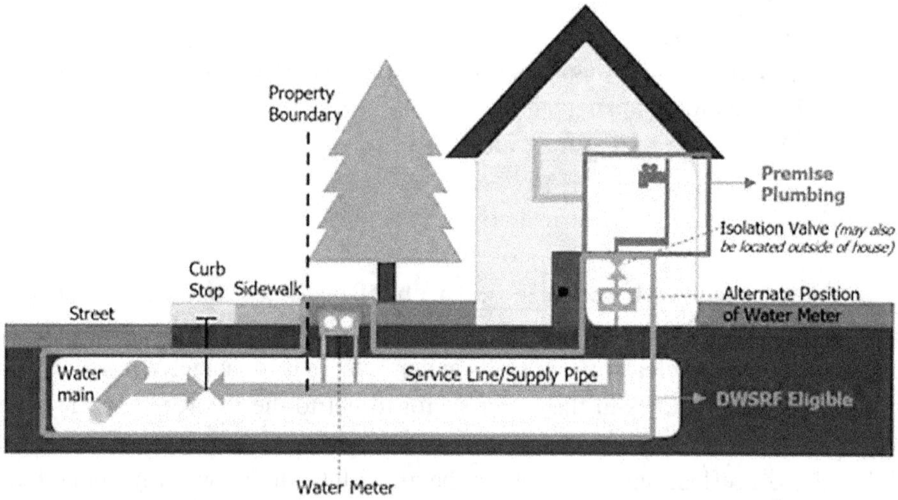

Traditional meter reading often involved a technician visiting the home meter monthly and collecting data on consumption. This historical monthly visit would cost the utility in terms of labor, equipment, vehicle maintenance, time, and could be error-prone since the readings were often handwritten into a ledger. Data could be difficult to obtain if the meter was obstructed such as by a dog or in a crowded basement. The utility would need to periodically review the manual ledger data and adjust the records to keep the readings as accurate as possible. Utilities would also need to send a technician to a home for the end of service or to begin new service after the home was sold.

In the late 1960s, a patent for "M2M" or machine-to-machine communication, whereby machines would automatically communicate with each other, was issued (Kennedy, 2016). This M2M concept became the forebear of automatic meter reading (AMR) technology which allows meters of all types to transmit to receivers without the need for handwritten data collection. With AMR, technicians could walk up to meters with a receiver and obtain the usage data. A subsequent evolutionary step in meter reading was advanced AMR drive-by meter reading that allowed the technician to stay in the vehicle and simply drive past the home as the meter transmitted a reading to a receiver in the vehicle.

Figure 3. AMR Drive-by Testing.
Image of one of the chapter authors at work with AMR equipment in 2012.

According to Scott (2003), by the end of 2002, 49.3 million AMR units were shipped to all utilities in North America (gas, electric, and water), and 9.5 million of those units (19.3%) were shipped in 2002 alone. Adding AMR, which often represented a significant capital investment for the utility as meters were changed out from old analog devices to the newer digital devices, was adopted by many water utilities in the 2000s and is still commonly used as of 2019.

Similarly, advanced metering infrastructure (AMI) for two-way communications between utility customers and the organizations who manage the area water resources has begun to grow in use as more cities seek to become "smart cities" (Talari et. al, 2017). A smart city initiative can involve the build-out of a grid of devices that supports all utilities, including water, gas, and electric monitors, and offers expanded services such as crisis management, demand response, timed pricing, and remote servicing. The AMI system is generally a fixed network that utilizes smart meters to transmit and receive data such that driving to the home location is no longer required except for initial installation or periodic system maintenance.

Figure 4. Example of Advanced Metering Infrastructure.
Reprinted from Guidance for Designing Communications Systems For Water Quality Surveillance and Response Systems by the United States Environmental Protection Agency. (2016, July). https://www.epa.gov/sites/default/files/2016-09/documents/srs_communications_guidance_081016.pdf

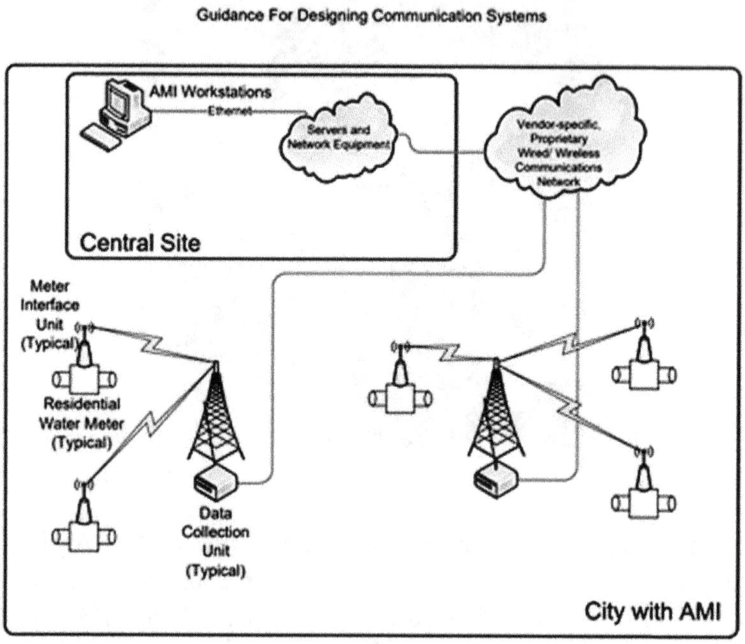

These AMI smart meters can provide real-time information on the working condition of the water meter, any alarm notices, and additional data about the status of the water system, as well as the state of any other utility or service that is on the grid (Mohapatra & Rath, 2020). On and off servicing can be done in a fully remote state. Smart metering technologies are also able to achieve supportive outcomes such as early detection of leaks (Monks et al., 2019). Smart meters and AMI can offer long-term cost savings due to the decreased need to drive to locations, the increased real-time system monitoring and early detection of any concern and can assist customers with usage management with real-time alerts on unusual consumption patterns. The systems clearly decrease the percentage of unobtainable data for utilities and provide more data for customers. A customer experiencing a water leak can be notified earlier and avoid a large water bill for water that was lost due to underground and invisible pipe defects between the meter and the home.

SCADA and Potential Cyber Security Concerns

Once the water crosses the meter and then is used in a home or business, much of it is returned to the utility via the pipes in the collection system. This used water will be cleaned and then recycled to the water supply, such as being returned to creeks, lakes, and streams, as a part of the urban water cycle. The process of recycling water must be closely observed and managed. A technology known as supervisory control and data acquisition (SCADA) is used when monitoring and controlling water treatment distribution centers, wastewater collection, and wastewater treatment facilities. Per Temido et al (2014), the information made available by SCADA systems helps to improve the system operation, by providing better management of storage capacity, optimization of the pumping stations, and pressure management across the system. SCADA is a part of a fully integrated control system and can be somewhat simplistic in small systems, but often runs to a level of high complexity for large urban areas. These systems offer from nascent to advanced Artificial Intelligence (AI) technology which assists facility operators in an analysis of system variables for data-driven decisions and has been deployed by utilities of all sizes. As noted by Russell (2019), 79% of water utilities have fully implemented SCADA.

Since SCADA systems in the past were generally environmentally autonomous and not connected to external world networks, they were often not seen as being threatened by cyber vulnerabilities. Security was largely due to the "gap" between the water service network of devices and the external world. However, cyber threats, as well as SCADA systems, have become more complex. Per Cherdantseva et. al (2016), escalating sophistication with modernization as well as real-time continuous operation of distributed, multi-component architecture underpins the growth of cyber threats to SCADA systems. SCADA systems are also exposed to a wide range of cyber threats because of the ongoing standardization of communication protocols, "internet of things" or IoT hardware components, and explosive interconnectivity across all types of digital devices including smartphones. Noteworthy as an example of cyber concerns, in 2006, a hacker penetrated the operation system of a water treatment facility in Harrisburg, USA (Cherdantseva, 2016). According to Zetter (2011), hackers gained remote access into the SCADA control system of the city water utility in Springfield, Illinois, and destroyed a pump. "The reality of today's environment is that there is no 100% safe and secure system in the world," says Luis Casado, the director of Gannett Fleming's water division in Russell's 2019 review of the internet of water. As can be seen, the simple purchase and installation of a single smart valve or pump could open a gateway entry point for malicious exploits if great care is not taken. Therefore, some cyber security concerns must be addressed to operate safely even in a "gap" enabled SCADA system.

The potential loss from a cyber-attack to a lifeline resource such as a water service may be so severe that the risk, which is calculated as a product of the loss from the attack and the probability of the attack, is estimated as a substantial risk even with a very low attack occurrence probability (Cherdantseva et. al, 2016). The loss of water service or a threat to water service is considered a very high risk for many communities. SCADA systems must be operational 24 hours per day, 7 days a week and there is often no test environment for real-time risk analysis that mirrors the exact same setup. It can be difficult to impossible to test out all the weaknesses and threat areas in a full simulation. A common approach to this risk analysis concern that is sometimes different by operational staff and/or regulations is for the water utility to conduct periodic and substantive cyber security risk assessment methods that analyze all people, products, and processes involved with the SCADA system from a perspective of potential vulnerabilities. In fact, on October 23, 2018, America's Water Infrastructure Act (AWIA) was signed into law in the United States as the most extensive set of changes to the Safe Drinking Water Act since the 1996 Amendments (EPA, 2021). Furthermore, AWIA Section 2013 requires community drinking water systems serving more than 3,300 people to develop or update risk assessments and emergency response plans (ERPs) at least every 5 years, including explicit requirements for cybersecurity.

The required planning exercises driven by AWIA, focusing on resiliency for all known and anticipated hazards, are initially often tabletop reviews that leverage available databases and subject matter expertise to evaluate threats and vulnerabilities. Water businesses must undertake a full system review every 5 years, but many also assess in an ongoing data-driven fashion to keep the full review information up to date during intervening years. In the additional ongoing risk analysis approach, each new device is reviewed prior to use in the system, and just as importantly each new employee is subjected to extensive and ongoing background checks to ensure that any possible weaknesses are identified. The resulting knowledge from periodic large-scale tabletop exercises and ongoing threat assessments is used to close any loopholes and to lobby stakeholders for the funding required to make essential improvements that will reduce threat exposure. In support of water utilities, the Environmental Protection Agency (EPA) in the US has provided a mechanism to conduct a drinking water or wastewater utility risk assessment called the Vulnerability Self-Assessment Tool (VSAT Web 3.0). This online free tool can be used to begin to estimate risks and vulnerabilities, including cyber risks, and evaluate potential improvements to increase security for critical utility assets during the 5-year review process (EPA, 2021). VSAT can also be used with another EPA tool, CREAT, which is used to specifically assess climate-related risks. CREAT is necessary because climate changes can pose significant risks to water utilities and the tool assists in activities including threat level cataloging of critical assets such as pumps in low-lying areas that may be subject to flooding during high rains or assets which may be endangered by extended drought periods causing dry pumps and pipes. To assess the long-term impacts of their natural disaster threat along with other vulnerabilities, CREAT allows water utilities to import VSAT results directly into CREAT (EPA, 2022).

Unfortunately, even with the extensive tools offered by EPA and others, there is some remaining risk assessment and response concern in the US for several small and very small drinking water suppliers who do not fall under the AWIA requirements. According to the Centers for Disease Control, most people in the US receive their water services from public systems, with public referring to public service, not public ownership. A recent estimate of the approximately 155,693 public water systems serving over 286 million people in the United States, 52,110 (33.5%) are community systems that provide water to the same population year-round. Another 103,583 (66.5%) are non-community systems, including 84,744 transient systems that provide water where people do not remain for long periods of time such as a campground supply and 18,839 nontransient systems which provide water to the same people at least six months a year, but not all the calendar year such as in a school (EPA, 2007). Originally enacted in 1974 and reauthorized in 1996, the Safe Drinking Water Act (SDWA) includes classifications by EPA for community systems based on the population size served. The use of this classification gradually shifted focus from 2000 to 2015 with increases in population in certain areas. As of 2015, very large to large community water systems, classified as serving over 100,000 people, delivered water to over 141 million people, while medium community water systems which serve 3,301 to 100,000, delivered water for over 139 million people, and small to very small systems supporting 3,300 or fewer people delivered water to over 23 million in the US population (EPA Office of Water, 2018). Note that if the original broader classification from the Safe Drinking Water Act is used, a small system would be any system serving 10,000 or fewer people, and a greater percentage of all systems, over 97%, are considered small versus large under the broader SDWA classification. The broader classification of small is not used as often by many since 2015 and the American Water Works Association (AWWA) 2019 State of the Water Industry Report, for instance, also uses the refined classification delineating small systems as serving 3,300 or fewer people (AWWA, 2019).

Thus, AWIA, a 2018 federal law that applies to systems serving more than 3,300 people, does not cover the public water systems of over 23 million people and these very small but numerous systems may need more awareness, support, oversight, and assistance with cybersecurity. It also does not cover vulnerabilities for those water services which are not a part of the SDWA, such as private wells which are used by approximately 13% of US households and are not federally regulated. When closely looking into small and very small water systems risk awareness, there are examples of how small system vulnerability can be exploited; one such event was recently investigated by the FBI in 2021. This investigation will be reviewed in the next section of this chapter for the period during COVID from 2019 to 2021.

As of 2017, to deepen the risk awareness, many state and local governments have stepped in to address concerns for smaller water utilities and have enacted various additional water quality accountability acts, such as in New York and New Jersey (Germano, 2019), which include cybersecurity as well as other areas for risk assessment as a part of an emergency preparedness blanket for all size systems. Similar to these local regulatory initiatives to add more threads for water safety, weaving together all hazards protection and preparedness for water systems continues by federal, state, and local governments in conjunction with supporting professional agencies. As can be seen, the COVID-19 pandemic has, undoubtedly, highlighted the importance of emergency preparedness in the water sector. Emergency planning allows for an improved response to crises such as pandemics and climate change. Sowby (2020) highlighted several policy statements by the American Water Works Association (AWWA) and the American Society of Civil Engineers (ASCE) aimed at improving water utility (physical and financial) resilience through ensuring power reliability, diversification of water portfolios/emergency interconnections, public involvement, customer communication, and maintaining cash reserves. While policy statements by industry professional associations do help guide public policy, they are ultimately voluntary.

Maintenance Management Data Analytics Including CMMS

Technology supports needed data analytics for data-driven decision making. Leveraging maintenance management data analytics by using a Computerized Maintenance Management System (CMMS) offers additional pieces of truth to help solve the challenges of maintaining aging water networks and improving system soundness. While SCADA is generally focused on real-time daily performance, CMMS technology tends to be focused on the system asset needs of maintenance, energy, and reliability. For water and wastewater facilities, energy consumption and maintenance costs can comprise over 30% of total operating costs. In fact, for many local governments, public drinking water and wastewater plants can often be the largest energy consumers, accounting for 30 to 40% of total energy consumed in that community (U.S. EPA, 2013). When one considers asset management, maintenance, and reliability, one may find that for many utility businesses the picture is largely foggy to very dim.

According to the American Society of Civil Engineers (ASCE) 2017 Infrastructure Report Card, significant new investment and increased efficiencies are needed as US filtration plants, pipes, and pumps age past their useful life (ASCE, 2017). Much of the current water infrastructure grew out of the water pollution concerns of the 1940s and escalated with the initial passing of the Federal Water Pollution Control Act of 1948, followed by the Clean Water Act in 1972 and the Safe Drinking Water Act in 1974. This infrastructure is thought to have a useful lifespan of 75 to 100 years. This means many components are beginning to reach the end of useful life for the older parts of the water system as of approximately 2000 to present, and that current water infrastructure will experience increasing failures if not renewed or replaced.

Strikingly, as of 2017, nearly six billion gallons of treated drinking water are lost daily due to leaking pipes, with an estimated 240,000 water main breaks occurring each year in this critical service (ASCE, 2017). To address the declining health of US water infrastructure, asset management supported by a CMMS can assist managers and stakeholders with business data on capital infrastructure and the timing of new essential investments.

Figure 5. Example Use of Maintenance Management Data Analytics.
Reprinted from Framework for Comparing Alternatives For Water Quality Surveillance and Response Systems by the United States Environmental Protection Agency. (2015, June). https://www.epa.gov/sites/default/files/2015-07/documents/framework_for_comparing_alternatives_for_water_quality_surveillance_and_response_systems.pdf

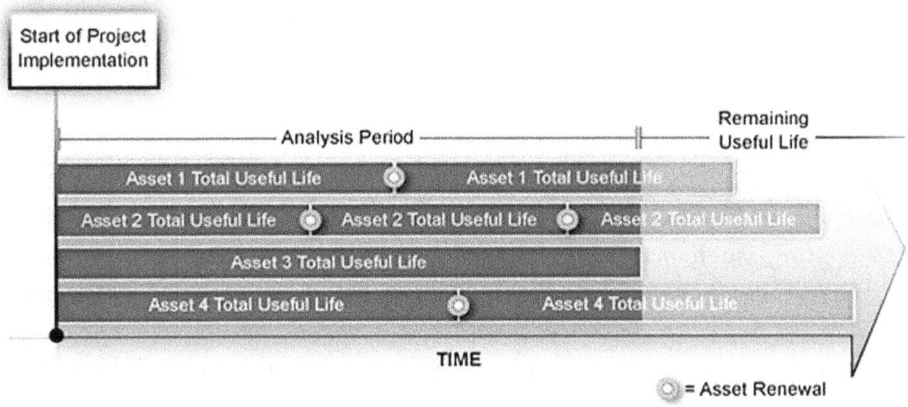

It must be noted that ongoing maintenance such as pipe linings to extend the life of old piping sections versus periodic investments such as new pipes are not the same financial undertaking and that both must be considered in business budgeting. A CMMS, when built and maintained with ongoing asset data, can support a critical assets inventory with data on consistent condition and performance, can inform planning to maintain, repair, and replace assets that are largely hidden underground, and can be foundational for funding essential business activities which support this vital community service. Moreover, the 2018 AWIA federal law specifically requires water utilities serving more than 3,300 people to assess the resilience of the pipes and constructed conveyances as well as the financial infrastructure of the system; thus, a CMMS at some level will be needed to collect and maintain the descriptive system data in the US.

Asset management is not just a critical infrastructure issue but can also be a public health issue with respect to water services. Maintaining an appropriate inventory and understanding of publicly owned assets will help locate solutions when a crisis occurs in the extended asset environment. For example, the 2014 crisis in Flint, Michigan was the result of a failure to apply corrosion inhibitors to the water after the water supply source was changed, thus causing lead from aging pipes to leach into the drinking water (Kennedy, 2016). A stronger asset management approach, along with better risk analysis, may have helped system managers more fully understand the true downstream consequences of changing water supply while not updating the water chemistry to maintain the pipe conditions needed to deliver safe drinking water. Because of either a willing or unwilling lack of asset condition knowledge, criminal charges were filed against current and former state employees who were seen as responsible for the Flint water crisis because of their allegedly misleading officials about Flint's water treatment plant, which was not in compliance with lead and copper rules (Kennedy, 2016). It is clear that meager asset management in water service can lead to overwhelmingly poor immediate and long-term business outcomes including erosion of public trust, costly settlements, and consumer health impacts. According to Robertson (2020):

"...after nearly $400 million in state and federal spending, Flint has secured a clean water source, distributed filters to all residents who want them, and laid modern, safe copper pipes to nearly every home in the city that needed them. Its water is as good as any city's in Michigan. And to compensate its just under 95,000 people for the damage they've suffered—economically, medically and psychologically—the city and state reached a settlement in August that will pay nearly $650 million to Flint residents."

Collaborative Technology Including GIS for Water System Surveillance

The use of collaborative software such as Geographical Information System (GIS) for integration, coordination, and analysis across various platforms and processes supports location-based data dashboards. As observed by Russell (2019), GIS is used by 64% of utilities to locate pipes and facilities, manage assets, and analyze system capabilities.

Figure 6. Collaborative Technology for Water Quality Surveillance and Response System.
Reprinted from Dashboard Design Guidance For Water Quality Surveillance and Response Systems by the United States Environmental Protection Agency. (2015, November). https://www.epa.gov/sites/default/files/2015-12/documents/srs_dashboard_guidance_112015.pdf

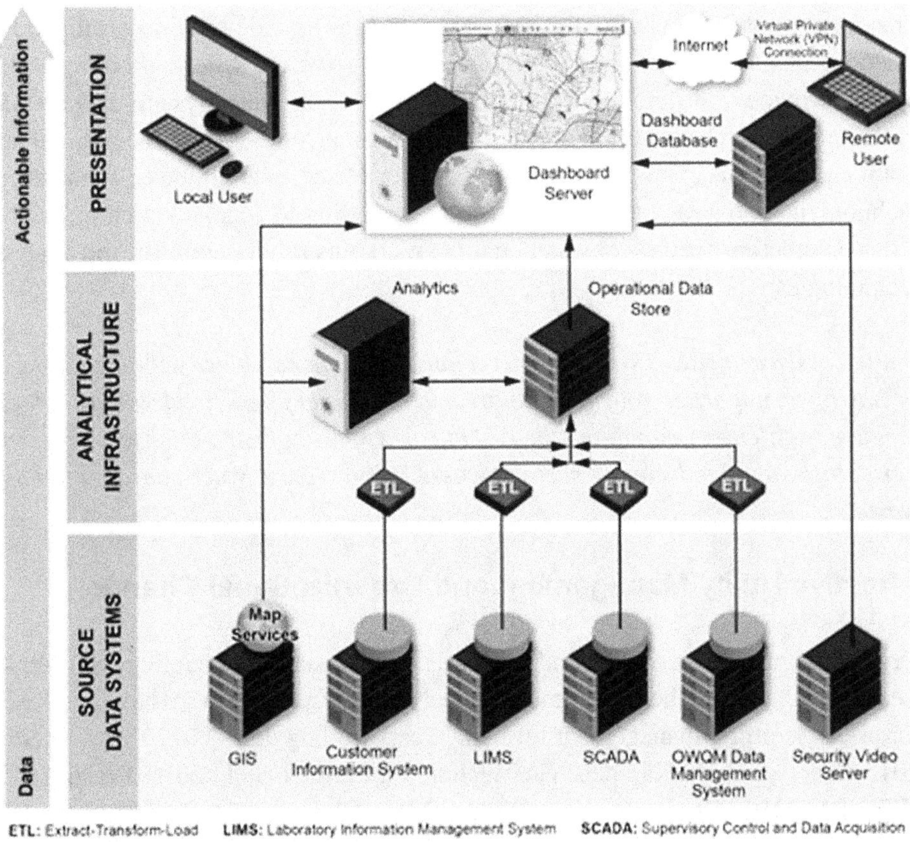

GIS technology offers positional clues that feed hydraulic modeling, pressure gradient insights, and locational information for tracking pressure complaints leading to a more cohesive understanding of system operation. As stated in the AWWA Computer modeling of water distribution systems manual (2011):

GIS stores and displays geographically referenced information, i.e., information that is easily understood through map display. Spatial relationships between entities are significant for most information stored in GIS. GIS has potential to store vast amounts of information useful for system analysis, including pipe assets, customer meter locations, zoning and land parcel data, aerial photography or other land bases, street locations, digital terrain models (DTMs), digital elevation models (DEMs), and jurisdictional boundaries.

GIS, CMMS, and SCADA have advanced to the point that the collaboration across data system platforms has led to many utilities having dashboards that present most of the data visually. Software packages provide file formats that other systems can read and feed into the analytical infrastructure. Information from each system provides a different but essential framing of the water system data and the collaborative dashboard often presents the clearest picture of system operation and maintenance. This cross-talk environment is expected to continue to enhance the technological prowess of the water sector. According to Manjakkal et. al (2021), when completing a review of water quality (WQ) sensors, deployment, and analysis technologies:

The use of sensor networks and Internet communications combined with GIS tools will be having an important role in the future and can be very beneficial to stakeholders in not only efficiently managing the WQ but also in water distribution management, agriculture, and landscaping sectors, where it can reduce water consumption and wastage.

Effective Utility Management and Organizational Change

Organizational change models and standards such as financial modeling, ISO 14001, Lean Six Sigma, and the Malcolm Baldrige National Quality Award have been used for effective utility management for many years leading up to 2019 (Mitre Media, 2015; Searcy, Morali, Karapetrovic, Wichuk, McCartney, McLeod, & Fraser, 2012; U.S. EPA, 2016). One water utility leadership model of significance is the Effective Utility Management (EUM) coalition which includes the U.S. EPA and a number of major water sector associations such as the American Water Works Association, National Association of Clean Water Agencies, Association of State Drinking Water Administrators and others since 2008 (EPA, 2021). The EUM coalition has a leadership group comprised of US utility leaders from various sized organizations as well as water professionals with engineering, operations, maintenance, administrative knowledge that fostered the creation of the EUM Primer. The Primer, first released in June 2008, is directly addressed to the "water leader" and offers those leaders guidance in the form of the "Ten Attributes of Effectively Managed Water Sector Utilities" (EUM, 2017).

The ten attributes are designed to support sustainable water infrastructure and operation. The attributes listed in the 2017 updated primer are not expressed as mere aspirational goals but instead are noted as a "set of opportunities for improving utility management and operations" (EUM, 2017). The ten attributes include: Product Quality, Customer Satisfaction, Stakeholder Understanding and Support, Financial Viability, Operational Optimization, Enterprise Resiliency, Employee and Leadership Development, Infrastructure Strategy and Performance, Community Sustainability, and Water Resource Sustainability. The EUM Primer offers instruction on actualization of the attributes at water utilities in the US to improve organization-wide performance and specifically speaks to the need for substantive leadership:

"Leadership" refers both to individuals who can be effective champions for improvement, and to teams that provide resilient, day-to-day management continuity and direction. Effective leadership establishes and communicates a long-term vision for the organization and embodies a commitment to cultivating the organization's culture, helping to ingrain methods to achieve the utility's vision into the organization's day-to-day operations.

Leadership, then, is a strong key attribute to resilience for water utilities. The water industry has to be ready to deliver their products come rain or shine, and extreme weather such as drought has to be an aspect of resilient planning. For example, one area of resilience is being ready to respond to large and extended weather events, including climate change. Climate readiness refers to adapting to and addressing climate change impacts on system infrastructure according to the *Drinking Water Infrastructure Needs Survey and Assessment, Fifth Report to Congress*, and US water utilities have increased efforts dedicated to anticipating and proactively addressing the potential effects of climate change at the water utility level (EPA, 2013). Thus, these utilities are demonstrating the key attribute of Enterprise Resiliency.

Furthermore, the EUM Primer offers insights for utilities that are faced with organizational change. The primer advises on this topic in the section called "Continual Improvement Management". There is a clear emphasis on self-assessment, measurement, and explicit change management process as the basis for sustainable organizational change. The EUM primer (2017) states:

Continual improvement management is usually implemented through a complete, start-to-finish management system, also referred to as a "Plan-Do-Check-Act" framework. Continual improvement plays a central role in effective utility management and is critical to making progress on the Attributes.

According to Cousar (2017), the implications of well-planned continuous improvement for the water sector are highlighted by one utility leader's direct observation:

>$5 Million hard-dollar savings, 34% reduction in total number of injuries, 8% improvement in top-box customer satisfaction. 6.5% reduction in operating expenses. Still working on improving employee engagement scores, reducing the customer outage hours, and reducing the number of lost-work-day injuries.

As can be seen in the preceding paragraphs, from the home meter to water sector leadership, utilities up to 2019 were operating and delivered service competently that was seldom disrupted on any large scale. The highest-profile water sector crisis in the decade up to 2020 was the asset management failure and subsequent public health crisis associated with lead piping in Flint, Michigan (Robertson, 2020). Addressing the underpinning issues that led to the Flint tragedy for water businesses and the communities they serve, specifically costly replacement of aging lead pipes, continues in 2022.

WATER SECTOR CONCERNS DURING COVID

Considering the time period of 2019 to 2022, the COVID-19 pandemic has emphasized the urgency of fully recognizing water access as a public health priority. In the United States, for example, "water poverty" is an increasing concern in the wake of the pandemic. Water poverty is the result of water customers' inability to pay for the business service which provides essential water for drinking and flushing. In response to pandemic-driven water poverty, several states have suspended water shutoffs, and more than 400 cities have placed moratoriums on water shutoffs (Renukappa et al. 2021). Recent water demand research throughout the pandemic identified a general relocation of water consumption, as one would expect, from public places to residential homes (Vidal, 2020). Vidal (2020) also noted intensive water use practices at the start of the pandemic, which decreased with time. Cook and Makin (2020) estimated that the increase in domestic water use was between 15-20%, while businesses saw a reduction between 30-50%. Utilities have had to adapt on many fronts to continue operations; for instance, in light of social-distancing measures necessitating limited possibilities for site visits and in-person meetings, implementation of phone calls/remote working, and collection of visual substitutions (via photo and video) has occurred. Many utilities are trialing new technologies and methods of communication, reconfiguring office species, and accommodating a hybrid or work-from-home (WFH) schedule for employees.

While COVID-19 has certainly caused unprecedented delays in project delivery, it has simultaneously accelerated a digital transformation in the US and abroad. For example, WatEner (2020) reported that a municipal water utility in Karlsruhe Germany "used a platform to support the operation and management of its drinking water distribution system. The platform operated a Demand Forecast System (DFS) that applies Artificial Intelligence and Pattern Recognition Techniques, adapting dynamically not only to meteorological parameters (such as temperature, rainfall, and humidity) but also to changes in consumer behavior" (Renukappa et al. 2021).

Moreover, technologies such as digital meter reading installation (smart meters) and improved digital channels for communication are on the rise. Likewise, the development of 'digital twins' (DTs) for utilities, a virtual replica of the physical assets or processes involved in urban water infrastructure, is also on the rise (Poch et al. 2020). DTs can be used to "remotely monitor actual processes and can apply AI techniques to conduct predictive maintenance of the assets...thus avoiding setbacks and production losses" (Poch et al. 2020). In the water sector, DTs are typically used to aid in the management of treatment processes and water networks and to anticipate (or simulate) failure events (Martínez et al. 2007; Poch et al. 2020).

Tools such as DTs help streamline utility operations while improving accuracy and operational resilience. The pandemic has revealed that the digitalization of utilities is a vital priority. Organizations are moving towards digital records, information, and tracking for both ease of communication during emergencies and greater data analytics capabilities. As expected, organizations that "invested in and implemented digital transformations experienced the benefits first-hand" when the pandemic struck; "[t]he current disruptions caused by the pandemic have single-handedly pushed digital technologies adoption and adaption to the top of the agenda" (Renukappa et al 2021). Moreover, as Renukappa states, "[t]he industry has gained understanding of interdependencies across the supply chain and the need to address digital adaptation and interoperability issues." Decentralized digital systems are more resilient, offer redundancy, and minimize personnel density– critical features in the case of epidemics/pandemics (Garrido-Baserba et al. 2018; Sun et al. 2020, Poch et al. 2020). Poch et al (2020) presents an interesting vision for utility digital revolutions, structured into three levels: (1) a technological level (adapting the tools needed to become 'Fit for Future'), (2) an economic level (contributing to the financial sustainability of utilities by supporting asset, resource, and energy management), and (3) a water-human-data level (to increase the confidence citizens have in public water services).

It is also interesting to consider the pandemic's impact on utility demand and growth projections. COVID-19 caused a significant shift in domestic vs. business demand. This shift may, in turn, impact the prioritization of projects in asset management programs (Renukappa et al 2021). Renukappa et al (2021) explored the impact of the pandemic on six water and sewage organizations throughout the UK. Through interview discussions, the authors found that the greater demand for domestic water caused a reassignment of resources to ensure water supply, which delayed the delivery of certain engineering and construction projects. This sentiment was echoed by several utilities throughout the UK.

COVID-19 induced delays on water sector construction projects has pushed the industry towards greater integration of Information and Communications Technology (ICT) at many phases of project delivery, from inception to cooperation (Renukappa et al. 2021). Many interviewees took advantage of cloud-based technologies (ie. Microsoft Suite, Google) for sharing project knowledge and managing information. These powerful collaboration platforms helped aid in the discussion of design and construction drawings, for instance. Moreover, tools such as Building Information Modeling (BIM) can be used to generate three-dimensional models of assets and construction plans; "isolated project teams could comment in real-time as design drawings were generated" and demoed (Renukappa et al. 2021). Many utilities shelved construction projects that were not started when the pandemic began, even though their detailed design was complete (Renukappa et al. 2021). Many construction phase risk-assessments had to be revisited to incorporate pandemic-related factors such as social distancing and additional personal protective equipment (such as face masks). Construction phasing was often reworked to limit the number of workers on-site at a given time; this impacted many project managers who had to accommodate this extra planning. As with many other industries, Renukappa et al (2021) found that interviewees also installed various handwashing and sanitation points throughout construction sites and offices and provided disposable gloves. Mandatory training sessions were also implemented by some utilities to "ensure awareness of... revised working practices" (Renukappa et al. 2021). There are many on-going conversations and lessons to be learned. Clients and contractors are revisiting contract agreements, and learning how to pivot in light of pandemic impacts on-site operations, cost, safety, program changes, etc. The authors claim that their interview study offers a tentative and exploratory first look at the pandemic's impact on water and wastewater utilities. While findings cannot necessarily be generalized across the board, it does offer some common themes for further investigation by utilities everywhere (Renukappa et al. 2021). At the conclusion of their study, Renukappa et al (2021) recommend that utilities follow "the sound practices set out for project delivery teams by the Institution of Civil Engineers and the Construction Leadership Council– ... stipulat[ing] a Roadmap to Recovery by Restarting, Resetting and Reinventing."

Lawson et al. (2021) conducted a similar interview-style study of executives from 10 UK water companies to investigate the organizational response to the pandemic and how this response impacted operational delivery. Similar themes emerged as with the prior studies discussed here, namely: observed changes in demand/customer behavior, and necessary changes to operational practices. With regards to lessons learned, respondents echoed three main lessons: 1) the failure to adequately consider and prepare for the scale of the threat, 2) the success of sector-level collaboration, and 3) a need to embrace new modes of working (Lawson et al. 2021). Lawson et al. (2021) noted that many organizations began cross-training employees to help address mass absenteeism as a result of COVID-19; another strategy was the recruitment of university students and contacting recently retired staff to assess their willingness to return to work if needed. Weeks prior to lockdowns in the UK, organizations also implemented operational interventions such as isolating operation teams with specific specialist skills and minimizing contact between different teams to reduce the possibility for transmission. Some organizations even ran mass WFH pilot tests to assess the capacity of IT networks (Lawson et al. 2021). Many respondents echoed the sentiment that adaptation measures were more reactive than proactive (Lawson et al. 2021).

The COVID-19 pandemic offered a lesson in the complexity and interdependencies of water systems, resulting in some organizations reevaluating other existing business continuity and response plans (Lawson et al. 2021). The Water Research Foundation (WRF) hosted a virtual research summit aimed at establishing best practices for environmental surveillance of SARS-CoV-2 on April 27, 2020. An important focus area was the sensitivity of analytical results as an indicator of changes in community prevalence of COVID-19 (WRF, 2020). The Water Environment Federation (WEF) quickly developed materials providing general information about SARS-CoV-2 and outlining guidance for the wastewater industry and continue to update online data the applicability of which translated across drinking water and wastewater reuse (WEF, 2021). One example of a WEF tool is *Key Safety Information for COVID-19 and Other Biohazards Fact Sheet for Wastewater Workers* (WEF, February 9, 2021). Compiling this data from around the world will help better understand the magnitude and spread of the virus, while also helping understand the efficacy of current treatment systems.

WATER SECTOR CONCERNS AFTER COVID

For the water sector as of 2022, a pressing real issue is: COVID may be here to stay, so now what? Water utilities must have adaptive business models that support sustainable outcomes considering unexpected challenges, including global pandemics. Water is necessary for life and communities rely on this shared infrastructure's continued existence before, during, and after a crisis. Where/what needs the most attention today including and outside of COVID/SARS-CoV-2 response? According to the Centers for Disease Control and Prevention (CDC), the virus that causes COVID-19 has been found in untreated wastewater; however, there is no current evidence that any disease has resulted from exposure while working at a wastewater facility, but wastewater practices and data will require ongoing review (CDC, 2020). Nonetheless, living with potentially infectious diseases necessitates ongoing technological investigation and expanding understanding. Exploiting data analytics for tracking COVID-19 in wastewater as an indicator of spread and potential forecast mapping is developing; quantitative data analysis and measurements in untreated sewage can provide information on changes in total infection in the community contributing to that wastewater treatment plant per the CDC (Larsen & Wigginton, 2020). Changes in water sector organizations and communities caused by COVID/SARS-CoV-2 that will likely stay in play include new data collection requirements: The CDC and the US Department of Health and Human Services (HHS), in collaboration with agencies throughout the federal government, are initiating the National Wastewater Surveillance System (NWSS) in response to the COVID-19 pandemic (CDC, 2021).

Pecson et al. (2020) offers a perspective on what is known, what is being learned, and how it will help to answer the central question of whether existing systems adequately protect public health considering emerging pathogens and concerns. Pecson et al. (2020) note that the water industry should be commended for their effective mobilization of resources to better understand SARS-CoV-2. For example, many countries (including the Netherlands, Frant, Australia, and the United States) implemented targeted SARS-CoV-2 monitoring campaigns to address critical knowledge gaps as early as February/March 2020, analyzing municipal wastewater to track the prevalence of COVID infections. Nevertheless, the collection and interpretation of data from wastewater is an emerging field requiring further development and refinement of analytical methods that work towards cross-laboratory comparability and quality assurance.

According to Ji et al. (2021), wastewater-based epidemiology (WBE) which investigates the source and survival characteristics of coronavirus in the aquatic environment, has gained global attention, along with studies of virus detection methodologies, water hygiene, and the impact of the pandemic on the water ecosystem. Collaborations and funding for such endeavors have materialized rapidly– with supporters in academics, local governments, utilities, and state and national research groups (Pecson et al 2020). One of the greatest concerns surrounding emerging wastewater-derived contaminants is their potential impact on potable reuse applications. Potable reuse, which allows the use of water more than once to expand a community's available water supply, is the system most likely to be impacted by any emerging contaminant (Pecson et al 2020). In order to evaluate whether current indirect potable reuse (IPR) log reduction value requirements need to be updated in response to SARS-CoV-2 (or any other emerging microbial contaminant), it's imperative to leverage technology to consider: (a) the concentration range of infectious pathogens in the source water (b) the effectiveness of existing treatment barriers (both natural and engineered) in removing– or inactivating– the pathogen and (c) the risk of infection from consumption of treated water. Pecson et al. (2020) refers to this process as Quantitative Microbial Risk Assessment (QRMA). This process has been used through the decades to inform regulatory frameworks including "U.S. EPA's Surface Water Treatment Rules, WHO's potable reuse guidelines, and California's IPR regulations" (Pecson et al 2020). Comparing SARS-CoV-2 outcomes to historical QRMA processes and outcomes can help determine whether there are unique characteristics about the virus that might necessitate updating existing regulations.

Pecson et al. (2020) notes that while a growing body of literature demonstrates that the genetic material of SARS-CoV-2 can be detected in raw wastewater, an important limitation of molecular methods used for detection testing is their inability to differentiate between infectious and inactivated viruses. Several studies have found a rapid inactivation of the virus in the fluids of the lower GI tract, suggesting that the proportion of infectious viruses in raw wastewater may be limited. Moreover, one of the defining features of SARS-CoV-2 (consistent with other respiratory viruses) is the lipid envelope that surrounds its protein capsid and genetic material; this lipid envelope is fragile and therefore coronaviruses are less persistent in the environment and more sensitive to treatment by chemical disinfectants (as compared to the non-enveloped enteric viruses often targeted in potable reuse) (Pecson et al 2020). Several studies suggest that SARS-CoV-2 is sensitive to UV disinfection as well– with inactivation doses orders of magnitude lower than what is currently used in UV-based advanced oxidation processes in California's potable reuse treatment trains (Pecson et al 2020).

Physical removal of SARS-CoV-2 (via activated sludge and secondary clarification, advanced treatment with membranes) is expected to be higher as compared to other enteric viruses due, in part, to size; SARS-CoV-2 has a diameter of about 100 nm while other enteric viruses have diameters of about ~30 nm (Pecson et al 2020, Poch et al. 2020). Note that in terms of conventional water treatment, many in academia and other water professionals agree that conventional treatment methods using filtration and disinfection processes can inactivate the virus (Poch et al. 2020). With regards to the risk of infection from consumption of treated water, however, SARS-CoV-2 may still pose a challenge as it has shown to be significantly more effective than viruses typically targeted by current regulations. Further study is needed to ascertain whether SARS- CoV-2 can be transmitted through the ingestion of drinking water and to what extent (Ji et al. 2021). Pecson et al (2020) concludes by stating the assumption that SARS- CoV-2 is "unlikely to drive new, stricter requirements for the treatment of potable water supplies," but that further study of the virus is warranted to confirm. The State Water Resources Control Board in California has allocated resources to better understand the impact of SARS- CoV-2 on treatment requirements in direct potable reuse applications; data on nucleic acid concentrations in raw wastewaters has been collected between 2019-2021 and will be used to determine regulatory directions Pecson et al (2020).

Up Next: Future Forward for the Water Sector

With so many challenges while working to provide uninterruptable service, what is signaling on the road ahead for the water sector as of 2022? One imperative signal for the water sector is the need to respond to an aging workforce. Per the American Society for Civil Engineers (ASCE) 2021 US infrastructure study: "[b]etween 2016 and 2026, an estimated 10.6% of water sector workers will retire or transfer each year, with some utilities expecting as much as half of their staff to retire in the next five to 10 years." Likewise, according to the ASCE's same infrastructure report, "there is a water main break every two minutes, and an estimated 6 billion gallons of treated water lost each day in the U.S. - enough to fill over 9,000 swimming pools". Sustainability of water infrastructure requires understanding technology's impact on staff, operations, leadership shifts, and considering how organizations may need to change to address infrastructure needs.

Updated regulatory requirements may be essential to push some infrastructure renewal since it will likely be costly. In 2021, EPA began the implementation of new lead and copper rules for drinking water called the Lead and Copper Rule Revisions (LCRR) which focus on replacing all lead service lines and compliance tap sampling amongst other requirements (EPA, 2021). Seachange regulatory requirements such as LCRR can spur significant new initiatives in AI including new technology for improved detection of lead service lines and software to support improved asset inventory with replacement schedules. Leaders will likely find the increased need to be agile in unpredictable landscapes as climate change becomes more pressing in different microclimates or new pandemic episodes occur. They may need to use more data analysis for strategic decision-making, whether that includes invoice process improvements to better support rapid changes in customers' economic situations, or water treatment plant design based on altered flow data. The power of forecasting, tracking spending, identifying trends (i.e., disinfection byproducts (DBP) concentrations in a system being seasonal and how that may impact water plant treatment response) has become vital to maintaining public trust. Mentorship of staff will need to move to the front burner to preserve data analysis capabilities and to encourage new skill sets including new remote work options and virtual remote-control technologies. Clearly, the future of the water sector offers vast technological and operational challenges as well as enormous opportunities.

REFERENCES

American Society of Civil Engineering. (2017). *Infrastructure Report Card: Drinking Water*. Retrieved from https://www.infrastructurereportcard.org/wp-content/uploads/2019/02/Full-2017-Report-Card-FINAL.pdf

American Society of Civil Engineering. (2021). *Infrastructure Report Card: Drinking Water*. Retrieved from https://infrastructurereportcard.org/cat-item/drinking-water/

American Water Works Association. (2011). *Computer modeling of water distribution systems*. American Water Works Association.

American Water Works Association. (2019). *State of the Water Industry*. Retrieved from https://www.awwa.org/Portals/0/AWWA/ETS/Resources/2019_STATE%20OF%20THE%20WATER%20INDUSTRY_post.pdf

Burke, W. W., & Litwin, G. H. (1992). A causal model of organizational performance and change. *Journal of Management*, *18*(3), 523–545. doi:10.1177/014920639201800306

Center for Disease Control and Prevention. (2020). *Water and COVID-19 FAQs*. Retrieved from https://www.cdc.gov/coronavirus/2019-ncov/php/water.html

Center for Disease Control and Prevention. (2021). *National Wastewater Surveillance System (NWSS): A new public health tool to understand COVID-19 spread in a community*. Retrieved from https://www.cdc.gov/healthywater/surveillance/wastewater-surveillance/wastewater-surveillance.html

Cherdantseva, Y., Burnap, P., Blyth, A., Eden, P., Jones, K., Soulsby, H., & Stoddart, K. (2016). A review of cyber security risk assessment methods for SCADA systems. *Computers & Security*, *56*, 1–27. doi:10.1016/j.cose.2015.09.009

Cotterill, S., Bunney, S., Lawson, E., Chisholm, A., Farmani, R., & Melville-Shreeve, P. (2020). COVID-19 and the water sector: Understanding impact, preparedness and resilience in the UK through a sector-wide survey. *Water and Environment Journal: the Journal / the Chartered Institution of Water and Environmental Management*, *34*(4), 715–728. doi:10.1111/wej.12649 PMID:33230399

Cousar, R.D. (2017). *Counting on change: A quantitative evaluation of the relationship of continuous improvement to organizational performance outcomes at United States public sector clean water utilities*. Academic Press.

EUM. (2017). *Effective Utility Management: A Primer for Water and Wastewater Utilities*. Retrieved from https://www.epa.gov/sites/default/files/2017-01/documents/eum_primer_final_508-january2017.pdf

Heflin, C., Jensen, J., & Miller, K. (2014). Understanding the economic impacts of disruptions in water service. *Evaluation and Program Planning*, *46*, 80–86. doi:10.1016/j.evalprogplan.2014.05.003 PMID:24950018

Ivleva, N. P. (2021). Chemical Analysis of Microplastics and Nanoplastics: Challenges, Advanced Methods, and Perspectives. *Chemical Reviews*, *121*(19), 11886–11936. doi:10.1021/acs.chemrev.1c00178 PMID:34436873

Ji, B., Zhao, Y., Wei, T., & Kang, P. (2021). Water science under the global epidemic of COVID-19: Bibliometric tracking on COVID-19 publication and further research needs. *Journal of Environmental Chemical Engineering*, *9*(4), 105357. doi:10.1016/j.jece.2021.105357 PMID:33747765

Kennedy, M. (2016, April 20). *Lead-Laced Water In Flint: A Step-By-Step Look At The Makings Of A Crisis*. NPR. Retrieved from https://www.npr.org/sections/thetwo-way/2016/04/20/465545378/lead-laced-water-in-flint-a-step-by-step-look-at-the-makings-of-a-crisis

Larsen, D. A., & Wigginton, K. R. (2020). Tracking COVID-19 with wastewater. *Nature Biotechnology*, *38*(10), 1151–1153. doi:10.103841587-020-0690-1 PMID:32958959

Lawson, E., Bunney, S., Cotterill, S., Farmani, R., Melville-Shreeve, P., & Butler, D. (2021). COVID-19 and the UK water sector: Exploring organisational responses through a resilience framework. *Water and Environment Journal: the Journal / the Chartered Institution of Water and Environmental Management.*

Manjakkal, L., Mitra, S., Petillot, Y., Shutler, J., Scott, E., Willander, M., & Dahiya, R. (2021). Connected Sensors, Innovative Sensor Deployment,and Intelligent Data Analysis for Online Water Quality Monitoring. *IEEE Internet of Things Journal*, *8*(18), 13802–13824. doi:10.1109/JIOT.2021.3081772

McKeever, A. (2021, September 21). COVID-19 surpasses 1918 flu as deadliest pandemic in U.S. history. *National Geographic*. Retrieved from https://www.nationalgeographic.com/history/article/covid-19-is-now-the-deadliest-pandemic-in-us-history

Mitre Media. (2015, September 11). Moody's Research Reports. *By Issuer* [data set]. Retrieved from http://www.municipalbonds.com/bonds/moodys_reports/

Mofijur, M., Fattah, I. R., Alam, M. A., Islam, A. S., Ong, H. C., Rahman, S. A., ... Mahlia, T. M. I. (2021). Impact of COVID-19 on the social, economic, environmental and energy domains: Lessons learnt from a global pandemic. *Sustainable Production and Consumption, 26*, 343-359.

Mohapatra, H., & Rath, A. (2020). IoT-based Smart Water. In IoT Technologies in Smart-Cities: From sensors to big data, security and trust (pp. 63-93). The Institution of Engineering and Technology. doi:10.1049/PBCE128E_ch3

Monks, I., Stewart, R. A., Sahin, O., & Keller, R. (2019). Revealing unreported benefits of digital water metering: Literature review and expert opinions. *Water (Basel)*, *11*(4), 838. doi:10.3390/w11040838

Mukarram, M. (2020). Impact of COVID-19 on the UN sustainable development goals (SDGs). *Strategic Analysis*, *44*(3), 253–258. doi:10.1080/09700161.2020.1788363

National Infrastructure Advisory Council. (2016). *Water Sector Resilience Final Report and Recommendations*. Retrieved from https://www.cisa.gov/sites/default/files/publications/niac-water-resilience-final-report-508.pdf

Pawson, R., & Tilley, N. (1997). *Realistic Evaluation*. Sage.

Pecson, B., Gerrity, D., Bibby, K., Drewes, J. E., Gerba, C., Gersberg, R., Gonzalez, R., Haas, C. N., Hamilton, K. A., Nelson, K. L., Olivieri, A., Rock, C., Rose, J., & Sobsey, M. (2020). Editorial Perspectives: Will SARS-CoV-2 reset public health requirements in the water industry? Integrating lessons of the past and emerging research. *Environmental Science. Water Research & Technology*, *6*(7), 1761–1764. doi:10.1039/D0EW90031A

Poch, M., Garrido-Baserba, M., Corominas, L., Perelló-Moragues, A., Monclús, H., Cermerón-Romero, M., Melitas, N., Jiang, S. C., & Rosso, D. (2020). When the fourth water and digital revolution encountered COVID-19. *The Science of the Total Environment*, *744*, 140980. doi:10.1016/j.scitotenv.2020.140980 PMID:32687996

Poister, T. H. (2010). The future of strategic planning in the public sector: Linking strategic management and performance. *Public Administration Review*, *70*, 246–254. doi:10.1111/j.1540-6210.2010.02284.x

Raftelis. (2020). *Financial Impact of the COVID-19 Crisis on U.S. Drinking Water Utilities*. Retrieved from https://www.awwa.org/Portals/0/AWWA/Communications/AWWA-AMWA-COVID-Report_2020-04.pdf

Renukappa, S., Kamunda, A., & Suresh, S. (2021). Impact of COVID-19 on water sector projects and practices. *Utilities Policy*, *70*, 101194. doi:10.1016/j.jup.2021.101194 PMID:34873382

Robertson, D. (2020, December 23). Flint Has Clean Water Now. Why Won't People Drink It? *POLITICO*. Retrieved from https://www.politico.com/news/magazine/2020/12/23/flint-water-crisis-2020-post-coronavirus-america-445459

Russell, P. (2019, September 2). *The Internet of Water*. ENR. Retrieved from https://www.enr.com/articles/47424-data-drives-smarter-water-systems-decisions

Scott, H. (2003, July 1). *Water AMR Market Continues Strong Growth*. WaterWorld. Retrieved from https://www.waterworld.com/home/article/16190548/water-amr-market-continues-strong-growth

Searcy, C., Morali, O., Karapetrovic, S., Wichuk, K., McCartney, D., McLeod, S., & Fraser, D. (2012). Challenges in implementing a functional ISO 14001 Environmental Management System. *International Journal of Quality & Reliability Management, 29*(7), 779–796. doi:10.1108/02656711211258526

Sowby, R. B. (2020). Emergency preparedness after COVID-19: A review of policy statements in the US water sector. *Utilities Policy, 64*, 101058. doi:10.1016/j.jup.2020.101058 PMID:32406402

Talari, S., Shafie-khah, M., Siano, P., Loia, V., Tommasetti, A., & Catalão, J. (2017). A Review of Smart Cities Based on the Internet of Things Concept. *Energies, 10*(421), 421. Advance online publication. doi:10.3390/en10040421

Temido, J., Sousa, J., & Malheiro, R. (2014). SCADA and smart metering systems in water companies. A perspective based on the value creation analysis. 12th International Conference on Computing and Control for the Water Industry, CCWI2013. *Procedia Engineering,* (70), 1629–1638. 10.1016/j.proeng.2014.02.180

U.S. EPA. (2004, September). *Primer for Municipal Wastewater Treatment Systems*. Retrieved from https://www3.epa.gov/npdes/pubs/primer.pdf

U.S. EPA. (2008). *Ensuring a Sustainable Future: An Energy Management Guidebook for Wastewater and Water Utilities*. Retrieved from https://nepis.epa.gov/Exe/ZyPDF.cgi/P1003Y1G.PDF?Dockey=P1003Y1G.PDF

U.S. EPA. (2013). *Drinking water infrastructure needs survey and assessment, fifth report to Congress* (EPA 816-R-13-006). Retrieved from https://water.epa.gov/grants_funding/dwsrf/upload/epa816r13006.pdf

U.S. EPA. (2015, June). *Framework for Comparing Alternatives For Water Quality Surveillance and Response Systems*. Retrieved from https://www.epa.gov/sites/default/files/2015-07/documents/framework_for_comparing_alternatives_for_water_quality_surveillance_and_response_systems.pdf

U.S. EPA. (2015, November). *Dashboard Design Guidance For Water Quality Surveillance and Response Systems*. Retrieved from https://www.epa.gov/sites/default/files/2015-12/documents/srs_dashboard_guidance_112015.pdf

U.S. EPA. (2016, July). *Guidance for Designing Communications Systems For Water Quality Surveillance and Response Systems*. Retrieved from https://www.epa.gov/sites/default/files/2016-09/documents/srs_communications_guidance_081016.pd

U.S. EPA. (2019, March). *Addressing Lead in Drinking Water with the Drinking Water State Revolving Fund.* Retrieved from https://nepis.epa.gov/Exe/ZyPDF.cgi/P100WD7W.PDF?Dockey=P100WD7W.PDF

U.S. EPA. (2021). *Proposed Revisions to the Lead and Copper Rule*. Retrieved from https://www.epa.gov/ground-water-and-drinking-water/proposed-revisions-lead-and-copper-rule

Usman, M., Farooq, M., Jellali, S., Charabi, Y., Al-Sadi, A. M., & Al-Badi, A. (2021). COVID-19: The Challenges and Opportunities for Water, Air, Agriculture and Energy Sectors. *International Journal of Agriculture and Biology, 25*(5), 1085–1095. doi:10.17957/IJAB/15.1767

Water Environment Federation. (2021). *Key Safety Information for Covid-19 and Other Biohazards*. Retrieved from https://www.wef.org/globalassets/assets-wef/news-hub/coronavirus/key-safety-information-for-covid-19-and-other-biohazards.pdf

Water Research Foundation. (2020). *Virtual International Water Research Summit on COVID-19*. Retrieved from https://www.waterrf.org/event/virtual-international-water-research-summit-covid-19

Chapter 3
Ethical Staffing in the COVID-19 Digital Age:
Are a New Set of Ethical Practices Needed to Guide Selection?

Kevin Jones
Indiana University-Purdue University Columbus, USA

Nazish Saba
Indiana University-Purdue University Columbus, USA

ABSTRACT

This chapter discusses ethical hiring practices in the digital age and argues for the necessity of ethical hiring processes. An examination of hiring before and during the COVID-19 pandemic describes the increased use of digital technologies to aid the ethical hiring process. A particular focus is on the increased use of artificial intelligence (AI) tools to aid in screening and selection. Unethical hiring practices are examined showing the potential adverse outcomes which may result. The chapter concludes with model and a set of guidelines to illustrate the components that enable ethical hiring processes and incorporate the use of AI.

INTRODUCTION

The Challenges of Hiring

Hiring employees is an essential component of a successful organization. Finding the best talent and gaining optimum performance from hires are the basic goals

DOI: 10.4018/978-1-6684-4358-3.ch003

of all employers. When choosing the best potential performer, it is important that organizations consider all available information about various candidates and objectively decide on the best potential performer for the organization. Ideally, hiring choices are the best match for the organization and the applicant.

The COVID-19 pandemic significantly disrupted the employment process and added complexity to the already challenging hiring process. Hiring processes involve handling large amounts of data, including sorting through applications, cover letters, resumes, and other information. This time-consuming activity is conducted by busy professionals who must allocate significant time to review data and interview candidates.

The rise of the COVID-19 pandemic had many consequences for organizations. Companies and the workforce faced many challenges at once. Work was restricted from office settings to hybrid and remote work. There was rising concern regarding employee safety due to COVID-19, and that fear caused an increase in demand to work remotely. During the peak of the COVID-19 pandemic, many businesses were temporarily or permanently shut down, whereas others continued operation despite the challenges (Institute for Business Value, 2020). Depending on the impact of the pandemic and the economic implications, some organizations have temporarily froze hiring, while others have increased hiring. Companies are adapting fast to new structures to maintain operations and retain existing talent (Pratt & Florentine, 2021).

The emergence of the pandemic accelerated the use of technological tools to enhance the hiring process. In early 2021, the Society for Human Resource Management (Maurer, 2021) predicted that organizations would increase the use of "virtual" technologies. Many organizations in various industries have incorporated digital technology to meet hiring needs. The pandemic hiring challenges have led more employers to choose artificial intelligence (AI) tools (Loten, 2022). This increased use of AI in hiring continues despite potential ethical concerns associated with AI-aided hiring (Dattner et al., 2019; Hauer et al., 2021). The reality is that organizations are transforming their hiring practices. In the post pandemic environment, the hiring practices of many organizations will increasingly rely on AI tools to support the efficiency and effectiveness of hiring, despite ethical concerns and potential preferences of applicants. Given the presence and expected increase in digital technology to enable the hiring process, the question is: Are a new set of hiring ethics needed?

HIRING AND ETHICS

Hiring is a continuous process, as organizations are always looking for top talent. In the United States, various federal and state laws create expectations for objectivity, fairness, nondiscrimination, equal opportunity, and privacy in the hiring process. To align with these legal aspects, many organizations have policies that give expectations for an equitable and objective hiring process.

Ethical hiring can be described as ensuring that applicants are objectively screened, and that candidate consideration is based on the qualifications specified in the position description. Additionally, ethical hiring involves selecting the best candidate based on the objective data recruiters collect during screening, interviewing, and selection decision-making. Hiring processes should be free of prejudice, discrimination, and unfounded biases.

The hiring process inherently has an unavoidable bias in the preferences of hiring managers and/or human resource (HR) professionals present when making hiring decisions. The goal is not to eliminate bias (this is not fully attainable), but to understand biases that supersede the collected data which should drive the employee selection. Ethical hiring involves understanding the components of the hiring process that can result in inequitable, discriminatory, or ineffective hiring, and subsequently designing processes to mitigate adverse potentials in the hiring process. Ethical hiring processes enable those in the hiring process to mitigate the adverse impacts of biased beliefs and perceptions that can hinder the ability to select the best potential performer.

ETHICAL HIRING CONCERNS

Simply stated, unethical hiring reduces the importance of qualifications, experience, and/or preparation and focuses primarily on the hiring managers personal preferences. When thinking about the ethicality of hiring, most applicants generally trust the hiring process sufficiently to believe that their qualifications will be considered equitably and fairly. In other words, applicants assume that those reviewing their application are following a set of hiring ethics. Unfortunately, in some cases, hiring managers do not follow ethical hiring practices. Unethical practices can lead to ineffective hires and can create potential for litigation when poorly executed. Concerns with hiring can be found in the core areas of hiring including job posting, applicant review and screening, and selection interviewing.

Job Posting: Posting a job is a visible step associated with hiring. It is the essential trigger for applications for the open positions. The simple goal for a job posting is to enable the most qualified applicants to submit their credentials for review. Describing position requirements and qualifications are essential components of job posting. Hiring managers are often excited when a posting is offered to fill a new or existing position. While position posting is seemingly a simple and straightforward process, unfortunately there are several areas which require training, experience, or guidance. Some areas of ethical concern with posting are:

- Does the posting allow for the most qualified applicants to apply?
- Is the posting narrowly written to target specific individuals?
- Is the position written to exclude various competencies and capabilities?
- Is the position posted in a timeframe that only allows a limited number of qualified applicants to apply?
- Does the position language discourage applicants from diverse groups of people (this may also be illegal)?

When seeking applicants, it seems logical that making a posting accessible to as many qualified people as possible is in the best interest of a hiring manager. Unfortunately, where and how postings are made available may limit the creation of the broadest qualified applicant pool. While the digital age has increased our ability to connect with a wide range of applicants through online application processes, are we still limiting applicant access due to:

- Software requirements which limit the ability of some applicants to submit their information.
- A set of non-intuitive application steps which can create barriers for qualified candidates to submit their applications.
- Accessibility barriers that limit the opportunity of those with disabilities to apply.

Postings can become unethical when they limit access to the posting, are written to exclude rather than to include, or do not share essential information for selection.

Applicant Review and Screening: One of the most important components of hiring is maintaining confidentiality. When applicant's information reviewed and screened, their information should be limited to a small group of hiring participants. Further, the evaluation of a person's qualifications during a screen must be based upon the stated criterion in either the posting and/or job description. Sharing information to non-essential people in the hiring process compromises the integrity of the search and can adversely affect an applicant. Some areas of applicant review and screening are clearly unethical:

- Discriminatory practices including screening out applicants on the basis of age, race, ethnicity gender, national origin, disability, religion, gender identity, etc. (Note, these practices may lead to litigation).

- Eliminating applicants based on non-performance or non-job critical issues (e.g., A person's name).
- Screening out candidates based primarily on personal opinions that are not supported by the posting/job description.

The goal should be to select the best candidates based on the expressed position requirements and qualifications. Selections based on deviations from the position information communicated to applicants can be deemed as unethical hiring.

Selection Interviewing: Interviewing candidates is a necessary component of the hiring process. At the same time, many potential unethical practices can emerge during selection interviewing. Moving from an applicant to a candidate typically requires an interview. An in-person or increasingly live online conversation enables the candidate to present their qualifications to the hiring participants. Selection interviewing helps the hiring participants to choose among qualified candidates. There are multiple ways to conduct selection interviewing ranging from behavioral based selection to competency focused selection. Subsequent to interviewing, decisions are made about the candidate's potential performance as an employee. Unfortunately, the selection interview is subject to bias and misperception as humans are evaluating other human's performance. Arguably, it is impossible to eliminate all bias from selection interviewing, but the goal should be to create the most objective process possible reducing the overt bias on unsubstantiated opinion. While there are a multitude of potential unethical behaviors in interviewing some overtly unethical issues found in selection interviewing are:

- Prejudgment of candidates before the opportunity to interview.
- Going off-script or asking irrelevant questions.
- Favoring candidates without supportive qualification information.
- Differential treatment of preferred candidates compared to less preferred candidates.
- Showing interest in some candidates and less interest in other candidates through demeanor and attention.
- Focusing on an undefined "fit" with the organization that can mask personal preferences for specific individuals regardless of qualification.

With the importance placed on selection interviewing, the act of interviewing should receive ethical scrutiny to ensure the best possible ethical outcomes for the participants in the process.

UNETHICAL HIRING CONSEQUENCES AND COSTS

When a job posting goes online, recruiters are flooded with applications. This situation may have gotten worse with the pandemic since companies are not limited to hiring geographically. Shortlisting candidates by reviewing CVs and interviewing them is challenging and time-consuming for recruiters. In these cases, potential employees' profiles may get overlooked. Another drawback is that it also takes longer than expected time for recruiters to get back to candidates with the updates. This gives candidates the feeling of being ghosted. An organization in this situation is not only hurting an applicant, but also hurting its image (Green, 2022). With the power of social media, candidates can be vocal in sharing their unpleasant experiences with an organization's hiring process.

All companies, small or big, private or government, and public or nonprofit, have default standards for ethical staffing that comply with federal and state law. However, despite the efforts made to keep personal biases aside, an ethical dilemma sometimes comes in the way of fair hiring (Corporate Finance Institute, n.d.). In the United States, the U.S. Equal Employment Opportunity Commission (EEOC) (2019) "prohibits discrimination of any type in employment relationships" (p.1). With laws and regulations in place, this is by far the most common and continually reported legal issue related to recruitment. Deliberate acts of discrimination or harassment, whether carried out by an organization or a staff member, are easy to notice. According to EEOC, "for example, a help-wanted ad that seeks 'females' or 'recent college graduates'" is an open refusal to hire men or people over 40. Often, overt intentions are not there, and discrimination occurs due to personal bias; for example, asking an applicant their gender or marital status, or asking a female applicant if she is pregnant.

The worst scenarios are when victims are unaware of being discriminated against. One of the EEOC settled cases is Dollar General, where this retail chain discriminated against an African American by offering lower wages than White applicants, based on criminal background. Upon investigation, it was discovered that Dollar General not only performs a broad criminal background, but also discourages convicts from applying. The result was that Dollar General would pay $6 million to settle a class race discrimination. These funds would be distributed to all those African Americans who had been discriminated against between 2004 and 2019 (EEOC, 2019).

Unethical practices of selection may lead to legal, financial, and reputation costs. When applicants do not perceive hiring as ethical, a variety of potential adverse consequences arise for organizations. Despite expressed legal requirements and various organizational policies, participants in a hiring process may engage in unethical hiring practices. For applicants, unethical hiring practices may lead to unfair judgment of their capabilities and loss of opportunity. Given the potential for adverse consequences, it is crucial to plan specific actions to reduce the possibility of unethical hiring practices.

COVID-19 AND DIGITAL TECHNOLOGY INFLUENCE ON HIRING

Traditional ways of interviewing candidates by inviting them to a facility are being replaced by software such as Zoom and Skype. The onboarding experience is increasingly provided remotely than in-person. The COVID-19 pandemic has introduced tele-hiring and teleworking and made a single job posting interesting to many more people as employers. Prospective employees are not limited by geography. There are now more job applications for a single open position than ever before.

In the COVID-19 era, the fast-moving change environment also pushed organizations to look at AI as the only socially acceptable tool for hiring. AI technology in the recruitment industry is evolving, and organizations are also moving toward it slowly. Following a pandemic reset, many companies need an increased workforce, which has created more interest in using AI for hiring.

The pandemic has brought certain challenges to the hiring industry (Arbix & Comin, 2020). The influence of the pandemic creates the need for a reset, where some companies are working toward maintaining their operations, while others are trying to resume the crisis.

Many organizations conclude that AI in the hiring process can increase efficiency. According to the Forbes Expert Panel and Forbes Coaches Council (2021), AI can also "screen for suitable talents" and "remove unconscious bias" (p. 1). With the increased use of digital technology in the hiring process during the COVID-19 pandemic, AI selection tools have increased the capability to aid in the hiring process.

ARTIFICIAL INTELLIGENCE TECHNOLOGY AND HIRING

In order to meet tasks on time, the overburdened recruitment staff needs AI in the organizations. The HR staff and AI have to work together to achieve effective results. AI tools are not there to replace humans, but to work with overburdened staff by using their expertise to optimize results. AI improves the process of pushing ads in front of targeted potential employees. Also, AI adds efficiencies in identifying qualified candidates from applicants and from screening and selection of candidates (based on interviews and data) to sending an offer and providing an onboarding experience to the new hire. In short, digital technology, especially AI, can help save time and valuable resources.

Surrounded by various technologies, the recruitment industry is also slowly adapting innovative approaches to cover shortfalls of the traditional hiring process. For example, with AI screening, the hiring managers handle large quantities of applications more efficiently and promptly. Several companies are offering AI-driven technologies to make the recruitment process quick, precise, and unbiased. The next-generation automation tools are available to meet firms' different needs for different stages of hiring to boost efficiency and fill vacant roles within the given timeframe.

Raju (n.d.) sheds a light on the AI technologies available in the market. The following are examples:

- **Job Ad Posting:** When it comes to job posting, AI can help in tailoring job descriptions, by focusing on highlights and articulating in a manner that attracts the right candidates. Words selection is crucial in the job posting. Tools such as Textio use an "augmented writing platform" which focuses on job content by recommending word choices that are gender-neutral for any given job posting. For example, a phrase such as "we are looking for 'recent grad'" does not appeal to a person in their 40s.
- **Candidate Sourcing:** After job ad posting, the next challenging step is to shortlist top candidates and engage them for the next step in hiring. Outsourcing also means seeking out passive candidates who perfectly match job requirements. Tools such as Entelo and Arya allow companies to scan top-tier talent across the open Web not limited to LinkedIn. These tools use application tracking system and candidate relationship management modules to match and rank candidates and track the application during the hiring process. These tools also help in engaging potential employees via emails and text messages.
- **Candidate Assessment:** After finding the right pool of candidates for their in-person interviews, recruiters conduct a preliminary assessment that includes SWOT analysis, background, and legal check. This is another time-consuming, yet important, step in hiring. There are many preemployment assessment tools commonly known as chatbots. Some of these tools are Mya, JobPal, and Harver. These automated assessment tools are not only designed to screen candidates for job technical and general skills, but also to keep them engaged by providing 24/7 virtual assistance.
- **Interviewing Process:** After prescreening and short-listing candidates, the next crucial step in the hiring process is interviewing. To provide a great and unbiased experience to the candidates, companies are widely using AI-powered tools such as Hirevue, and Talocity. These tools perform live video interviews with candidates asking them predetermined questions and

recording their answers for analysis. AI tools analyze facial expressions, intonation, and word choice to evaluate if a candidate is a good fit for the role.

- **Onboarding Process:** Once the selection is done based on assessment and interview, AI is also leading in the event of the onboarding experience. Tools such as Enboarder and Apical make it easy to review paperwork and learn about policies.

ETHICAL CONCERNS WITH ARTIFICIAL INTELLIGENCE INFLUENCED HIRING

While technologies such as AI have great utility, certain challenges require attention before adopting them as part of the hiring process. Unfortunately, challenging questions that have become crucial during the pandemic are: What are ethical practices in the use of digital technologies in the hiring process? How can recruiters ensure that AI makes unbiased decisions?

The biggest concern with AI technologies is whether they comply with legal regulations. Hiring managers need to use this technology with care. A general observation is, for AI to perform at its best, it needs more data, and that data can be provided by hiring staff from their experience and expertise. Poorly used or ineffective AI hiring systems can negatively affect an organization.

The Amazon case study is one of the most discussed cases in AI hiring failures. Dastin (2018) reported that, since 2014, Amazon had been using its in-house AI-powered recruiting program. This program learned from the company's resume data pattern of 10 years, which was provided to it for learning purposes. By using that knowledge, it started filtering out female candidates. It downgraded and assigned lower scores to resumes that include the word "women's," hence giving preference for hiring men. When discovered, this program was put to rest in 2017.

Although ethical concerns about staffing processes have been present for a long time, the COVID-19 pandemic raises new potential ethical concerns regarding the staffing processes, due to the increased use of digital technologies in the hiring process. For example, with the presence of social media and the availability of a seemingly unlimited supply of information online about each applicant, issues of privacy and appropriateness of information used in a staffing process are emerging without a clear set of legal or ethical guidelines.

A potential ethical concern is the ways that applicant information is handled. Information shared about applicants through email, texts, Tweets, and calendar postings can be difficult to manage, and, once delivered, are difficult to rescind. While some organizations have sought to address ethical concerns with social media use and other Internet data through policies, there is not a commonly held set of ethical standards when using readily available Internet data sources when hiring.

Recruiters' burden has been shared by AI, since COVID-19. Part of the shared responsibility is preemployment screening. It is a very useful tool, as it helps in filtering out candidates based on their skills, traits, emotional intelligence, and physical ability for a position.

However, the use of AI as a time-efficient and resourceful medium also requires its careful, legal, and job-related use. AI uses algorithms and collects massive amounts of data by scanning hundreds and thousands of Web sites. It does not only look at CVs, but it also looks beyond to find a "good" fit for the role. AI searches through social media profiles, blogs, meetups, chat rooms, and other miscellaneous online media, and collects a handful of information that may or may not be related to a job posting. It can provide relevant information such as an overview of candidates' professional profiles along with other unnecessary information, such as political affiliation, the organization they support, volunteer activities or religious beliefs.

Social media provide platforms such as LinkedIn, Monster, and Indeed for hire. On the other hand, these platforms raise ethical concerns for employers when using them for screening (White, 2021). With the prevalence of social media, where it seems that almost everyone has an online presence, there is a need to set legal guidelines for use of social media screening for hiring purposes. Scanning applicants' social media presence can raise a concern about invading their privacy and can also lead to discrimination in hiring. A relevant example is a scenario where a candidate's application gets rejected at initial screening because the recruiter misjudged an applicant by accessing someone else's social media profile who has the same name as the applicant.

The pandemic has caused various shifts in hiring, including more emphasis on video hiring using tools such as Zoom to interview employees. Potential ethical issues emerge regarding the ways employees are interviewed on a video screen vs. an in-person interview. A question that can be asked is: Are biases enhanced when viewing someone on video instead of in person? Another question that can be asked is: Does using online video create an inadvertent digital divide between those who possess technologies and training to present themselves better on camera giving an edge compared to those with lesser training and technology? Also, does a video call favor certain forms of nonverbal behaviors and/or personalities? These questions and others emerge, raising potential ethical hiring concerns.

While AI-driven screening tools enable some employers to reduce the challenge of sorting through applications, these tools create new ethical issues. Much has been written about algorithmic bias in AI. Unfortunately, this bias also may be found in AI-driven screening tools. Can AI tools account for the range of variables that can make a good hire? Can those with strong capabilities be excluded from hiring opportunities due to variables the AI does not recognize? Can the use of AI screening tools lead to unethical outcomes? Organizations must consider the power of AI tools to create efficiencies in the hiring process, but it is also necessary they reflect on the potential adverse outcomes when using these tools.

USING ARTIFICIAL INTELLIGENCE TO ENHANCE ETHICAL HIRING

A Proposed Model

Presently, humans are still the best at identifying talent to perform in the organization. Humans' ability to critically analyze candidate variables to determine their future performance in a specified position is arguably more art than science. Indeed, in most cases, the choices of those to hire are strongly influenced by individual hiring manager preferences. While data collected during the hiring process should be the primary influence on decisions, many hiring decisions are made with limited use of data and rely more on the human perception of the candidate. While relying on human perception is not necessarily the wrong approach, all humans are subject to bias in their decision-making. Unfortunately, these biases can result in biased decisions based on incorrect variables or, in the worst-case, based on prejudices, stereotypes or overt discriminatory practices.

It is easy to say that there will always be unethical (e.g., bias) variables in the hiring process. A cynic can conclude that unethical hiring practices are normal and will always be part of hiring. On the other hand, an argument can be made that AI-infused hiring allows to reduce unethical hiring by supplementing human judgment. Arguably, digital technology can reduce the adverse effects of bias, so enabling a more objective, hence ethical, selection process.

Digital Age Ethical Hiring Model

To give guidance on ways to have humans work in tandem with AI, Figure 1 presents a staged approach for using AI in the hiring process.

Figure 1.

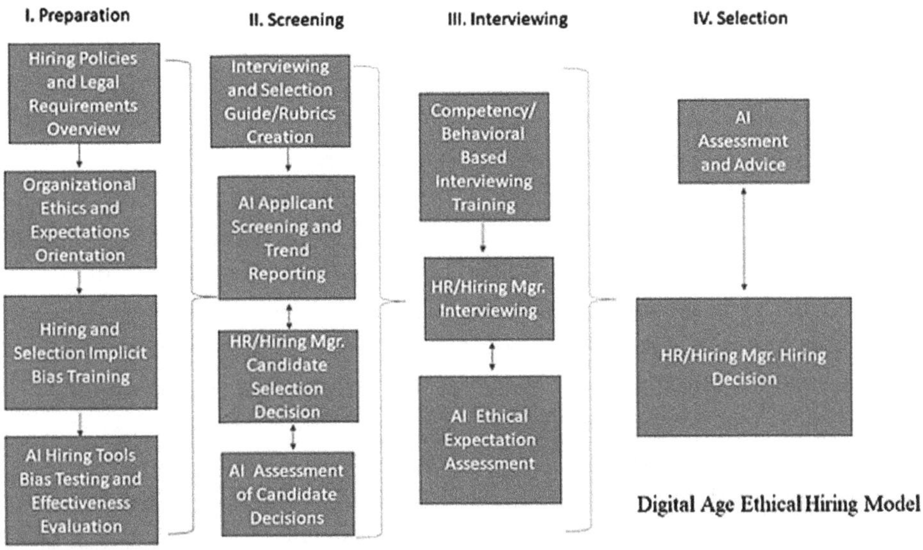

1. **Preparation Phase:** In this phase, good hiring practices involve preparing those who will engage in the hiring process to follow legal requirements, organizational policies, and ethical expectations of the organization. Additionally, technology tools such as AI must be thoroughly tested to ensure that outcomes that involve AI align with the legal, policy, and ethical requirements for the hiring process. It implies a thorough planning process involving HR to ensure the policies and the legal requirements are met. Additionally, organizational leadership should specify the ethical expectations for the hiring process. This assumes that there is an identifiable ethical code of conduct that has been communicated widely within the organization. This is an essential component of ensuring ethical hiring practices. Finally, the information technology specialist should work closely with HR to test the selected AI technology and ensure the desired performance aligns with the organizational policy and ethical expectations. It is during this phase that all participants are fully trained to ensure that the hiring practices are consistently executed across the organization.
2. **Screening Phase:** This phase involves selecting from among applicants to identify candidates and determine the final employee selection. The screening process has always been a time-consuming activity, and AI screening tools have created efficiencies in moving qualified applicants to become candidates for the position. Unfortunately, the AI screening process, like the human screening process, can result in unethical (e.g., unmerited preferential treatment) and/or illegal practices (e.g., discrimination). This is why it is essential that, in the

Preparation Phase, the AI tool is thoroughly vetted to reduce the potential of bias emanating from the use of technology. The screening phase includes the use of hiring selection guides or rubrics. Building these selection guides and rubrics into the AI screening tool enables the consistent hiring practices that organizations desire. The consistency of the hiring process helps to reduce the legal exposure from candidates who feel they were unfairly excluded from candidate selection. Data should be compiled to identify trends to report on hiring objectives such as increasing the number of culturally diverse candidates. Using the data generated from the AI screening, humans review the results and make recommendations. A final step involves resending the information through the AI to evaluate the candidate selections. This parallel evaluation of data evaluation processes reduces the probability of unethical candidate progression of applicants.

3. **Interviewing Phase:** In this phase, humans lead the interview using the selection guide and rubrics to make consistent decisions. Interviews should be based on a competency/behavioral-based approach to focus on a candidate's actual performance and acquired capability. The hiring manager and HR representative should work together to ensure ethical and legal compliance during the interviews, including questions, data collected, and the correct and consistent use of the selection guide or rubric. After the interviews, AI can be used to evaluate the data collected and give recommendations on the candidate interviews to provide advice for the final selection process. The inclusion of AI in this stage is not turning the decision to AI, but using a data-based analysis to inform the human decision-making process.
4. **Selection Phase:** In this final phase of the hiring process, the decision on the candidate to choose is made. Historically, this process has been influenced by many subjective factors even when using objective measures. While data are often used to hire, arguably more decisions are made based on a "feeling" or "sense" about the candidate. While some will argue that this is the only way that humans can select, the true goal should be to make selections based upon the data received through the selection process. Compiling the screening data, the interview data, along with the application, resume, and cover letter, are, in the best circumstances, difficult to interpret objectively. The argument made here is that the AI system should make the recommendation on who to hire based on the data. Undoubtedly, the final decision should be made by humans comparing their analysis of the data with that of the AI system. The goal is to ensure the most objective decision by using a more objective source of decision making (i.e., the AI system) to help ensure that decision is made based on the collected data.

The model proposed focuses on including AI in all stages of the hiring process. Importantly, the effectiveness of the model is contingent on using AI tools that have been thoroughly tested to avoid adding additional biases in the hiring process emanating from the AI tool. The authors are not suggesting that AI should replace human judgment. On the other hand, using human capability supported by AI capability can result in more effective and ethical hiring processes.

A NEW SET OF ETHICAL HIRING PRACTICES

As organizations think through the hiring process in the digital age and especially as they move beyond the COVID-19 pandemic, they can work to ensure that the processes for hiring meet desired ethical and legal standards. Hiring during the post-COVID pandemic has become especially challenging with shortages of workers, changing attitudes and expectations of workers, and the so-called "great resignation," where many are simply choosing not to work or leaving employers with whom they are not satisfied. These challenges, added to the normal challenge of hiring, suggest a need to redesign hiring approaches.

Another outcome of the pandemic is a heightened expectation of fairness and equity in hiring. The commitment to ensuring equity, fairness, and opportunity with a focus on hiring the best candidates are essential goals in the post pandemic era. While many commonly accepted ethical practices exist for hiring, considerations of lessons learned and reflection about opportunities to enhance the commitment to ethical hiring are needed. The following recommendations address ethical hiring:

- Ethical practices can be attained.
- Conversation and training should take place before the selection process on ethical practices.
- Clear policies are necessary on staffing and selection ethics.
- More communication about ethical selection policies and practices is needed.
- Agreement on ethical actions should align with policy.
- AI can be used to enhance the hiring process.

Some practices that support an effective hiring process include comprehensive training to develop competencies in ethical hiring. The training should emphasize the significance of ethics and the potential consequences of unethical hiring. Further, the training should address employment laws and labor rules. An important point of emphasis is to ensure diversity, equity, and inclusion in the recruitment process to enable the best candidates to be selected regardless of their identity. Moreover, those who conduct ethical hiring practices should be rewarded and those who disregard standards should be penalized.

Arguably, the use of AI technology can enhance the hiring process. The COVID-19 pandemic has increased the use of AI and other digital technology in the hiring process (Guha, 2021). Given this increased usage, it is critical to articulate a set of desired ethical practices. Consideration of desired ethical practices must precede the implementation and use of digital technologies in the hiring process:

- Hiring processes are determined before choosing a digital technology to aid in the selection.
- Clear goals need to be determined for the use of the digital technology.
- Training is delivered to all users of the digital technology.
- Data safeguards and pathway tracking of collected data should be evaluated and determined.
- When using AI to aid in screening, a continuous audit of the results should be conducted to determine potential algorithmic bias or adverse outcomes for applicants.
- Humans should be the final judge of screening results, instead of full reliance on digital technology AI decision-making.
- All applicants should be made aware of the use of digital technology.

AI can be used to create efficiencies in the hiring process. In many ways, the power of AI can improve the overall outcome of hiring. Unfortunately, AI can potentially add bias to the hiring process. AI must be used in tandem with human input and evaluation to ensure optimum fairness. In other words, even though an organization should use AI tools to increase efficiency and consistency, human evaluation should also play a role in the determination of candidate selection using data generated to ensure fairness, equity, and opportunity for all applicants.

SUMMARY

The COVID-19 pandemic has created both opportunities and challenges for using digital technologies. The use of digital technologies in the hiring process will only increase, making the need for ethical practices more urgent. Ethical practices, while always necessary, are especially important in future post pandemic hiring, where the use of digital technologies will surely increase as hiring accelerates.

As employers move to integrate hiring processes that were used during the pandemic and incorporate these processes into the post pandemic activity, the goal will be to gain greater proficiency and effectiveness in the use of the newer processes. As digital hiring tools begin to be normalized, organizations should work to develop, train, and reinforce that ethical hiring is good for both the organization as well as the prospective hire.

Perhaps, a new set of ethics are required to enable hiring managers to work effectively with AI tools to ensure fair treatment of all applicants and candidates for a position. Organizations must plan for ethics to be at the center of all hiring processes to enable the best potential outcomes from hiring processes.

REFERENCES

Arbix, G., & Comin, A. A. (2020). The pandemic, technology, and work are at the crossroads. *Internet Sectoral Overview, 12*(4), 11-17. https://www.cetic.br/media/docs/publicacoes/6/20210115080952/internet_sectoral_overview_year-12_n_4_artificial_intelligence_and_work.pdf

Corporate Finance Institute. (n.d.). *Ethical dilemma*. https://corporatefinanceinstitute.com/resources/knowledge/other/ethical-dilemma/

Dastin, J. (2018, October 11). *Amazon scraps secret AI recruiting tool that showed bias against women*. https://www.reuters.com/article/us-amazon-com-jobs-automation-insight-idUSKCN1MK08G

Dattner, B., Chamorro-Premuzic, T., Buchband, R., & Schettler, L. (2019, April 25). *The legal and ethical implications of using AI in hiring*. Harvard Business Review. https://hbr.org/2019/04/the-legal-and-ethical-implications-of-using-ai-in-hiring

Forbes Expert Panel and Forbes Coaches Council AI. (2021, August 27). *Key ways artificial intelligence can improve recruiting in the hiring process*. Forbes. https://www.forbes.com/sites/forbescoachescouncil/2021/08/27/key-ways-artificial-intelligence-can-improve-recruiting-in-the-hiring-process/?sh=78908a412c52

Green, S. (2022, February 2). *What is candidate experience & why you should be paying more attention to it*. https://www.innoflow.io/2022/02/02/what-is-candidate-experience-why-you-should-be-paying-more-attention-to-it/

Guha, T. (2021, February 4) *AI is transforming how and where we work*. Accenture. https://www.accenture.com/us-en/blogs/business-functions-blog/how-ai-is-transforming-work

Hauer, M. P., Kevekordes, J., & Haeri, M. A. (2021). Legal perspective on possible fairness measures: A legal discussion using the example of hiring decisions. *Computer Law & Security Review*, *42*, 105583. doi:10.1016/j.clsr.2021.105583

Institute for Business Value. (2020). *COVID-19 and the future of business.* IBM Corporation. https://www.ibm.com/thought-leadership/institute-business-value/report/covid-19-future-business

Loten, A. (2022, January 7). *Employers, investors take notice of ai tools to speed job recruitment.* The Wall Street Journal. https://www.wsj.com/articles/employers-investors-take-notice-of-ai-tools-to-speed-job-recruitment-11641599629

Maurer, R. (2021, March 26). *The pros and cons of virtual and in-person interviews.* https://www.shrm.org/resourcesandtools/hr-topics/talent-acquisition/pages/pros-and-cons-virtual-in-person-interviews.aspx

Pratt, M. K., & Florentine, S. (2021, August 20). *Employee retention: 10 strategies for retaining top talent.* https://www.cio.com/article/251060/how-to-improve-employee-retention.html

Raju, V. (n.d.). *30 AI-driven tools to optimize your talent acquisition process.* https://www.aihr.com/blog/artificial-intelligence-talent-acquisition/

U.S. Equal Employment Opportunity Commission. (2019, November 18). *Dollar General to pay $6 million to settle EEOC class race discrimination suit.* https://www.eeoc.gov/newsroom/dollar-general-pay-6-million-settle-eeoc-class-race-discrimination-suit

U.S. Equal Employment Opportunity Commission. (n.d.). *Prohibited employment policies/practices.* https://www.eeoc.gov/prohibited-employment-policiespractices

White, S. K. (2021, September 17). *AI in hiring might do more harm than good.* https://www.cio.com/article/189212/ai-in-hiring-might-do-more-harm-than-good.html

ADDITIONAL READING

Bock, L. (2015). *Work rules! Insights from inside Google that will transform how you live and lead.* Twelve.

Brockman, J. (Ed.). (2020). *Possible minds: Twenty-five ways of looking at AI.* Penguin.

Diamandis, P. H., & Kotler, S. (2020). *The future is faster than you think: How converging technologies are transforming business, industries, and our lives.* Simon & Schuster.

Kelly, K. (2017). *The inevitable: Understanding the 12 technological forces that will shape our future*. Penguin.

Klinvex, K. C., O'Connell, M. S., & Klinvex, C. P. (1999). *Hiring great people*. Briefcase Books.

Lee, K. F., & Chen, Q. (2021). *AI 2041: Ten visions for our future*. Currency.

Merisotis, J. (2015). *America needs talent: Attracting, educating, & deploying the 21st-century workforce*. RosettaBooks.

Schwab, K., & Malleret, T. (2020). *COVID-19: The great reset*. World Economic Forum.

Shell, E. R. (2018). *The job: Work and its future in a time of radical change*. Currency.

Smith, B., & Browne, C. A. (2021). *Tools and weapons: The promise and the peril of the digital age*. Penguin.

Sucher, S. J., & Gupta, S. (2021). *The power of trust: How companies build it, lose it, regain it*. Public Affairs.

KEY TERMS AND DEFINITIONS

Applicant: A person that submits a request for consideration for position using an application, resume/CV, letter, or other required information.

Artificial Intelligence (AI): Algorithmic applications that enable computers to simulate human actions and thinking using data.

Candidate: A person who has applied and is determined to have the qualifications to be considered for hiring.

COVID-19 Pandemic: The title of the global spread of an airborne and surface contact infectious disease attributed to the SARS-Co-V2 virus.

Ethical Hiring: Hiring processes that ensure that applicants are objectively screened, and candidate consideration are based on the qualifications specified in the position description. Further, applications and candidates are processes in accordance with applicable policies and laws ensuring fairness, equity, transparency, and objectivity.

Hiring Ethical Issues: Hiring process or actions that are inconsistent with legal requirements, organizational policy expectations or actions that create inequitable or unfair selection outcomes.

Hiring Processes: Activities that enable the recruitment, screening, interviewing, and selection of applicants for organizational work roles.

Chapter 4
Leadership Using Internal Controls for a Decentralized (Remote) Workforce

Lisa Bushur-Harris
Western Governors University, USA

ABSTRACT

This chapter covers internal controls and segregation of duties needed to protect company assets when a decentralized (remote) workforce is utilized. Company assets focused on in this chapter include cash, customer billings, inventory, computers, data, expense report reimbursements, payroll checks, and time engagement. Also discussed is the value of company reputation in terms of customer service levels, social media presence, and corporate social responsibility. Managers and leaders must be knowledgeable on how to monitor their subordinates in a remote workplace.

LEADERSHIP USING INTERNAL CONTROLS FOR A DECENTRALIZED (REMOTE) WORKFORCE

During the post-Covid environment, more employees work remotely than ever before. Leaders will want to be aware of areas of risk related to protection of their entity's assets, reputations, and societal goodwill, that are more vulnerable with a decentralized (remote) workforce. An entity's leaders cannot make first-hand observations of hardly any employee activity in a remote environment without the proper internal controls. With this new trend of a remote workforce comes new challenges for leadership to effectively manage their entities.

DOI: 10.4018/978-1-6684-4358-3.ch004

Backing up in time, in a pre-Covid environment when most workers performed their service on company property, it was significantly easier for supervisors and managers to observe employee performance. Managers could physically see work being performed, or lack of work being performed (if employees are goofing off, making blunders, or botching their work product). Additionally, communication was different in the pre-Covid environment when most employees were working on-site, in that office discussions were more easily held and did not have to be intentional, where a comment could sometimes point a manager in the direction of a potential issue or trend. In the pre-Covid environment, managers could also more closely observe how much time their employees spent on many tasks.

The company assets discussed in this chapter with respect to protecting them in a decentralized work model include cash, customer billings, inventory, computers and similar equipment, data, expense report reimbursements, payroll checks, time and engagement, and company reputations (as affected by customer service, social media, and corporate social responsibility). All of these assets require specific protections and the prescribed controls for each are a bit different from each other. The ideas included in this chapter may not be all inclusive.

While some may not think of a company's reputation as an asset, indeed it is if you think of companies that have suffered a downfall due to a poor reputation. Reputationsciences.com (2021) states that corporate culture, social responsibility, company leadership, financial performance, customer sentiment, and products & service all make up a company's reputation. Reputationsciences.com lists these companies has having bad reputations at the beginning of 2021:

1. Phillip Morris-Tobacco companies have faltered as smoking has been determined to cause cancer.
2. Sears-Store closings and falling customer interest in their brand has hurt Sears.
3. Comcast-Being saddled with many lawsuits has caused Comcast to struggle.
4. Facebook-Accusations of not managing customer privacy and allowing manipulation of the 2016 U.S. presential election have lessened Facebook's reputation, although Facebook still seems to have a strong foothold.
5. Goldman Sachs-Reputation damage due to the 2008 financial crisis in the U.S. still haunts Goldman Sachs.

This chapter first discusses very general information about controls, and then moves to the more specific.

Fraud Triangle- The fraud triangle helps explain why employees may make the decision to commit fraud and theft against their employer. As Dorminey, Fleming, Kranacher, and Riley (2010) explain, the fraud triangle can explain why employees can be motivated to commit fraud against their employer, and these motivations are still very much (or more) at play in a remote work environment. The three sides of the fraud triangle include *perceived financial pressure*, *rationalization (moral justification)*, and *perceived opportunity* to commit and conceal a theft. See Figure 1 for the three sides of the fraud triangle.

Figure 1. The Fraud Triangle

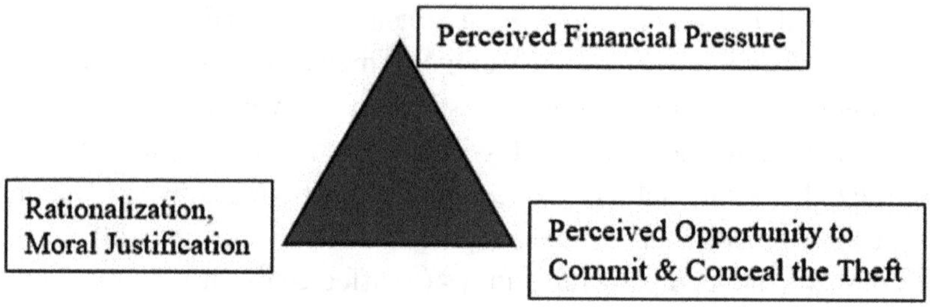

Dorminey et all also describe a fraud diamond, that adds a fourth variable called "*fraudster's capabilities*". See Figure 2 for the Fraud Diamond.

Figure 2. The Fraud Diamond

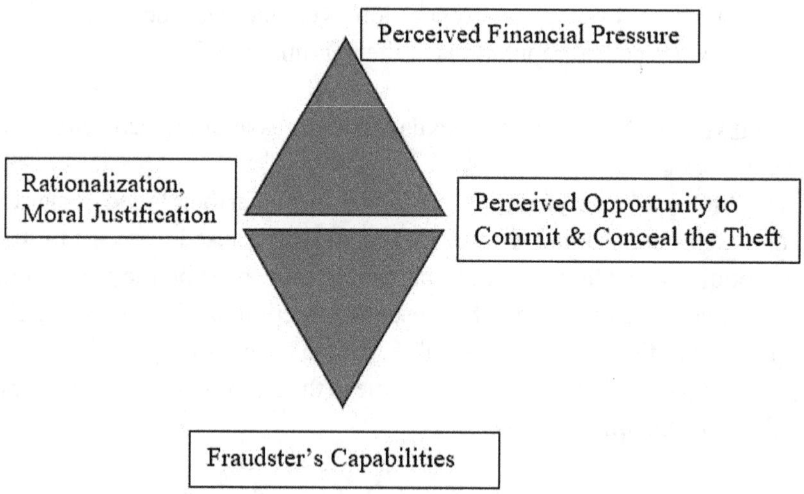

Think of a situation where an employee perceives financial pressure (such as their child needs a life-saving cancer treatment), and opportunity (the employee handles all cash transfers and general journal entries for their employer), and rationalization (the employee feels they deserve the money they steal because the employer's health insurance is inadequate to cover the needed medical treatment). And then if we add the fourth variable, the fraudster's capabilities (it would not be too difficult for the employee to set up an account and make the wire transfer into an account they own). Segregation of duties between those employees that can initiate a cash transfer, approve the transfer, record the cash transfer in the accounting records, and reconcile the cash accounts each month end, may very likely prevent such a theft. More about theft of cash is discussed later in this chapter.

Now think about another scenario related to stealing an intangible asset, for example, time. An employee feels pressure because they have two online full-time jobs, and management at each employer does not know about the other job that the employee keeps. The pressure presents because the employee feels they need both jobs in order to financially make ends meet and to make the house payment on their large two story, three car garage home. The employee sees the opportunity, in that neither employer will suspect they are actually managing two full-time jobs, since both are remote. The employee's rationalization is that they feel they are not fairly compensated at either job, and the employee feels capable that they can manage the responsibilities of both positions. What is not fair to the employers is the employee is not fully able to engage in either position, since they both require a full-time commitment.

In the one instance where this author observed this scenario, the employee was caught, and also had become burned out and quit the lower paying job, with no notice, because she needed to give immediate attention to the job she was keeping. This employee was caught because she was listed on the website for one of the employers as one of the key employees, and coworkers of the second employer noticed her inclusion on the other employer's website and spoke up to their bosses.

Fraud can happen whether a workforce is on-site or working remotely, but with a remote workforce being more prevalent than ever, the internal controls must become more sophisticated. Additionally, the increased use and complexity of computer systems makes implementation of effective internal controls more challenging than ever.

Segregation of duties can be a very important policy, even in a remote workforce. This is a list of duties that should be separate and independent of each other to protect a company:

- Shipping & forwarding shipping documents to billing
- Billing customers
- Posting cash receipts
- Following up requests for credit memos to customer accounts
- Bank reconciliations

- General ledger postings

Of the above list, only the first position, shipping, is generally not able to be performed remotely. In a larger company, segregation of duties is easier to carry out because there are more employees available over which to spread the responsibilities. In smaller companies, a segregation of duties may need to be supplemented with the owner of the company, or their delegates, reviewing transactions, if there are not enough members in the entity to have a satisfactory segregation of duties.

Another relatively simple internal control is to implement a rotation of duties, without advance notice to employees. (This is also a good idea for cross training purposes, in case someone leaves their position unexpectedly, or to cover vacation times.) See the above list of duties related to segregation of duties. It would be quite difficult for employees in these areas to cover any malfeasance if their duty was periodically and unexpectedly moved to someone else. It has been noted that when a new person takes over a given position, that has often been found to be the timeframe when a previous fraud is discovered. Managers should be very weary of an employee who refuses to take vacation time, or who is very territorial over their work duties.

Interestingly, it may actually increase internal controls to have a remote workforce in the accounting area because it would reduce the chance of collusion because the employees are less likely to know each other well, or develop close relationships, or develop a joint plan to defraud the company. Collusion can occur. This chapter mentions segregation of duties, but if there is collusion among employees, the segregation of duties is not an effective safeguard by itself.

The Sarbanes Oxley Act includes whistleblower protection for employees who report fraud in a publicly traded company (whistleblowers.gov/statutes/sox_amended). Employees may report fraud related to mailings, banking, securities, or any rule covered under the Securities and Exchange Commission. Additionally, employees may seek remedies for any retaliation related to their reporting (whistleblowing) such as reinstatement of their job and position, payback of expenses incurred, and compensation for damages (whistleblowers.gov/statutes/sox_amended). The Whistleblower program is administered by the United States Department of Labor, Occupational Safety and Health Administration.

One employee received a six-million-dollar award in a Whistleblower case (Zuckerman, 2021). The employee was an accounting executive who lost her job in retaliation for reporting accrual of unauthorized bonuses for executives (Zuckerman). This employee won her case because she reported a circumvention of an internal control related to having executive bonuses approved by the board before accrual (and related payment) (Zuckerman).

Now we'll move to more specific assets that need protections:

Cash-The protection of the entity's cash is obvious, but there are new risks in this area. Many expenses are paid by electronic transfer of funds, but there still needs to be a segregation of duties related to paying expenses, approval of the payment of same, and review and reconciliation of bank accounts. Many banks offer a service where two approvals are needed for a cash transfer to be processed, so that no one person can initiate a transfer by themselves.

Companies should make use of a lockbox at a bank for all cash receipts from customers where the bank receives all checks from customers, the bank deposits all of the checks in the company's account, and then the bank sends a copy of all checks and the remittance advices to the company for posting of the cash receipts in the accounts receivable system. By using a lockbox, it makes a remote accounting system possible because the remittance advices can simply be emailed to the accounting personnel to enter, and it keeps employees' hands off company cash receipts.

This author owns a small CPA practice and one of the business clients was recently victim to theft of cash by complicated trickery. The head bookkeeper received an email that was seemingly from the majority shareholder and CEO of the company that said something like this, "Hi Margie, please transfer $68,000 to Smith Enterprises at Bank of America, routing number xxx, account number xxx. I'll bring the documentation for this equipment purchase when I come back into town next Monday. Thanks, John" So many things looked reasonable to Margie, such as:

- The email came from John's company email address,
- John did often make the equipment purchase decisions,
- It was common for John to email Margie with similar requests, and
- The amount was reasonable based on the type of equipment the company often purchased

It was astounding how much relevant information the perpetrator had been able to collect to pull off such a crime. By the time John made it into the office the following Monday, the funds had already been moved out of the Smith Enterprises bank account, the investigating FBI agent stated.

Ransom attacks on organizations is a relatively new risk area for organizations. This is when malicious attackers lock an organization's data, which can cripple them. The attacker requests a ransom payment (usually in the form of bitcoin) in order to unlock the organization's data. Data that organizations can lose access to include customer data such as the customer database, accounts receivable, sales orders, estimates previously issued, and more. Other areas could relate to educational records being locked from a university, architecture plans for a construction developer, CAD files for an engineer, tax return data for a tax preparer. The list could include every occupation and industry possible. It's important that data security is set at a sufficiently high level to protect organizations from ransom attacks.

Orcutt (2020) discusses how electronic currency (such as bitcoin) works, and opines it may never completely replace cash. With the inability to track digital currencies, this is good news from the standpoint of internal controls.

Chen, Li, and Zhang (2021) make the point, however, that too much overzealousness in areas of cash internal controls can backfire because it can decrease an entity's operational efficiencies. Chen et al further explained that an entity's success can be hampered if the over-enthusiastic internal controls hinder employee creativity, stifling management attention and energy, robbing employee enthusiasm and trust. Chen et al further state that a cost/benefit analysis should be considered before implementing excessive internal controls over cash, and such an analysis should be considered from every possible aspect (management, operations, employees, marketing, research, and more). This author does not disagree with Chen, Li, and Zhang, but does caution that it is difficult for an entity to grow while also safeguarding assets, without some measure of systematic internal controls.

This author's intention is not to teach about how to kite cash, but this scenario could be maliciously carried out by an employee: an employee with cash management responsibilities could create a fictitious company similarly named to the company they work for, and then wire transfer $10,000 from one of the many employer cash accounts to their own account, then in a few days, transfer $10,000 from a second employer cash account to the first, and then in a few days, transfer $10,000 from a third employer cash account to the second. (The amount of the cash wire transfer is not relevant, only that it's sufficiently immaterial to the company, to not raise many eyebrows.) After numerous transfers are made to cover previous transfers out, then finally a general journal entry is made in the general ledger to write off the missing $10,000, and the description would likely be entered as "Miscellaneous adjustment required by auditors", or "Miscellaneous adjustment per John Smith, CFO". The employee removes the $10,000 from the fictious account, and continues to work at their place of employment, unless questions arise, in which case they would quit. They may have even taken the job with an assumed name and social security number. There are many variations of the kiting scheme. A segregation of duties between employees' responsible for cash transfers, bank reconciliations, and general ledger entries may have prevented this theft.

This is completely "old school", but an organization's controller, CFO, comptroller, or similar key employee should be reviewing a company's general ledger for propriety, to ensure all postings are for reasonable amounts, and all vendors and customers listed are proper. With the fast-paced data is moved in today's environment, the review process is often skipped or shortened, but this is still a very necessary step in the protection of any company's assets. This step is likely even more important in a remote work environment, so that the controller, CFO, or comptroller can be more aware of the types of transactions occurring in the organization, because they do not have face to face interactions with accounting personnel.

Positive Pay is a relatively new banking product that many banks offer, where the customer provides their bank with the check number and dollar amounts of checks issued, and the bank will only allow checks on that list to clear their customer's account (positivepay.net). This is especially helpful to protect against a company's checks being duplicated and issued as if bona fide company checks. When checks are presented to the bank that are not on a company's Positive Pay list they provided to the bank, then an image of that check is sent to the company, and they can approve it, or have it returned, and then they also have the document in order to follow up unauthorized checks (positivepay.net) with law enforcement.

Some banks employ a system related to wire transfers where two employees have to separately log into the banking system to approve a wire transfer, and each has a fob (similar to a key fob for an automobile) from the bank that has a security code that changes every few minutes. Each of the employees must enter the correct security code in order for the wire transfer to be initiated. This process is helpful because it requires two employees' approval, and those employees must also have physical custody of the fob with the ever-changing security code.

Fraud related to billings to customers- Employees with responsibility for customer billings can manipulate invoices after goods are shipped, and then have customers pay them personally for a discounted amount. Alternatively, employees could potentially not bill favored customers at all. Depending on the scenario, an employee could also falsely bill customers to overstate revenues in order to meet a bonus goal, and then write off the excess billings after the bonus is safely received.

Ring (2019) reinforces internal controls, including checking the name of customers on invoices to ensure they are bona fide, reviewing purchase orders for goods, matching goods received against shipping records, and monitoring the billing process. For a satisfactory segregation of duties in the billing and accounts receivable area, differing employees (and ideally, differing departments for larger companies) should be responsible for:

- ordering goods from vendors,
- shipping goods to customers,
- billing customers,
- receiving and applying payments from customers, and
- preparing bank reconciliations.

Probably the largest billing fraud in the US known at this time is Medicare fraud. Coustasse (2021) discussed Medicare fraud, and how health providers have bilked the Social Security system out of billions of U.S. dollars by billing for medical services that were more detailed and extreme than they actually were. The medicare.gov website includes information about how Medicare recipients can report suspected Medicare fraud for medical claims that appear fraudulent by way of being embellished, or not being for services received at all.

And not to make this writing about how to kite receivables (either), but a malicious employee with cash transfer responsibilities could create a fictitious company similarly named to the company they work for, and then as a large customer (that has lots of activity with the employer) pays their invoices to the company via wire transfer, they could post the receipt for only portion of the cash wire transfer from the customer, and then have a portion of that cash transferred to their own account. In the very next transfer from the affected customer, that cash could be posted to the first invoice, to cover the first shortage. And then the third transfer could cover the shortage on the second invoice. This type of kiting is particularly tricky because it just appears that the customer is paying slowly, but no cash will be easily identified as missing. (This type of kiting is often caught when the malicious employee forgets to keep up with their kiting activity, and with no cover up, an employee from the collections department calls the customer for payment, and realizes the invoice HAS been paid, and investigates further.)

Misappropriation of inventory-shipments of an entity's inventory can be "accidentally" shipped to an incorrect address where an employee can pick up the inventory, and a company will generally reship the inventory to the bona fide customer upon customer request. Controls are needed to ensure all shipments are being delivered to bona fide customers.

This author ordered a carpet shampooer from Amazon.com during the holiday season last year, and it never arrived, and so this author filed a claim with Amazon.com, and they promptly sent another carpet shampooer. After the holidays, the neighbor sheepishly knocked at the door with the original carpet shampooer in hand. He said that his mother had shipped so many Christmas gifts for his children, that he didn't even look at who the packages were addressed to until Christmas morning, and that's when they realized they erroneously possessed the carpet shampooer. Now this author has two carpet shampooers. This was an innocent mistake, but see how readily Amazon.com reshipped the lost item? Most reputable companies operate in the same manner, but this should be monitored for abuse.

Bassamboo, Moreno, and Stamatopoulos (2020) wrote about their "optimal polynomial-time algorithm" that audits inventory levels, in their attempt to combat misstated inventory levels due to employee theft (and customer shoplifting, spoilage, and damages). Their work is also important because their model helps safeguard against replenishing inventory too soon, which can also waste company resources by not efficiently deploying assets until needed. (Not having enough inventory on hand is an issue, when a customer needs the inventory immediately, and this could possibly lead to a customer finding a new supplier. Having too much inventory on hand is an issue because it ties up excess cash.)

Jenks (2019) explains that inventory internal controls are important because:

- A company will know when to reorder (or to manage the manufacturing process so that more goods are available when needed);
- A company will know when they have adequate levels of inventory, and thus preserve cash and not have it tied up in excess inventory;
- The entity will have more dependable data on inventory costs to include on their financial statements;
- Sales analyses will be more accurate, so that a company can make good decisions on which products to promote and which ones to potentially enhance or drop; and
- Shrinkage will be more easily identified, and with a good system, one may be able to determine which products have been misappropriated.

A proper segregation of duties is necessary concerning ordering inventory, determining manufacturing goals, shipping inventory, and billing customers. If the same persons or departments are responsible for more than one of these areas, then misappropriation may occur. Collusion is more likely to occur if employees within the same department have responsibilities that would allow for a coverup of a misappropriation. For example, the shipping department forwards bills of lading to the billing department so that invoices can be prepared, but if the shipping department employee ships the goods to themselves, family, or friends, they will not want an invoice prepared or sent, and if the shipping employee can prevent an invoice from being prepared, that employee has just stolen inventory. With a segregation of duties between shipping and invoicing, this erroneous shipment would be much less likely to occur.

Computers & similar equipment-Computer replacement should be monitored for frequency and validity, and all broken or obsolete computers should be returned to the entity for verification of the condition of the computer or equipment. This same procedure should be applicable for printers, scanners, cell phones, tools, and all other company assets in the possession of employees. Deceitful employees (which is likely rare) could ask for a new laptop or other equipment each year, and then supply the extras to family members and friends. Some computers and equipment can be repaired and sent out again for employee use. An entity should also set firm policies regarding safeguarding against theft of laptop computers, such as no storing company laptops in vehicles or hotel rooms. Managing the safety and replacement of computers and similar equipment can be a significant savings for entities. Thieves who steal laptops also can gain access to valuable and confidential data, discussed in the next paragraph.

Data-Laptops, desktops, and complete computer systems all can be susceptible to data theft. An entity should have policies regarding keeping sensitive data encrypted and policies against opening suspicious email attachments in order to prevent malicious attacks on entity systems and theft of data. Many organizations make use of a virtual private network, so that outsiders have more difficulty in hacking their way into a system. It goes without saying that a thorough password policy should be part of any entity's security system.

Many IT departments are instructed to only give employees access to the areas they need within a company's computer systems, in order to reduce the chance of unauthorized use. If you work in Accounts Receivable, you don't need access to any other areas, such as Payroll, or Accounts Payable. If you're in Sales, you don't need access to be able to change the amount of invoices in the Accounts Receivable system. If you're in Purchasing, you don't need access to the Accounts Payable system that would allow a change in amounts that are due to vendors.

Another reason for the requirement for employees to return company computers when they are done with them is so that the hard drive can be wiped clean of any confidential information on customers (including their credit card information for payments), patients (including their personal medical history, other HIPPA-protected information, and their credit card information for payments), students (including their grades, or other FERPA-protected information), or other stakeholders. Old computers should not be "handed down" to children or others, but returned to the issuing organization for proper disposal of data. Embarrassing data breaches are often reported in the media, and you don't want that to be your entity on which those reports are based. The computers should also be swiped to protect any proprietary information.

Some of the more common companies reporting the largest data breaches for 2021 according to Bluefin.com (2021) include:

- Facebook
- Instagram
- LinkedIn
- Kroger
- Volkswagon/Audi

Loayza (2020) stated, "For at least the last decade, the United States has been in a state of low-level cyber war. Foreign nations have hacked and infiltrated American companies to steal intellectual property and gain competitive advantages." Both

China and Russia have been accused of trying to steal U.S. proprietary information via the web.

Expense report reimbursements-employees can fake expense reports, and the related receipts. It's a good policy to require advance approval of expenditures and have expense reports reviewed by the direct supervisor of employees, since they will be most familiar with what expenditures are appropriate.

Similar to expense report abuse is abuse of a company credit card. In this author's private practice, clients are warned when there are numerous expense entries labeled, "Amazon.com". A person can buy ANYTHING at Amazon, so don't you think upon audit, an auditor is going to ask for documentation what the Amazon.com purchases are on a company credit card and what was the business purpose? In addition to Amazon, this author has observed employees inappropriately charge personal car repairs and personal cell phone bills to company credit cards. The most laughable credit card purchases this author once refused to deduct on a client income tax return were charged on February 14th and were charges to Victoria's Secret and Godiva Chocolates.

Heroux and Goupil (2020) studied and wrote about the sociocognitive determinants related to supervisor approvals of subordinate's expense reports, for purposes of understanding these determinants in order to learn best practices for saving organizations money on expense report reimbursements. Heroux and Goupil's concluded that when manager's feel that the expense report policy helps protect organizational assets, they are more likely to adhere to the expense report approval procedures. They also concluded that when manager's feel they are entirely and exclusively responsible for the approval of expense reports, managers are more likely to adhere to the expense report approval procedures. These lessons of getting the buy in of managers related to expense report procedures should be noted by all organizations.

Supervisors of employees with access to company assets should be aware to be on the lookout for employees living way beyond their means. Does your employee have a new boat and condo at the lake, their spouse doesn't work, and you have no idea how the employee could have afforded those niceties? Does your employee seem to be losing a lot of cash at a casino, and you don't know where that cash is coming from? Do you get tagged in social media posts showing the employee in international vacation spots on a frequent basis? Perhaps supervisors should consider whether the employee has happened upon additional funds at their organization's expense.

Payroll scheme-This malfeasance could be undertaken by an employee in a payroll department. Let's say Briana Smith is a payroll supervisor and notices that an employee, Brian Smith, quits his job. Instead of terminating Brian Smith in the payroll system, Briana could instead have Brian's bi-weekly payroll check deposited to her bank account by entering her banking information into the payroll system.

Because it's a remote workforce, the Manager doesn't realize "Brian" is still getting a payroll check.

Two overall internal controls come to mind that would have prevented the above payroll scam: 1.) managers should be given a profit & loss statement for their department on a monthly basis for review, and these managers should follow up on unusual and unreasonable items. 2.) internal audit should on a test basis, make sure that employees are truly bona fide people by talking with them, checking to see if they have social media accounts, and if they are real employees.

Another suggestion would be for Human Resources departments to have termination checklists that would include:

- enter terminations in the payroll system, so that the payroll employees cannot continue processing a payroll check for terminated employees
- collect any badges or security code fobs from terminated employees
- change the passwords to all affected systems
- collect their laptop (as already explained previously)

A manager should sign off on the completed termination checklist, verifying the proper steps have been taken. (Of course, the termination checklist would include COBRA insurance elections and other information about benefits, but this chapter is only covering internal controls needed in a remote workforce environment.)

One control that companies perform is a match against payroll records and vendor records, to see if any of the names, addresses, or bank account numbers match. This control is making sure no employees have set themselves up as a vendor to receive fictitious payments. When a match is found between an employee and a vendor in a name or address or bank account number, then that's identified as an exception and should be followed up by a manager. (Often the exception turns out to be that the employee turned in a bona fide expense report that was legitimate and properly approved for payment. But if the vendor payment has no backup support, it would be time to investigate that vendor and employee further.)

Time and engagement-Don't we all believe the more time spent on an activity, that more output will result, or the quality of the output will increase? But what about the stories of employees sleeping on the job, or playing or shopping on the internet for hours? An employee at a car manufacturer plant once bragged and others that he's able to read the daily newspaper every work day from front to back, including every article and ad, while on duty on the assembly line because according to the employee, the union had got the line to move so slowly, that the employee had massive amounts of idle time between cars landing in his work area for him to complete his task on the car. Many employees shop online while they are supposed to be working. In 2018, employees spent on average 45 minutes per day doing online shopping says readwrite.com (2020).

Frankly, use of time will need to be monitored. And time is a company asset, if the employer is paying for it. Employees may spend time on personal endeavors instead of working. Lost time is quite expensive for an entity. Some employers have programs that can monitor the number of computer key strokes per shift, and similar types of engagement measures. Others have goals or KPIs that remote employees need to meet, such as number of calls handled, amount of time on the phone, etc.

Now a warning, over-monitoring of remote employees can be invasive and show a level of distrust of the employee, which will be counterproductive to a company's goals and mission. Nguyen (2020) explains that managers should make sure not just to monitor engagement data, but to measure the outcomes of employee efforts; the data is not the goal, but rather the goal is to achieve desired outcomes. Nguyen also suggests using the least invasive way available to monitor employee engagement.

Many organizations use Key Performance Indicators (KPIs) as a way to measure effectiveness of an employee. Perhaps the KPI is number of patient appointments held by phone, number of passing students, number of orders taken, number of prescriptions filled, number of widgets made and delivered. With the use of KPIs, focus on hours spent can be de-emphasized, and KPIs can be a more important focus.

Van Fleet (2020) makes a very good point about efficiency and effectiveness of employee time, and stresses "doing the right job" more than "doing the job right". Van Fleet recommends that managers remain focused on the main company goals, and that will lead to the proper use of time. This author was once in a networking group with a car salesperson, and he was also an officer of the networking group. This person spent so much excessive time on making the networking group better (organizing extra events, building props, and preparing financial reports), that one had no idea how he had time to sell cars and make a living. He was really good at doing the wrong job.

In summary, leaders need to be vigilant in a way that is effective related to monitoring a remote workforce, yet not overly intrusive to employees. Management needs to get what they are paying for in terms of labor, and employees need to feel trusted and respected; that balancing act likely needs to be considered.

Customer service-The level of an entity's customer service should be constantly monitored, and this can be done through the use of recordings (with appropriate permissions) and/or customer surveys. Some customer service specialists may let their tone of voice, speed, quality of services, or quantity of calls deteriorate if they are not monitored. Employee time is an asset that should be protected. Also, leaders are able to give feedback when they are able to monitor the customer service specialist's contact with customers.

Positive experiences with an entity's customer service team can make for a positive company image, and the reverse can also be true of a negative experience leading to a negative company image.

Some firms don't like to hear complaints from their customers. Ironically those who complain and have their problems resolved will tend to be more satisfied, have higher repeat purchase behavior, and generate more positive word-of-mouth publicity than customers whose complaints weren't resolved or who did not complain in the first place. Purity (2021)

Why don't most organizations ever make formal apologies? The most common answer is because then it may open the organization up to lawsuits, and the apology can be interpreted as an admission of guilt. However, Patel & Reinsch (2003) opined that when an entity apologizes to consumers or to the public, this act is <u>not</u> an admission of guilt, and so in a time when corporations want to and are expected to be socially conscious, this lesson likely needs to be remembered. If apologies are not an admission of guilt, then it could only help if an entity expresses regret for a mishap or misdeed.

Social Media-Employees' social media accounts can be damaging to the employee's reputation, but also for their employer's reputation. Employees may share information on social media that is unprofessional, racists, sexist, adverse to the entity's mission, or other inappropriate posts. If the employee shares their employer name, this can be quite damaging. Setting up policies related to behavior when using social media are vital. Rauf (2021) states that social media should be monitored for its negative effects and for ethical issues. Rauf also explains there seems to be a "new sociology of morality" regarding social media. It does appear that almost anything is allowable to be posted on social media. Social media is a relatively new platform that must be considered by any organization or company.

Corporate Social Responsibility-Entities often desire to increase their social image by participating in corporate social responsibility endeavors, which is a focus of an entity on societal and environmental concerns, including sustainability. In a 2021 study, El Samea & Rashed show that positive perceptions of an entity's corporate social responsibility activities impacts the purchase decisions of consumers who are interested in their products. Corporate social responsibility can be broken down into these four categories: economic responsibility, moral responsibility, legal responsibility, and philanthropic responsibility (El Samea & Rashed), and so when considering these four categories, this gives entities many opportunities to focus on corporate social responsibility.

However, negative information about a company quickly erases the good feelings that have developed about an entity's corporate social responsibility activities (Bogel,

2019). Also, entities that have been victim to negative press, can overcome it by participating in corporate social responsibility (Bogel, 2019).

Tesla, Inc. displays social responsibility in that it produces electric vehicles, which are a greener alternative to vehicles that use gasoline. But then when there is negative publicity, it counterbalances the positive influence of the social responsibility. Examples of this are when there is a vehicle accident of one of the Tesla self-driving cars with the driver asleep at the wheel, or when the founder, Elon Musk, is accused of making a public gaffe, such as making remarks on Twitter about Telsa stock performance or insider information that have concerned the SEC.

Managers may want to remember that mishaps in the customer service and social media arenas can quickly have a negative impact on an entity's corporate social responsibility campaign and overall reputation. Because customer service and social media professionals often work remotely, this is an area that needs constant monitoring by management.

Advantages of the additional internal controls mentioned in this chapter- Besides the obvious protection of company assets and company reputation, there's another advantage to implementation of internal controls of the areas mentioned in this chapter. These additional internal controls may also lend themselves nicely to company growth. With the added measures, an entity's business model may be more scalable, and growth may be easier.

CONCLUSION

The post-Covid environment has led to a decentralized (remote) workforce which leads to new responsibilities for employers and their management teams to protect company assets (including cash, customer billings, inventory, computers, data, expense report reimbursements, payroll checks, time engagement, and company reputation. Company reputation is discussed in terms of customer service levels, social media presence, and corporate social responsibility.

The fraud triangle explains perceived financial pressure, perceived opportunity to commit and conceal a theft, and rationalization, from the perspective of the perpetrator to a crime, and the fraud diamond, which adds the fraudster's capabilities. Generally, all parts of the fraud triangle (and diamond) are present when fraud has occurred.

Internal controls are an important key to safeguarding company assets, including a segregation of duties. There should be separate personnel responsible for initiating transactions, general ledger recording, approving entries, finalizing transactions, and reviewing transactions. Cash responsibilities should be separate from custody of inventory, posting of customer cash receipts, and accounts payable functions. Cash internal controls also include Positive Pay, and security code fobs. Rotation of job duties, and a whistleblower policy can also play an important role in a company's internal control system.

Employee time and engagement should also be monitored, but it's important to monitor the desired outcome, not the hours worked, so as to not offend employees with invasive monitoring measures, as those may erode trust between employees and their employer.

In summary, leaders need to be vigilant in a way that is effective related to monitoring a remote workforce, and company assets, including company reputation. Of course strong leadership can lend greatly to decreased fraud and unethical behaviors, but this chapter focused on internal controls. Effective leadership in a decentralized work environment will be more fully discussed elsewhere.

REFERENCES

Bassamboo, A., Moreno, A., & Stamatopoulos, I. (2020). Inventory Auditing and Replenishment Using Point-of-Sales Data. *Production and Operations Management, 29*(5), 1219–1231. doi:10.1111/poms.13153

Bluefin.com. (n.d.). https://www.bluefin.com/bluefin-news/2021-biggest-data-breaches-so-far/

Bögel, P. M. (2019). Company reputation and its influence on consumer trust in response to ongoing CSR communication. *Journal of Marketing Communications, 25*(2), 115–136. doi:10.1080/13527266.2016.1166146

Burity, J. (2021). The Importance of Logistics Efficiency on Customer Satisfaction. *Journal of Marketing Development and Competitiveness, 15*(3), 26–35. doi:10.33423/jmdc.v15i3.4537

Chen, H., Li, T., & Zhang, C. (2021). Going Too Far Is as Bad as Not Going Far Enough: An Inverted U-Shaped Relationship between Internal Controls and Operational Efficiency. *Journal of International Accounting Research, 20*(2), 25–50. doi:10.2308/JIAR-17-571

Coustasse, A. (2021). Upcoding Medicare: Is Healthcare Fraud and Abuse Increasing? *Perspectives in Health Information Management, 18*(4), 1–14. PMID:34975355

Dorminey, J. W., Fleming, A. S., Kranacher, M.-J., & Riley, R. A. (2010). Beyond the Fraud Triangle. *The CPA Journal, 80*(7), 16–23.

El Samea, D. A., & Rashed, A. (2021). The Impact of Customer's Perception of the Practices of Corporate Social Responsibility on Purchase Intention. *Journal of Accounting, Business & Management, 28*(2), 83–103.

Héroux, S., Fortin, A., & Goupil, C. (2020). Adherence to expense report approval control: An application of the theory of planned behavior. *Journal of Applied Accounting Research, 21*(3), 397–413. doi:10.1108/JAAR-12-2018-0198

Jenks, L. (2019). The Importance of Inventory Control. *Skin Inc, 31*(4), 58–63.

Loayza, A. (2020). Cyber Privateers: Protecting American Intellectual Property from Cyber Theft. *Landslide, 12*(5), 27–32.

Nguyen, A. (2020). On the Clock and at Home: Post-COVID-19 Employee Monitoring in the Workplace. *People & Strategy, 43*(3), 30–33.

Orcutt, M. (2020). An Elegy for Cash: The Technology We Might Never Replace. *MIT Technology Review, 123*(2), 32–37.

Patel, A., & Reinsch, L. (2003). Companies Can Apologize: Corporate Apologies and Legal Liability. *Business Communication Quarterly, 66*(1), 9–25. doi:10.1177/108056990306600103

Rauf, A. A. (2021). New Moralities for New Media? Assessing the Role of Social Media in Acts of Terror and Providing Points of Deliberation for Business Ethics. *Journal of Business Ethics, 170*(2), 229–251. doi:10.100710551-020-04635-w

Readwrite.com. (n.d.). *Why You Should Allow Employees to do Online Shopping at Work*. https://www.reputationsciences.com/companies-with-the-best-and-worst-reputations-at-the-start-of-2021/

Ring, C. (2019). Construction Corner: Don't Let Invoice Fraud Take Your Company Down. *Business Credit, 121*(5), 31.

Van Fleet. (2020). *Quality Time: Productivity Through Time Management*. Information Age Publishing.

ZuckermanJ. (2021). Zuckermanlaw.com/jury-awards

KEY TERMS AND DEFINITIONS

Corporate Social Responsibility: An undertaking of a private business to make themselves socially accountable and to make decisions consistent with taking care of people in need, society, and the earth. Corporate social responsibility endeavors generally focus on economic, environmental, and societal projects.

Fraud Diamond: A four-sided depiction of the determinants of fraud that include the three determinants of the fraud triangle, plus a fourth side, the fraudster's capabilities.

Fraud Triangle: A three-sided depiction of the determinants of fraud: perceived financial pressure, perceived opportunity to commit and conceal a theft, and rationalization (moral justification).

Internal Controls: Procedures designed to protect an entity's assets, the most common of which is a proper segregation of duties.

Sarbanes Oxley Act: Passed in 2002, this act covers required practices by publicly traded companies for financial and accounting purposes, and is intended to safeguard investors (shareholders), employees, and other stakeholders from losses due to financial statement fraud by company executives, and is administered by the Securities & Exchange Commission.

Segregation of Duties: A best practice of ensuring that processes surrounding company assets have more than one employee involved in a transaction, to reduce the risk of fraud or theft. A proper segregation of duties includes separate personnel responsible for initiating, general ledger recording, approving, finalizing, and reviewing transactions.

Societal Goodwill: How members of society feel about a company or organization in terms of how well that company or organization acts as a corporate citizen, regarding how they treat their customers, society, and their environment.

Chapter 5
Leadership and Strategic Decision Making Under Exogenous Shocks Such as COVID-19

Jim A. McCleskey
Western Governors University, USA

ABSTRACT

This manuscript addresses three topics relevant to the field of leadership and strategy. The topics include the effect of individual and group decision processes on leadership decision-making, the newest directions and developments in the strategic leadership literature, and the effect of exogenous shocks such as COVID-19 on leadership decision-making. It includes a thematic literature review on individual and group processes in leadership decision-making. It addresses affective and cognitive processes, cognitive biases and heuristics, group decision-making, and affective and cognitive processes in individual decision-making. It describes the latest literature on strategic leadership decisions under uncertainty and artificial intelligence as an aid to decision-making. It includes a discussion of exogenous shocks such as COVID-19 on strategic leader decision-making. It also describes the impact of exogenous shocks on small and medium-sized enterprises, crisis management, stress on decision-makers, effective decision-makers, and leader behaviors that impact crisis decision-making.

DOI: 10.4018/978-1-6684-4358-3.ch005

INTRODUCTION

This chapter aims to illuminate three specific research topics related to strategy by utilizing an extensive review of the extant literature. The first is how individual and group decision processes help or hinder leadership decision-making. The second topic addressed is the new conceptual and theoretical developments and research findings in strategy. The final topic presented in the chapter is the impact of exogenous shocks such as COVID-19 on strategic decision-making. The chapter begins with an overview of business decision-making and the strategy formulation process.

BACKGROUND

Leadership Decision-Making

No topic is as essential to success in leadership as effective decision-making. Unfortunately, leadership decision-making is a complex, fuzzy, and challenging topic for practitioners and scholars. Decision-making occurs inside the *black box* of organizational leaders' and members' minds and often within a vast array of corporate boundaries and communication links (Kleinbaum & Stuart, 2014). Despite the challenging terrain of organizational settings, people carry out decision-making, and decision-makers are simply individuals or small groups. The distinctive ability of humans to make good decisions and choose appropriate means to achieve relevant goals has resulted in unprecedented historical advancement and success (Hastie & Dawes, 2010). Multiple group and individual processes affect decision-making, and they can be divided into categories based on whether the processes are affective or cognitive (Vinokur, 1971). Some recent research publications in strategy development and execution discussed cognitive biases, strategic decision-making, and the context of uncertainty. Other published work examined the use of simulations to approximate decision-making situations, artificial intelligence (AI) as an aid to decision-making, and the quest for temporary competitive advantage (Acciarini et al. 2020).

The COVID-19 pandemic created a profound exogenous shock to societal and economic structures. This shock landed on all organizations but appeared to create an incredibly challenging situation for small and medium-sized firms due to their unique characteristics. The new reality created by the pandemic ushered in a severe crisis and the need for novel approaches to business decision-making. The correct lens for examining the outcomes involves an application of *crisis management* and leadership *decision-making under crisis*. The challenges associated with COVID-19 and the pandemic profoundly impact all areas of organizational and social life, and leadership decision-making is no exception.

MAIN FOCUS OF THE CHAPTER

Strategy Formulation Process

Leadership strategy formation is a coordinated set of activities undertaken by leadership and designed to help firms outperform their competition to achieve above-average returns and profitability (Gamble et al., 2021). Good strategy focuses on outcomes. Strategy formulation is an important reason some organizations fail while others succeed (Ott & Eisenhardt, 2020). Borrowing Rumelt's (2011) terminology, a *suitable* strategy formulation accepts and acknowledges the challenges faced by the organization, provides a realistic approach to overcoming those challenges, and focuses and coordinates all efforts toward the achievement of a decisive countermove or dramatic problem-solving effect for the organization.

To outperform other firms, organizational leaders must create a distinctive strategy and execute a unique competitive approach (Gamble et al., 2021). The combination of strategic objectives and specific operating methods utilized by the organization constitutes a business model, and the purpose of the business model is to assist the organization in the successful development and implementation of the strategy designed to achieve sustainable competitive advantage using a series of moves and counter moves (Gamble et al., 2021). Models simplify information, help summarize what is known, create visual images, and provide additional questions rather than answers (Krogerus & Tschäppeler, 2017). One of the most influential scholars in the field of strategy noted that strategy "is a path, not a fixed location," indicating the dynamic nature and process-driven quality of strategy formulation (Porter, 1980, p. xiv). In the 1990s, another classic author on strategy stated that revolutionary technological advancements, new communication and information technologies, flexible manufacturing, lower-cost labor, and new business methods conspired to heat dynamic markets across the globe (D'Aveni, 1994). These environmental forces are profoundly more relevant today, and organizations must account for these challenges while formulating and executing their strategies.

Leaders, both individually and in small groups, dynamically affect the success of the strategy formulation and implementation process. Once the leader formulates a strategy, careful evaluation is required. Not every strategy is a winning strategy, and some strategic decisions suffer from significant flaws. Organizational fit, positioning relative to internal and external environmental factors, sustainability, resource allocation, efficiency, and bottom-line financial performance all aid the decision-maker in evaluating successful strategies or flawed attempts (Gamble et al., 2021). Decision-making and strategy formulation are about people. While many factors, processes, and characteristics might be relevant to the strategy formulation and execution process, prior research suggested that the cognitive ability of the organizational leader plays a role in enacting novel strategic approaches, analyzing the external environment successfully, taking advantage of competitive opportunities, and executing strategic plans efficiently (Helfat & Peteraf, 2015; Ott & Eisenhardt, 2020).

HOW INDIVIDUAL AND GROUP DECISION PROCESSES AID OR IMPEDE LEADERHSIP DECISION MAKING

While group and individual processes may help or hinder business decision-making, prior research and conventional wisdom suggest that groups of leaders make better decisions than individual leaders (Stasser & Stewart, 1992). Various processes impact group and individual leader decision-making, including affective and cognitive processes (Vinokur, 1971). Affective refers to decision-making related to moods, feelings, attitudes, and affective processes, and organizational members frequently make decisions based on an emotional understanding of issues (Netz et al., 2020). Research suggests that when unforeseen events or exogenous shocks affect the operation of the firm, affective processes serve to focus cognitive resources, send necessary social signals regarding the high stakes nature of the situation, and may aid in the exchange of essential information (Netz et al., 2020; van Kleef, 2014). By contrast, cognitive processes typically involve general cognitive ability, attention, memory, and executive control (Gruszka et al., 2010).

The Wisdom of Crowds

Conventional wisdom suggests that groups of leaders make better decisions than individual leaders. It is counterintuitive to suggest otherwise, and popular literature states that "it is obvious that teams outperform individuals" (Katzenbach & Smith, 1999, p. 1). However, that conventional wisdom may not hold up to research and scrutiny in the realm of decision-making processes. Groups of leaders typically make better decisions than individual leaders when complex decisions are needed (Kao & Couzin, 2014; Stasser & Stewart, 1992) when rapid judgment is required (Gürçay et al., 2014), and whenever the group task requires augmented cognitive capabilities (Valeriani et al., 2017). However, the advantage of groups disappears if the group of leaders fails to communicate and collaborate effectively (Kerr & Tindale, 2004), withholds information from its members (Stasser & Titus, 1985), portrays a more aggressive or defensive style of interaction (Branson et al., 2010), enacts judgment biases, or is composed of weak leaders (Kerr & Tinedale, 2004; Kerr et al., 1996). For these reasons, previous researchers suggested that group leadership decision-making may be less efficacious than individual leader decision-making (Bahrami et al., 2010; King et al., 2012), particularly when the decision is time sensitive (Valeriani et al., 2017).

Other scholars posited that when group members are aware of the decisions of fellow group members, individual confidence increases and sometimes creates lower quality decision-making in the group on average (Lorenz et al., 2011; Valeriani et al., 2017). Group decision-making can also worsen over time due to increased group interaction and communication levels. Group decision-making is less effective than individual decision-making when the group members are geographically dispersed (Carneiro et al., 2019). Overall, the research on group decision-making processes has produced mixed results (Valeriani et al., 2017). Group decision-making may be better than decision-making by individuals depending upon the context, differences in group communications, group style, group composition, and the presence of solid emergent leadership.

Group Leadership Decision-making

Group decision-making typically involves establishing a consensus. Consensus requires discussion, deliberation, and sometimes debate between the group members to arrive at a decision that reflects the opinions of the entire group and is acceptable to all group members (Pérez et al., 2018). Achieving consensus is a group process that involves many individual processes. Groups accomplish consensus via a multistage iterative negotiation process that ends once the level of consensus reaches a minimum threshold requirement. The consensus process is dynamic because real-world consensus involves an ever-changing context, and new elements appear in the situation as the consensus is negotiated and achieved. Involving more people in the group decision process theoretically leads to better decisions, increased buy-in from the group, and higher quality decision-making. Ideally, group decisions would be unanimous; however, this is not always possible in organizational settings, particularly with leadership teams that may include members with high levels of ego involvement. Accordingly, the goal is a minimally acceptable consensus threshold (Pérez et al., 2018). As affective and cognitive processes play out within the group, they affect the consensus-building process.

Group Affective Processes

Group affective processes that impact decision-making include diffusion of responsibility and familiarization (Vinokur, 1971). Diffusion of Responsibility (DoR) suggests that people become less likely to decide if they are aware of others who can make the decision instead (Choo et al., 2019; Darley & Latane, 1968; Fischer et al., 2011). Choo et al. (2019) examined the whistleblower's decision-making context in an experimental design with 48 participants and found that in a high-stakes decision-making environment, DoR substantially reduced the willingness of participants to decide. Decision-makers were willing to allow others to take the risk associated with deciding, provided they knew that those others existed and could choose in their place (Choo et al., 2019). Prior research showed that the DoR effect reduced the likelihood that witnesses to a potential emergency would attempt to contact authorities or help. The presence of unseen others was adequate to substantially reduce decision-making on the part of witnesses faced with a high-stakes, potentially life-threatening situation (Darley & Latane, 1968). The DoR helps explain the well-documented *bystander effect*. In a meta-analytic study of 105 samples totaling 7700 participants, Fischer et al. (2011) explored the role of context, situational influences, degree of danger, presence of aggressors, and other factors on how the DoR plays out within the context of the bystander effect. Whenever others are present who can make difficult decisions, decision-makers are far less likely to act or encourage others to act. Applying this thinking to group decision-making processes reveals a troubling tendency for decision-makers operating within groups to avoid difficult and complex decisions while waiting for other group members to step forward. This hesitance is similar to the concept of social loafing but, in this case, falls under the more accurate heading of DoR.

Familiarization suggests that before a group member feels comfortable and familiar with the other group members, they will be cautious in their decision-making (Vinokur, 1971). After becoming familiar and comfortable with the group, the decision-making becomes less cautious and risk-tolerant. Representative bias, also known as familiarity bias, may be prompted by a lack of information and triggers cognitive shortcuts in decision-making (Jain et al., 2020). Prior studies suggested that effective group decision-making is contingent on having the right combination of group members and utilizing an optimal decision-making strategy (Kerr & Tinedale, 2004; Zhu et al., 2021). During the *forming* stage (Ilgen et al., 2005) described by Vinokur (1971) as pre-familiarization, team decision-making is primarily influenced by individual decision-making processes and styles until the group establishes a performance record. After that, the valence of past team performance solidifies as meaningful feedback and alters the team process of decision-making over time (Zhu et al., 2021).

Group Cognitive Processes

Group cognitive processes for decision-making include the planning perspective, rational decision-making, and brainstorming. The planning perspective refers to group decision-making using rational approaches based on preliminary plans and utilizing contingency frameworks (Netz et al., 2020). Group decisions are more likely to result in high-quality decision-making when group members have access to a larger supply and broader range of information (Brodbeck et al., 2007). Additionally, making group decisions in high-stakes strategic environments or due to crises requires high-speed cognitive processing, collaborative group communications, and the confidence to act (Netz et al., 2020). Specific group, cognitive process decision-making approaches include the enactment of "multi-criteria decision-making methods" to support complex decisions based upon a list of potential preferences, the use of weighted criteria, evaluation of alternatives, ranking of possible solutions, and the selection of an appropriate decision solution (Ardebili & Padoano, 2020, p. 1).

Group cognitive processes suffer from a variety of potential shortcomings. Groups tend to make decisions less effectively when they focus on negotiation rather than information exchange or when they enact biases, including sampling bias and repetition bias during the discussions (already shared information within the group is discussed more often than unshared information). Further, groups make decisions less effectively when the members give way to individual-level processes, including various types of evaluation bias such as ownership bias, social validation, and the consistency effect (Brodbeck et al., 2007). Group biases can stem from group size, prejudgment by individual group members, bias among individuals, and the group-judgment process (Kerr et al., 1996). Additionally, the likelihood of group decision-making bias increases as the group size increases. Group members favor shared information and social validation over new unshared information and unfavorable evaluation from peers. These processes can lead to less effective decision-making, groupthink, and missed opportunities.

Individual Decision-Making

Individual decision-making processes broadly fall into two categories based on the level of rationality. A rational decision-making process meets four specific criteria. It is based on an evaluation of the decision-makers' current resources, takes into account the consequences of the decision, evaluates the likelihood of outcomes based upon the level of uncertainty, and is adaptive to the constraints of the probability of results, range of choices, likely consequences, and those aforementioned resources (Hastie & Dawes, 2010). Rational decision-making processes are synonymous with cognitive processes. The second category of decision-making processes is not directly related to rational decision-making criteria and includes three types of affective decision-making processes: habits, conformity, and religious or cultural mandates (Hastie & Dawes, 2010). During affective decision-making processes, decision-makers do what they have done before, do what other people do or have done, or do what their cultural upbringing suggests they should do. Affective decision-making processes are common and powerful approaches to leadership decision-making because they operate quickly and efficiently on mental *System One* (Tversky and Kahneman, 1992). System One and System Two thinking are addressed in greater detail later in the chapter.

Individual Affective Processes

While cognitive biases receive generous treatment in the extant literature, a relative paucity of studies addressed individual affective processes based upon bias. Motivational biases include conscious and subconscious judgments based upon self-interest, external social pressure, or the context of the organizational problem (Montibeller & von Winterfeldt, 2015). The relative desirability of events, outcomes, or decision-making choices influences motivational biases. Examples may include overly optimistic forecasts, underestimating project costs, poor estimates of the time it takes to complete an organizational project or implementation, or any other situation or accurate judgment required by the decision-maker but withheld for personally motivated reasons. Motivational biases differ from cognitive biases. They are not based on errors in the decision maker's rational thinking and differ from heuristics because they do not represent mental shortcuts or attempts to streamline decision-making when facing complex or fuzzy problems. Motivational biases can be influenced by affective processes when a decision-maker holds an emotional predisposition for or against a specific decision, position, or outcome, whether consciously or subconsciously (Rottenstreich & Hsee, 2001; Siegrist & Sütterlin, 2014). Examples of motivational biases include the overconfidence bias (previously discussed), the desirability bias, the affect influenced bias, and the desirability of options bias. The desirability bias occurs when decision-makers assign higher probabilities and likelihoods to outcomes they already prefer to encounter. The affect-influenced bias occurs when potential outcomes of the decision can trigger an emotional reaction in the decision-maker that causes them to misestimate the probability of its occurrence. The desirability of options bias can lead decision-makers to incorrectly estimate probabilities in either direction that favors their preferred alternatives (Montibeller & von Winterfeldt, 2015).

Tool-based mental shortcuts are another example of individual affective processes that impact decision-making (Netz et al., 2020). These tools and techniques refer to affective reactions as real-world practices and mental shortcuts that managers use to perform strategic work, including decision-making. Cognitive biases, decision-making heuristics, and decision-maker intuition are examples of mental shortcuts that can arise from individual affective processes or individual cognitive processes. The next section of the chapter addresses these mental shortcuts in greater detail.

Individual Cognitive Processes

One of the critical challenges associated with leadership decision-making is uncertainty. In many situations, the decision-maker is unaware of all the possible outcomes, inputs, consequences, and likelihoods associated with a decision (Artinger et al., 2015). Uncertainty is one of the critical weaknesses of the popular *rational choice model* for decision-making. Rational choice modeling assumes that the likelihood of outcomes is known and that calculating an optimal solution is possible, which is typically not the case with real-world, dynamic, complex decision-making (Artinger et al., 2015).

Another potential danger associated with individual cognitive processes as an aid to decision-making is the utilization of cognitive biases by the decision-maker. A cognitive bias refers to irrational beliefs that can influence the ability to decide based on a rational examination of facts and evidence (Schwenk, 1986). Cognitive biases represent an ever-present element of individual strategic decision-making (Das & Teng, 1999).

Cognitive Biases. Jain et al. (2020) described various cognitive biases that can negatively impact decision-making, including the overconfidence bias, representative bias, anchoring bias, availability bias, regret aversion bias, loss aversion bias, mental accounting bias, and the herding bias. Overconfidence bias is the tendency for people to overestimate their capabilities and underestimate future uncertainty or risk. Representative bias, also known as familiarity bias, occurs during a lack of information and triggers cognitive shortcuts in decision-making. Anchoring bias occurs when the decision-maker becomes cognitively attached to a specific piece of quantitative data unrelated to the decision alternatives.

The availability bias occurs when decision-makers heavily rely on readily available knowledge rather than seeking alternative information to aid in the decision. Regret aversion bias is an affective bias created by excessively focusing on regret after making an incorrect decision. Loss aversion bias is primarily an affective process bias wherein decision-makers focus on potential losses and their fear of loss rather than potential gains. Mental accounting bias, also known as the Two-Pocket Theory, occurs when decision-makers treat each aspect of the decision as an independent decision rather than seeing a holistic decision and potential outcome. Herding bias involves decision-makers imitating the actions of other decision-makers without thinking independently about whether those decisions were rational and appropriate (Jain et al., 2020). Krogerus and Tschäppeler (2017) describe other cognitive biases, including confirmation and fast/slow errors. The confirmation error involves interpreting information about the decision based upon the decision makers' already existing assumptions or convictions. The confirmation error includes blocking out information that contradicts the decision maker's current opinion.

Tversky and Kahneman (1992) situated the fast/slow error in *System One* and *System Two* thinking. System One thinking is fast and intuitive and capable of solving well-understood or straightforward problems efficiently. System two thinking is intentional—addressing complex, fuzzy, or poorly understood problems. The cognitive bias occurs when decision-makers utilize their System One thinking to solve System Two problems (Krogerus & Tschäppeler, 2017). This list of cognitive and other biases is far from exhaustive, and McRaney (2011) published a list of forty-eight different biases. A discussion of all 48 cognitive and other biases is beyond the scope of this chapter. People often discuss cognitive biases and heuristics together; however, grouping these concepts together may be inappropriate since research suggests several positive outcomes associated with heuristics (Gigerenzer & Brighton, 2009).

Heuristics. Heuristics are cognitive shortcuts in decision-making that occur when information, processing capacity, and time are in short supply (Bingham & Eisenhardt, 2011). In contrast to cognitive and other biases, research suggests that heuristics can be more accurate than complex analytical procedures and are helpful for successfully exploiting cognitive abilities and environmental challenges (Gigerenzer & Brighton, 2009). Gigerenzer and Brighton (2009) posited that heuristics represent a form of pattern recognition, similar to machine learning concepts, based upon analytical shortcuts that allow decision-makers to predict more accurately than longer form rational decision-making strategies in many cases. Sometimes less information is more powerful and more accurate.

Heuristics are "ecologically rational," are accurate compared to the relative structure of the environment, and genuinely describe how decision-making functions in organizational settings (Pieterse & de Vries, 2013, para. 11). A few affective heuristics include the recognition heuristic, the fluency heuristic, the take-the-best heuristic, and the unit-weight linear model heuristic. Heuristics result in greater accuracy when the decision-making criteria are uncertain, the sample size is minimal, or the need to avoid variance is greater than the need to avoid bias. In these cases, bias can operate through simple heuristics and can be more efficient and accurate than a complex decision-making approach (Gigerenzer & Brighton, 2009). Pieterse and de Vries (2013) suggested that human decision-making depends on mental shortcuts and intuitive strategies such as relying on emotional responses, deciding after a brief wait, or heuristics. These mental shortcuts often result in better judgments and decision-making than extensive information gathering and evaluative processes.

THE NEWEST DIRECTIONS AND DEVELOPMENTS IN STRATEGY DEVELOPMENT AND EXECUTION

Recent research in strategy development and execution includes making strategic decisions in uncertain environments, using simulations to aid decision-making, utilizing AI to aid human organizational decision-makers, and the quest for temporary competitive advantages.

Strategic Decisions Under Uncertainty

Motlaghzadeh et al. (2020) studied decision-making in a multi-agent format utilizing game theory to examine strategic interactions among the decision-makers. This study surveyed the theoretical landscape based on cognitive decision- and rational decision-making models. The results showed that the authors successfully tested and evaluated a new decision-making methodology utilizing game theory. The authors noted that perfectly rational decisions are often not feasible in the real world because of the intractability of problems and the necessity of a hierarchical decision-making process (Motlaghzadeh et al., 2020).

Roelich and Giesekam (2019) studied decision-making under uncertainty in climate change action. The authors suggest that climate change mitigation is fraught with a high level of uncertainty based on the insufficiency of models, the need to set boundaries, the lack of accurate measurements, and several other systematic issues that make the construction of knowledge difficult if not impossible (Butler et al., 2015; Roelich & Giedekman, 2019). Citing several decision-making approaches designed to help decision-makers understand and manage uncertainty, the authors recommended decision-support approaches with multiple distinct drivers of decision-making, including behavioral, financial, and strategic, while highlighting the need to account for different attempts to mitigate risk and deal with uncertainty (Grubb et al., 2014; Roelich & Giedekman, 2019).

Schröder and Gilboa Freedman (2020) noted that while economists use economic preference and rational decision-making in utility theory to study decision-making under uncertainty, psychologists use personality traits, behaviors, and responses to contextual situations to analyze decision-making under uncertainty. The authors produced a study that examined both the economic concept of ambiguity and the personality trait of ambiguity intolerance in the same survey. This design is equivalent to studying both affective decision processes and cognitive decision processes, where the affective processes represent the psychological traits, and cognitive processes encompass the rational decision-making of the economists' model. The authors concluded that the relationship between economic and psychological measures to assess decision-making under uncertainty reveals a complex picture of decision-making characteristics that may not be measuring the same inputs or outcomes. Ambiguity aversion and ambiguity intolerance are distinct concepts, and economic preferences obtained from self-assessment survey questions appear to align better with the personality traits of ambiguity intolerance (Schröder & Gilboa Freedman, 2020). The authors note that the outcome may not be surprising since economic preferences measure human decision-making along narrowly defined models while personality traits attempt to measure more general patterns of thoughts and behaviors. Additional research in this area may prove fruitful.

An and Dorofeev (2019) examined foreign currency trading decision-making based on expert polls rather than the historical computer simulation models. The study revealed that historical simulation modeling was more accurate than expert polling for predicting the future value of foreign exchange currency pairs. Historical modeling was superior to the judgment of experts. Interestingly, expert polls were superior for predicting values beyond six months into the future compared to historical modeling. Short-term trends were more accessible from a historical modeling algorithm (An & Dorofeev, 2019). The use of algorithms in decision-making is only the beginning. Organizations also use machine learning and big data to aid decision-making in various applications.

Artificial Intelligence as an Aide to Decision Making

Huntingford et al. (2019) posited that AI could assist climate change research and decision-making about weather patterns and future dangers related to climate change. Climate change lends itself to an AI solution because of the complexity and fuzzy problem definition inherent in understanding weather and climate patterns. Stefan and Carutasu (2020) proposed using AI to aid ethical decision-making in organizations. The authors stress the need for a system with built-in reliability, safety, and transparency of all artificial intelligence-based decisions. Wong and Shayegan (2019) suggested that machine learning is not a new technology and has been increasingly applied to fields from economics to medicine to diagnostics. The authors argued that AI and machine learning could directly assist patient care by improving the accuracy of the prediction of diseases, the outcomes and likely prognosis for patients, and the analysis of potential treatments and solutions for in-patient care. Despite significant advancements in decision-making in various applications based upon AI and machine learning, ethical concerns have also sprung up around the use of machine learning.

Allen and Masters (2020) describe how utilizing AI and machine learning algorithms to make automated decisions can give rise to discrimination and promote inequality. Implementing automated decision-making in Europe has proceeded faster than implementing AI applications in the US. Automated decision-making involves drawing conclusions without the aid of human involvement. This type of automation involves using a specific, limited algorithm created by humans. However, automated decision-making can include utilizing complex AI algorithms supported by machine learning. Unfortunately, these systems can display discriminatory behavior through learning processes (Allen & Masters, 2020). Because decision-making in a machine language-based algorithm is hidden, significant concerns about the potential for discrimination along the lines of race, gender, ethnicity, or other factors have arisen alongside the technology. Opponents of the technology have raised concerns about the fact that there is not a credible mechanism to allow people to understand how decision-making occurs in the *black box* of artificially intelligent machine learning, resulting in a spectacular lack of transparency about the decision-making guided by the technology (Allen & Masters, 2020).

Eriksson et al. (2020) explored how organizations can use AI to formulate marketing strategies. Information has become a critical organizational asset, and the ability to analyze, visualize, and draw conclusions from data has become an essential organizational competency (Gartner, 2019). AI may provide decision-makers with a systematic ability to process and interpret relevant data and the ability to achieve specific objectives through adaption. Today, organizations use AI to translate big data into manageable information and knowledge for decision-makers. AI is becoming embedded in everyday life and garnering ever more attention regarding its utilization in business and commerce (Eriksson et al., 2020). After conducting 13 interviews on AI and marketing strategy formulation, the authors characterized the information across five distinct themes. Respondents saw opportunities to mitigate the inherent complexity of consumer behavior analysis, viewed AI as playing an operational role in ongoing data analytics, and believed AI could support strategic management decisions alongside traditional strategic decision-making activities. Respondents also thought that the organizational culture must be ready for AI and felt that AI influences how human resources are utilized and can help guide change in the business's organizational structure (Eriksson et al., 2020). The authors suggested that AI can "augment, replace, or complement the human decision-making process" (Eriksson et al., 2020, p. 809) and the future of marketing strategy creation (Jarrahi, 2018).

Temporary Competitive Advantage

Dagnino et al. (2021) posited that traditional methods of achieving competitive advantage disappeared quickly in many industries. Accordingly, organizational leaders must continually reformulate their strategic direction, rethink their plans, and acquire new sources of competitive advantage. A rapid pace of new product introductions, technological innovation, competitive actions, and increasingly fast moves and counter moves through innovation and adaption has resulted in a state of Hypercompetition in many industries. (D'Aveni, 1994). This observation led some scholars and practitioners to argue that today only a temporary competitive advantage (TCA) exists for organizations (Dagnino et al., 2021; McGrath, 2013; Sirmon et al., 2011). The challenge of TCA requires organizations to understand the antecedents of quickly evaporating competitive advantage, new methods of managing resources, and the competitive actions required to execute strategy in a TCA environment. While previous literature discussed the concept of sustainable competitive advantage, current literature focuses on the realities of TCA. Conceptually, TCA is not a novel theory, but rather the realization that the old theory did not fit the facts.

Traditional sources of competitive advantage have repeatedly failed to achieve competitive success, shown themselves to be insufficient for effective competition in the marketplace, failed to anticipate rapidly changing environments, and disappeared seemingly overnight. TCA emphasizes that advantage erodes routinely due to the dynamic interplay between competitors and the naturally occurring rivalry within industries (Dagnino et al., 2021). TCA is characterized by rapidly diminishing profits as competitors that previously enjoyed competitive advantage approached the mean profitability within the industry (Pacheco-de-Almeida, 2010). The antecedents of TCA include the process of globalization, increased demand uncertainty, high levels of technological advancement, and intense and rapid industry competitive actions based upon high levels of rivalry (Dagnino et al., 2021; D'Aveni et al., 2010). The requisite coping behaviors required of organizational managers and leaders given the realities of TCA include tighter management of resources, a better understanding of opportunities and threats, the ability to seize opportunities aggressively, and the skills to enhance, combine, protect, and reconfigure organizational assets as needed. Additional adaptations include high-speed recognition of competitive actions, high-speed decision-making, excellent execution of short product lifecycles, decreased time to markets, and the ability to project outcomes accurately. Organizational leaders must have outstanding pattern recognition and must be able to spot competitive sequences as they unfold in the marketplace (Dagnino et al., 2021; D'Aveni et al., 2010). The predictable consequences of the TCA environment include volatility of competitive performance, sudden and sometimes unexpected market share erosion, the loss of position for industry leaders, sudden opportunities to scale up production, and the "Red Queen effect" (Dagnino et al., 2021, p. 90). The *Red Queen Effect* refers to a situation where the challenging firm's aggressive actions and innovations improve its performance and it can dethrone the market leader; however, the challenging firm's actions also escalate the level of rivalry in the marketplace, which negatively affects the performance for all players in the market including the challenging firm. Accordingly, all competitors must increase the number, speed, and effectiveness of competitive actions, and all firms end up working harder and potentially realizing lower margins and less profitable operations (Dagnino et al., 2021).

Because of the volatility of performance associated with a TCA industry, conventional evaluations of profitability and performance become somewhat ineffective at capturing firm success or failure. When subjected to this volatility level, the financial performance analysis should be adjusted to reflect the presence of TCA and the variances between different markets, countries, and geographical marketplaces (Dagnino et al., 2021). In a TCA environment, product quality, performance, rate of product introduction, product lifecycle, and other key performance indicators related to speed and successful execution of the marketplace make ideal measures of organizational success rather than simply relying on financial performance ratios (Parker et al., 2017). The extant literature around TCA is growing, and the recognition that all competitive advantage is ultimately temporary is a highly intuitive observation of the practitioner concerned with strategy formulation and execution.

THE EFFECT OF EXOGENOUS SHOCKS SUCH AS COVID-19 ON LEADERSHIP DECISION MAKING

The COVID-19 pandemic has been categorized as a "metaphorical black swan event" (Kuckertz et al., 2020, p. 2), indicating a surprising and unpredictable event that triggers highly significant, severe consequences that impact the political and economic environment. The resulting lockdown of large portions of society and corresponding economic fallout constitute an acute crisis, and the consequences persist. While research suggests that a variety of different types of firms and organizations have been able to tap into their adaptive capacity and mitigate the effects of the crisis, small and medium-sized firms, particularly family-owned firms, may feel the effects of exogenous shocks more profoundly (Rodrigues et al., 2021; Soluk et al., 2021). These firms represent the economic backbone of industrialized nations. The more profound effects result from resource scarcity, unique forms of innovativeness, entrepreneurial orientations, and a tendency to prioritize economic and non-economic objectives (Soluk et al., 2021). Indeed, the very characteristics of these firms make them more susceptible to this type of exogenous shock (Rodrigues et al., 2021). The impacts on management and leadership decision-making have been and continue to be a concern in the post-COVID-19 economic reality. The type of decision-making required in a crisis, such as the one created by severe exogenous shocks to the economic and political system (e.g., COVID-19), is called *crisis management.*

Crisis Management

According to Al-Helou (2011), crisis management is an administrative process aimed at addressing crisis conditions with an emphasis on preparation and planning for the crisis and is primarily dependent on the organization's ability to predict crises and develop accurate scenarios and plans to mitigate those crises once they occur. Previously, scholars posited theoretically and empirically that effective decision-making is a critical responsibility for managers in crises (Aguilera, 2000; Hadley et al., 2011). Decision-making during times of crisis involves sensitive and vital decisions and typically includes a time limit, dynamic circumstances, and an increased need for definitive action (Al-Dabbagh, 2020). The decision-making process depends, at least in part, on the creativity and ability of the decision-maker, and previous authors argued that a dialectical and influential relationship exists between rational decision-making and decision-makers in crises (Al-Dabbagh, 2020; Youssef, 2017).

In addition to emphasizing rationality, scholars suggest that decision-making in a crisis should be deliberative, inclusive, transparent, and highly accountable to the various stakeholders involved in the decision (Norheim et al., 2021). Often, it may be challenging to accomplish this type of decision-making based on the high level of stress involved in crisis decisions and the compressed timeframe provided by crises. Other scholars developed models of leader efficacy for crisis decision-making (Hadley et al., 2011). The authors posited that the consequences of poor leadership in a crisis are costly and deadly. Their research specifically focused on public health crises since these types of crises are enormously significant for human health and safety. Significant crises typically involve cause, effect, and resolution ambiguity in addition to a high level of urgency. Crises create additional stress for the decision-maker, which exacerbates risk-taking and negatively impacts decision quality (Phillips-Wren & Ayda, 2020). Decision-makers under stress make riskier decisions, commit more errors of cognition, resort to stereotypes more often, and ignore context and situational factors during decision-making compared to decision-makers under lower levels of stress. These outcomes make it more difficult for decision-makers to operate effectively.

Effective decision-making requirements during a crisis include clarifying the situation to external stakeholders, maintaining a message of hope and resoluteness, providing coordination to the staff involved in effecting a resolution, and motivating individuals working towards the solution (Hadley et al., 2011). Two aspects characterize quality decision-making: the decision-making process and the decision-making outcome, and both the process and the outcome should produce a solution that maximizes the utility given the situational context (Phillips-Wren & Ayda, 2020). The successful resolution of a crisis manifests through a series of judgments and decisions (Aguilera, 2000; Hadley et al., 2011), and for that reason, information analysis and effective decision making constitute the essential core behaviors of the decision-maker.

Information assessment includes procedural and structural aspects of information flow and assessment, such as avoiding biases, minimizing errors, identifying necessary data, and collecting information (Hadley et al., 2011). Assessment of information must include evaluating the type and quantity of information available and may require access to multiple unique data sources. The quantity is significant because either too little or too much information leads to lower quality decision-making in a crisis (Phillips-Wren & Ayda, 2020). If the decision-maker receives more information than can be effectively processed, information overload results, and the excessively high volume of information operates akin to background noise, resulting in stress, distraction, and judgment errors. Hadley et al. (2011) posited that analyzing potential decisions in a crisis context requires generating multiple response options, using objective criteria to assess the preferred course of action, making logical recommendations, and taking appropriate actions based on the results.

Decision-making in crises often involves carrying out evaluation and assessment under significant physical and psychological demands, and the characteristic elements that define a crisis, including high-stakes outcomes, urgency, and a high level of ambiguity, severely constrain managers' ability to make effective decisions and to assess information accurately (Hadley et al., 2011). The time constraint is another significant factor impacting decision-maker effectiveness. Although the perception of time pressure may positively affect decision-maker focus, it is more often associated with lower quality decision-making and reduced performance quality as time-pressured decision-makers engage in a reduced number of cognitive tasks and process data with greater error rates (Phillips-Wren & Ayda, 2020). Decision-making under crisis labors from the inability on the part of the decision-maker to acquire a complete understanding of the underlying causes, mechanisms of influence, and likely consequences emanating from the crisis context (Hadley et al., 2020). Further, managers and leaders in crisis contexts experience severe time pressures and may not be able to acquire, analyze, and process information as effectively. Accordingly, a crisis constitutes a severe constraint on individual leaders' ability to assess information and successfully make decisions.

LEADER BEHAVIORS THAT IMPACT DECISION MAKING UNDER CRISIS

The COVID-19 crisis reminded everyone of the essential role that leaders play in mitigating unexpected and threatening events worldwide (Bundy et al., 2017; Lagowska et al., 2020). Accordingly, leaders received an increasing amount of attention and scrutiny during the crisis because followers expect leaders to provide comfort, hope, guidance, and meaningful information during times of uncertainty (Lagowska et al., 2020). Given the importance of effective leader decision-making under crisis, it is worth examining what traits or behaviors may increase leader efficacy in crises. Prior research examined the importance of ethical behavior during times of crisis (Potocan & Nedleko, 2021), the role of state and trait anxiety (McCleskey & Gruda, 2021), the impact of a decision-maker, and organizational resilience (Kuckertz et al., 2020; Rodrigues et al., 2021), the importance of effective communication, and the impact of increasingly complex networks (Jong et al., 2015). A review by Jong et al. (2015) sorted crisis management behaviors into five categories, including sense-making, meaning-making, decision-making, learning, and terminating. Sense-making involves interpreting signals and parsing out the correct message from the background. Meaning-making includes communicating the broader impact of the crisis to relevant stakeholders. Decision-making is self-explanatory and the central theme of this chapter. Learning includes evaluating the crisis context and appropriately extracting the lessons learned. Terminating includes taking appropriate measures to exit the crisis mode and return to a baseline of normal behavior.

All five of these crisis management behaviors are critical to successfully handling a crisis created by an exogenous shock. Mather (2020) argued that organizational leaders must act with urgency during a crisis to take responsibility, respond to mistakes, adapt and change constantly, communicate with transparency and clarity, remain aligned with the organizational purpose and values, practice agile decision-making, and tear down organizational silos. Large-scale exogenous shocks create significant shifts in the business and societal landscape, and leaders have an essential role in the required transformation of their organizations, governments, and processes to adapt to the new normal created in the post-shock environment (Mather, 2020).

FUTURE RESEARCH DIRECTIONS

This chapter described three specific research topics related to strategy by utilizing an extensive review of the extant literature. They include how individual and group decision processes help or hinder leadership decision-making, the new conceptual and theoretical developments and research findings in strategy, and the impact of exogenous shocks such as COVID-19 on strategic decision-making. While a thorough thematic literature review is included, additional research conducted inside organizational settings that considers specific leadership contexts and the effect of exogenous shocks including but not limited to COVID-19 is needed to establish the effects of COVID-19 specifically on leadership decision-making. This chapter is not an endpoint. Rather, it is a launching point for additional research to fully consider the effects of exogenous shocks including COVID-19 on leadership decision-making given the effects of group and individual processes.

CONCLUSION

Few topics are as interesting from a scholarly standpoint or as critical to the organization's success from the practitioner's standpoint as leadership decision-making and strategy. This chapter presented a thematic literature review designed to address three specific questions related to the extant literature. The first dealt with individual and group leadership decision-making processes, and the themes presented included affective processes, cognitive processes, cognitive biases, and the use of heuristics. The second question addressed new directions and developments in the literature on leadership strategy development and execution. The specific themes presented included leadership decision-making under uncertainty and AI as an aid to decision-making. The third topic involved the effect of exogenous shocks on leadership decision-making, and the themes in this section included the nature of exogenous shocks, the impact on small and medium-sized firms, the role of crisis management, stressors on leadership decision-makers, effective decision-making, and behaviors and traits that impact leadership decision-making under exogenous shocks.

ACKNOWLEDGMENT

This research received no specific grant from any funding agency in the public, commercial, or not-for-profit sectors.

REFERENCES

Acciarini, C., Brunetta, F., & Boccardelli, P. (2020). Cognitive biases and decision-making strategies in times of change: A systematic literature review. *Management Decision*. doi:10.1108/MD-07-2019-1006

Aguilera, D. C. (2000). *Crisis intervention: Theory and methodology* (8th ed.). Mosby.

Al-Dabbagh, Z. S. (2020). The role of Decision-maker in crisis management: A qualitative study using grounded theory (COVID-19 pandemic crisis as A model). *Journal of Public Affairs*, *20*(4). Advance online publication. doi:10.1002/pa.2186 PMID:32904918

Al-Helou, G. (2011). School crises in government secondary schools in the northern west bank directorates. *An-Najah University Journal, 24*(1), 223–262.

Alhawamdeh, H. M., & Alsmairat, M. A. K. (2019). Strategic decision making and organization performance: A literature review. *International Review of Management and Marketing*, *9*(4), 95–99. doi:10.32479/irmm.8161

Allen, R., & Masters, D. (2020). Artificial intelligence: The right to protection from discrimination caused by algorithms, machine learning, and automated decision-making. *ERA-Forum, 20*(4), 585-598. 10.100712027-019-00582-w

An, J., & Dorofeev, M. (2019). Short-term foreign exchange forecasting: Decision-making based on expert polls. *Investment Management & Financial Innovations*, *16*(4), 215–228. doi:10.21511/imfi.16(4).2019.19

Ardebili, A. A., & Padoano, E. (2020). A literature review of the concepts of resilience and sustainability in group decision-making. *Sustainability (Basel, Switzerland)*, *12*(7), 2602. doi:10.3390u12072602

Artinger, F., Petersen, M., Gigerenzer, G., & Weibler, J. (2015). Heuristics as adaptive decision strategies in management. *Journal of Organizational Behavior*, *36*(S1), S33–S52. doi:10.1002/job.1950

Bahrami, B., Olsen, K., Latham, P. E., Roepstorff, A., Rees, G., & Frith, C. D. (2010). Optimally interacting minds. *Science*, *329*(5995), 1081–1085. doi:10.1126cience.1185718 PMID:20798320

Bingham, C. B., & Eisenhardt, K. M. (2011). Rational heuristics: The 'simple rules' that strategists learn from process experience. *Strategic Management Journal*, *32*(13), 1437–1464. doi:10.1002mj.965

Branson, L., Steele, N. L., & Sung, C.-H. (2010). When two heads are worse than one: Impact of group style and information type on performance evaluation. *Journal of Business and Behavioral Sciences*, *22*, 75–84.

Brodbeck, F. C., Kerschreiter, R., Mojzisch, A., & Schulz-Hardt, S. (2007). Group decision making under conditions of distributed knowledge: The information asymmetries model. *Academy of Management Review*, *32*(2), 459–479. doi:10.5465/amr.2007.24351441

Bundy, J., Pfarrer, M. D., Short, C. E., & Coombs, W. T. (2017). Crises and crisis management: Integration, interpretation, and research development. *Journal of Management*, *43*(6), 1661–1692. doi:10.1177/0149206316680030

Butler, C., Demski, C., Parkhill, K., Pidgeon, N., & Spence, A. (2015). Public values for energy futures: Framing, indeterminacy, and policymaking. *Energy Policy, 87*, 665–672. doi:10.1016/j.enpol.2015.01.035

Carneiro, J., Martinho, D., Marreiros, G., & Novais, P. (2019). How cognitive and affective aspects can influence the outcome of the group decision-making process. *Expert Systems, 36*(3), e12257. doi:10.1111/exsy.12257

Choo, L., Grimm, V., Horváth, G., & Nitta, K. (2019). Whistleblowing and diffusion of responsibility: An experiment. *European Economic Review, 119*, 287–301. doi:10.1016/j.euroecorev.2019.07.010

D'Aveni, R. A. (1994). *Hypercompetition: Managing the Dynamics of Strategic Maneuvering*. The Free Press.

D'Aveni, R. A., Dagnino, G. B., & Smith, K. G. (2010). The age of temporary advantage. *Strategic Management Journal, 31*(13), 1371–1385. doi:10.1002mj.897

Dagnino, G. B., Picone, P. M., & Ferrigno, G. (2021). Temporary competitive advantage: A State-of-the-Art literature review and research directions. *International Journal of Management Reviews: IJMR, 23*(1), 85–115. doi:10.1111/ijmr.12242

Darley, J. M., & Latane, B. (1968). Bystander intervention in emergencies: Diffusion of responsibility. *Journal of Personality and Social Psychology, 8*(4), 377-383. doi:10.1037/h0025589

Das, T. K., & Teng, B. (1999). Cognitive biases and strategic decision processes: An integrative perspective. *Journal of Management Studies, 36*(6), 757–778. doi:10.1111/1467-6486.00157

Elbanna, S., Thanos, I. C., & Jansen, R. J. G. (2020). A literature review of the strategic decision-making context: A synthesis of previous mixed findings and an agenda for the way forward. *Management, 23*(2), 42–60. doi:10.37725/mgmt.v23i2.4621

Eriksson, T., Bigi, A., & Bonera, M. (2020). Think with me, or think for me? on the future role of artificial intelligence in marketing strategy formulation. *The TQM Journal, 32*(4), 795–814. doi:10.1108/TQM-12-2019-0303

Fischer, P., Krueger, J. I., Greitemeyer, T., Vogrincic, C., Kastenmuller, A., Frey, D., Heene, M., Wicher, M., & Kainbacher, M. (2011). The bystander-effect: A meta-analytic review on bystander intervention in dangerous and non-dangerous emergencies. *Psychological Bulletin, 137*(4), 517–537. doi:10.1037/a0023304 PMID:21534650

Gamble, J. E., Peteraf, M. A., & Thompson, A. A. (2021). *Essentials of Strategic Management: The quest for competitive advantage* (7th ed.). McGraw-Hill Education.

Gartner. (2019). Available at: https://www.gartner.com/en/doc/374107-predicts-2019-

Gigerenzer, G., & Brighton, H. (2009). Homo heuristics: Why biased minds make better inferences. *Topics in Cognitive Science*, *1*(1), 107–143. https://doi.org/10.1111/j.1756-8765.2008.01006.x

Grubb, M., Hourcade, J.-C., & Neuhoff, K. (2014). *Planetary Economics: Energy, climate change, and the three domains of sustainable development*. Routledge.

Gruszka, A., Matthews, G., & Szymura, B. (2010). *Handbook of Individual Differences in Cognition: Attention, Memory, and Executive Control*. Springer.

Gürçay, B., Mellers, B. A., & Baron, J. (2014). The Power of Social Influence on Estimation Accuracy. *Journal of Behavioral Decision Making*, *28*, 250–261. doi:10.1002/bdm.1843

Hadley, C. N., Pittinsky, T. L., Sommer, S. A., & Zhu, W. (2011). Measuring the efficacy of leaders to assess information and make decisions in a crisis: The C-LEAD scale. *The Leadership Quarterly*, *22*(4), 633–648. https://doi.org/10.1016/j.leaqua.2011.05.005

Hastie, R., & Dawes, R. M. (2010). *Rational Choice in an Uncertain World: The Psychology of Judgment and Decision Making* (2nd ed.). Sage.

Hauser, A., Eggers, F., & Güldenberg, S. (2020). Strategic decision-making in SMEs: Effectuation, causation, and the absence of strategy. *Small Business Economics*, *54*(3), 775–790. https://doi.org/10.1007/s11187-019-00152-x

Helfat, C. E., & Peteraf, M. A. (2015). Managerial cognitive capabilities and the microfoundations of dynamic capabilities. *Strategic Management Journal*, *36*(6), 831–850. https://doi.org/10.1002/smj.2247

Huntingford, C., Jeffers, E. S., Bonsall, M. B., Christensen, H. M., Lees, T., & Yang, H. (2019). Machine learning and artificial intelligence to aid climate change research and preparedness. *Environmental Research Letters*, *14*(12), 124007. https://doi.org/10.1088/1748-9326/ab4e55

Ilgen, D. R., Hollenbeck, J. R., Johnson, M., & Jundt, D. (2005). Teams in organizations: From input-process-output models to IMOI models. *Annual Review of Psychology*, *56*, 517–543.

Jain, J., Walia, N., & Gupta, S. (2020). Evaluation of behavioral biases affecting investment decision-making of individual equity investors by fuzzy analytic hierarchy process. *Review of Behavioral Finance, 12*(3), 297–314. https://doi.org/10.1108/RBF-03-2019-0044

Jarrahi, M. H. (2018). Artificial intelligence and the future of work: human-AI symbiosis in organizational decision making. *Business Horizons, 61*(4), 577–586.

Jong, W., Dückers, M. L. A., & van der Velden, P. G. (2016). Leadership of mayors and governors during crises: A systematic review on tasks and effectiveness. *Journal of Contingencies and Crisis Management, 24*(1), 46-58. doi:10.1111/1468-5973.12091

Kao, A. B., & Couzin, I. D. (2014). Decision accuracy in complex environments is often maximized by small group sizes. *Proceedings. Biological Sciences, 281*, 1–8. doi:10.1098/rspb.2013.3305

Katzenbach, J. R., & Smith, D. K. (1999). *The Wisdom of Teams: Creating High-Performance Organizations*. HarperBusiness.

Kerr, N. L., Maccoun, R. J., & Kramer, G. P. (1996). Bias in judgment: Comparing individuals and groups. *Psychological Review, 103*, 687–719.

Kerr, N. L., & Tindale, R. S. (2004). Group Performance and Decision Making. *Annual Review of Psychology, 55*, 623–655. doi:10.1146/annurev.psych.55.090902.142009

King, A. J., Cheng, L., Starke, S. D., & Myatt, J. P. (2012). Is the true 'wisdom of the crowd' to copy successful individuals? *Biology Letters, 8*, 197–200. doi:10.1098/rsbl.2011.0795

Kleinbaum, A. M., & Stuart, T. E. (2014). Inside the black box of the corporate staff: Social networks and the implementation of corporate strategy. *Strategic Management Journal, 35*(1), 24–47. https://doi.org/10.1002/smj.2090

Kowalczyk, M. (2017). *The Support of Decision Processes with Business Intelligence and Analytics: Insights on the Roles of Ambidexterity, Information Processing, and Advice*. Springer.

Krogerus, M., & Tschäppeler, R. (2017). *The Decision Book: 50 Models for Strategic Thinking*. WW Norton & Company.

Kuckertz, A., Brändle, L., Gaudig, A., Hinderer, S., Morales Reyes, C. A., Prochotta, A., Steinbrink, K. M., & Berger, E. S. C. (2020). Startups in times of crisis – A rapid response to the COVID-19 pandemic. *Journal of Business Venturing Insights, 13*, e00169. https://doi.org/10.1016/j.jbvi.2020.e00169

Lagowska, U., Sobral, F., & Furtado, L. M. G. P. (2020). Leadership under crises: A research agenda for the post-COVID-19 era. *Brazilian Administration Review, 17*(2), 1-5. doi:10.1590/1807-7692bar2020200062

Lorenz, J., Rauhut, H., Schweitzer, F., & Helbing, D. (2011). How social influence can undermine the wisdom of crowd effect. *Proceedings of the National Academy of Sciences of the United States of America, 108*, 9020–9025. doi:10.1073/pnas.1008636108

Mather, P. (2020). Leadership and governance in a crisis: Some reflections on COVID-19. *Journal of Accounting & Organizational Change, 16*(4), 579–585. https://doi.org/10.1108/JAOC-08-2020-0123

McCleskey, J., & Gruda, D. (2021). Risk-taking, resilience, and state anxiety during the COVID-19 pandemic: A coming of (old) age story. *Personality and Individual Differences, 170.* doi:10.1016/j.paid.2020.110485

McGrath, R. G. (2013). Six signs that your innovation program is broken. *Ivey Business Journal.*

McRaney, D. (2011). *You are not so smart.* Gotham Books.

Montibeller, G., & von Winterfeldt, D. (2015). Cognitive and motivational biases in decision and risk analysis. *Risk Analysis, 35*(7), 1230–1251. https://doi.org/10.1111/risa.12360

Motlaghzadeh, K., Kerachian, R., & Tavvafi, A. (2020). An evidential reasoning-based leader-follower game for hierarchical multi-agent decision making under uncertainty. *Journal of Hydrology (Amsterdam), 591*, 125294. https://doi.org/10.1016/j.jhydrol.2020.125294

Netz, J., Svensson, M., & Brundin, E. (2020). Business disruptions and affective reactions: A strategy-as-practice perspective on fast strategic decision making. *Long Range Planning, 53*(5), 101910. https://doi.org/10.1016/j.lrp.2019.101910

Norheim, O. F., Abi-Rached, J. M., Bright, L. K., Baeroe, K., Ferraz, O. L. M., Gloppen, S., & Voorhoeve, A. (2021). Difficult trade-offs in response to COVID-19: The case for open and inclusive decision making. *Nature Medicine, 27*, 10–13. https://doi.org/10.1038/s41591-020-01204-6

Ott, T. E., & Eisenhardt, K. M. (2020). Decision weaving: Forming novel, complex strategy in entrepreneurial settings. *Strategic Management Journal, 41*(12), 2275–2314. https://doi.org/10.1002/smj.3189

Pacheco-de-Almeida, G. (2010). Erosion, time compression, and self-displacement of leaders in hypercompetitive environments. *Strategic Management Journal, 31*, 1498–1526.

Pérez, I. J., Cabrerizo, F. J., Alonso, S., Dong, Y. C., Chiclana, F., & Herrera-Viedma, E. (2018). On dynamic consensus processes in group decision-making problems. *Information Sciences, 459*, 20–35. https://doi.org/10.1016/j.ins.2018.05.017

Phillips-Wren, G., & Adya, M. (2020). Decision making under stress: The role of information overload, time pressure, complexity, and uncertainty. *Journal of Decision Systems,* 1-13. doi:10.1080/12460125.2020.1768680

Pieterse, A. H., & de Vries, M. (2013). On the suitability of fast and frugal heuristics for designing values clarification methods in patient decision aids: A critical analysis. *Health Expectations, 16*(3), e73–e79. https://doi.org/10.1111/j.1369-7625.2011.00720.x

Porter, M. E. (1980). *Competitive strategy: Techniques for analyzing industries and competitors*. Simon & Schuster.

Potocan, V., & Nedelko, Z. (2021). The Behavior of Organization in Economic Crisis: Integration, Interpretation, and Research Development. *Journal of Business Ethics,* 1–19. doi:10.1007/s10551-021-04928-8

Rodrigues, M., Franco, M., Sousa, N., & Silva, R. (2021). COVID-19 and the business management crisis: An empirical study in SMEs. *Sustainability, 13*(11), 5912. doi:10.3390/su13115912

Roelich, K., & Giesekam, J. (2019). Decision making under uncertainty in climate change mitigation: Introducing multiple actor motivations, agency and influence. *Climate Policy, 19*(2), 175–188. https://doi.org/10.1080/14693062.2018.1479238

Rottenstreich, Y., & Hsee, C. K. (2001). Money, kisses, and electric shocks: On the affective psychology of risk. *Psychological Science, 12*(3), 185–190.

Rumelt, R. (2011). *Good Strategy Bad Strategy: The difference and why it matters*. Profile Books.

Schröder, D., & Gilboa Freedman, G. (2020). Decision making under uncertainty: The relation between economic preferences and psychological personality traits. *Theory and Decision, 89*(1), 61–83. https://doi.org/10.1007/s11238-019-09742-3

Schwenk, C. H. (1986). Information, cognitive biases, and commitment to a course of action. *Academy of Management Review, 11*(2), 298–310.

Siegrist, M., & Sütterlin, B. (2014). Human and nature-caused hazards: The affect heuristic causes biased decisions. *Risk Analysis*, *34*(8), 1482–1494.

Sirmon, D. G., Hitt, M. A., Ireland, R. D., & Gilbert, B. A. (2011). Resource orchestration to create competitive advantage: Breadth, depth, and life cycle effects. *Journal of Management*, *37*(5), 1390–1412. https://doi.org/10.1177/0149206310385695

Soluk, J., Kammerlander, N., & De Massis, A. (2021). Exogenous shocks and the adaptive capacity of family firms: Exploring behavioral changes and digital technologies in the COVID-19 pandemic. *R & D Management*, *51*(4), 364–380. https://doi.org/10.1111/radm.12471

Stasser, G., & Stewart, D. (1992). Discovery of hidden profiles by decision-making groups: Solving a problem versus making a judgment. *Journal of Personality and Social Psychology*, *63*, 426–434. doi:10.1037/0022-3514.63.3.426

Stasser, G., & Titus, W. (1985). Pooling of Unshared Information in Group Decision Making: Biased Information Sampling During Discussion. *Journal of Personality and Social Psychology*, *48*, 1467–1478. doi:10.1037/0022-3514.48.6.1467

Stefan, R., & Carutasu, G. (2020). Explainable machine learning for ethical artificial intelligence based decisions. *Journal of Information Systems & Operations Management,* 151-161.

Sull, D., Turconi, S., Sull, C., & Yoder, J. (2018). How to develop strategy for execution. *MIT Sloan Management Review*, *59*(2), 130–135.

Tversky, A., & Kahneman, D. (1992). Advances in prospect theory: Cumulative representation of uncertainty. *Journal of Risk and Uncertainty*, *5*(4), 297–323. doi:10.1007/bf00122574

Valeriani, D., Cinel, C., & Poli, R. (2017). Group Augmentation in Realistic Visual-Search Decisions via a Hybrid Brain-Computer Interface. *Scientific Reports*, *7*(1), 7772. https://doi.org/10.1038/s41598-017-08265-7

van Kleef, G. A. (2014). Understanding the positive and negative effects of emotional expressions in organizations: EASI does it. *Human Relations, 67*(9), 1145-1164. doi:10.1177/0018726713510329

Vinokur, A. (1971). Review and theoretical analysis of the effects of group processes upon individual and group decisions involving risk. *Psychological Bulletin*, *76*(4), 231–250. https://doi.org/10.1037/h0031568

Wong, N. C., & Shayegan, B. (2019). Patient centered care for prostate cancer-how can artificial intelligence and machine learning help make the right decision for the right patient? *Annals of Translational Medicine*, *7*(Suppl 1), S1–S1. https://doi.org/10.21037/atm.2019.01.13

Youssef, N. (2017). Requirements of decision-makers in confronting crises. *Journal of Legal Studies and Research*, *4*, 109–110.

Zhu, X. S., Wolfson, M. A., Dalal, D. K., & Mathieu, J. E. (2021). Team decision making: The dynamic effects of team decision style composition and performance via decision strategy. *Journal of Management*, *47*(5), 1281–1304. https://doi.org/10.1177/0149206320916232

KEY TERMS AND DEFINITIONS

Affect: Is emotion or desire, especially as influencing behavior, or action.

Affective: Is relating to moods, feelings, and attitudes.

Affective Bias: Is the tendency to differentially prioritize the processing of negative relative to positive events and is commonly observed in clinical and non-clinical populations.

Ambiguity: Is the quality of being open to more than one interpretation, i.e., inexactness.

Ambiguity Aversion: Is a preference for known risks over unknown risks. An ambiguity-averse individual would rather choose an alternative where the probability distribution of the outcomes is known over one where the probabilities are unknown.

Ambiguity Intolerance: Is a psychological construct that describes when individuals view ambiguous stimuli or events as a threat.

Artificial Intelligence (AI): Is the theory and development of computer systems able to perform tasks that normally require human intelligence, such as visual perception, speech recognition, decision-making, and translation between languages.

Availability Bias: Is the human tendency to think that examples of things that come readily to mind are more representative than is actually the case.

Bias: Is prejudice in favor of or against one thing, person, or group compared with another, usually in a way considered to be unfair.

Black Swan Event: Is an unpredictable occurrence that is beyond what is normally expected of a situation and has potentially severe consequences.

Brainstorming: Is a group discussion designed to produce ideas or solve problems.

Cognitive Bias: Is a systematic error in thinking that occurs when people are processing and interpreting information in the world around them and affects the decisions and judgments that they make.

Competitive Advantage: Is a condition or circumstance that puts a company in a favorable or superior business position.

Conformity: Is behavior in accordance with socially accepted conventions or standards.

Consensus: Is a general agreement or unanimity.

Consistency Effect/Consistency Bias: Is people's tendency to judge their own interpersonal behavior in a given situation in accordance with their general self-images, even if their actual behavior in the situation is partialled out.

COVID-19: Is defined as illness caused by a novel coronavirus called severe acute respiratory syndrome coronavirus 2 (SARS-CoV-2; formerly called 2019-nCoV), which was first identified amid an outbreak of respiratory illness cases in Wuhan City, Hubei Province, China.

Crisis: Is a time of intense difficulty, trouble, or danger.

Crisis Management: Is the process by which a business or other organization deals with a sudden emergency situation.

Decision-Making: Is the process of making choices by identifying a decision, gathering information, and assessing alternative resolutions.

Desirability of Options Bias: Is when decision-makers incorrectly estimate probabilities in either direction that favors their preferred alternatives.

Diffusion of Responsibility: Is a sociopsychological phenomenon whereby a person is less likely to take responsibility for action or inaction when other bystanders or witnesses are present.

Exogenous Shocks: Are unexpected or unpredictable events that occur outside an industry or country but can have a dramatic effect on the performance or markets within an industry or country.

Familiarity Bias: Is the idea that when people are faced with a choice between two gambles, they will pick the one that is more familiar to them.

Familiarization: Is the branch of mathematics concerned with the analysis of strategies for dealing with competitive situations where the outcome of a participant's choice of action depends critically on the actions of other participants.

Group Processes: Are the behaviors of the members of small working groups (usually between three and twelve members) as they engage in decision-making and task performance.

Groupthink: Is the practice of thinking or making decisions as a group in a way that discourages creativity or individual responsibility.

Habits: Are a settled or regular tendency or practice, especially one that is hard to give up.

Herding Bias: Is a tendency to follow what others are doing or thinking, i.e., following the rest of the herd without thinking.

Heuristics: Are simple, efficient rules, either learned or inculcated by evolutionary processes. These psychological heuristics have been proposed to explain how people make decisions, come to judgements, and solve problems.

Individual Processes: Are psychological processes that occur within individuals that cannot be seen whose existence can be inferred on the basis of people's behavior.

Leadership: Is the action of leading a group of people or an organization.

Leadership Strategy Formation: Is a coordinated set of activities undertaken by leadership and designed to help firms outperform their competition to achieve above-average returns and profitability.

Loss Aversion Bias: Is when a real or potential loss is perceived by individuals as psychologically or emotionally more severe than an equivalent gain.

Machine Learning: Is the use and development of computer systems that are able to learn and adapt without following explicit instructions, by using algorithms and statistical models to analyze and draw inferences from patterns in data.

Management: Is getting things done through other people utilizing the skills of planning, organizing, leading, and controlling.

Meaning-Making: Is the process by which people interpret situations, events, objects, or discourses, in the light of their previous knowledge and experience.

Mental Accounting Bias: Is when people treat assets as less fungible than they really are and then become susceptible to biased thinking, also known as the "two-pocket theory."

Overconfidence Bias: Is a tendency to hold a false and misleading assessment of our skills, intellect, or talent.

Ownership Bias: Is a decision-making bias that leads to an individual's tendency to prefer their own ideas over others' during the design process.

Pattern Recognition: Is the ability to detect arrangements of characteristics or data that yield information about a given system or data set.

Rational Choice Modeling: Is when individuals use rational calculations to make rational choices and achieve outcomes that are aligned with their own personal objectives.

Regret Aversion Bias: Is when the decision-maker refuses to make any decision because of the fear that the decision will turn out to be wrong and may later lead to feelings of regret.

Representativeness Bias: Is a mental shortcut that we use when estimating probabilities. When we're trying to assess how likely a certain event is, we often make our decision by assessing how similar it is to an existing mental prototype.

Resilience: Is the ability to withstand adversity and bounce back from difficult life events.

Sense-Making: Is the action or process of making sense of or giving meaning to something, especially new developments and experiences.

Social Validation: Is a psychological phenomenon where one or more passive individuals follow or conform to the actions of others within a group.

Strategy: Is an action that managers take to attain one or more of the organization's goals.

System One: Is our faster, automatic, intuitive, and emotional mode of thinking.

System Two: Is our slower, more effortful, and deliberate mode of thinking.

Two-Pocket Theory: Is when people treat assets as less fungible than they really are and then become susceptible to biased thinking, also known as the "mental accounting theory."

Chapter 6
Tailored Leadership as a Post-COVID-19 Opportunity for Enhanced Performance

Rebecca M. Melton
Western Governors University, USA

Kimberly B. Brooks
Western Governors University, USA

ABSTRACT

It is recognized that COVID-19 and its aftermath continue to present great challenges for many leaders. First, the chapter presents a review of leadership styles and their main tenets. Next, a framework by which leaders may tailor their approach to a task, issue, or employee challenge is suggested. Fifteen considerations are discussed with a user-friendly set of five overarching facets beginning with the letter P: purpose, person, place, project, and pacing. The authors propose that utilizing this framework may enhance the performance of the organization through a leader's consideration of the context of the scenario and its interaction with an employee or team. With each facet, the chapter discusses a critical connection to servant leadership. Further, questions for reflection are offered through which a leader may recognize strengths and glean actionable insights. Last, considerations for future research are presented.

DOI: 10.4018/978-1-6684-4358-3.ch006

INTRODUCTION

Post-COVID, employees face unprecedented challenges, both at home and work. On the family front, they may have been home-schooling children while balancing full-time work from the home office. People are dealing with the grief of losing loved ones to the virus. Others are managing the lasting physical consequences of the illness either physically or emotionally. Further, 30-40% of COVID survivors experience anxiety, depression, sleep abnormalities and post-traumatic stress (Mayo Clinic, 2021; Radraf, et al., 2021). Overall, Americans express fear and insecurity about the current vocational and economic crisis including, but not limited to rising inflation which as of this writing is up 7% the largest 12-month gain since June of 1982. In addition, employee resignations are at an all-time high, the highest levels reported in 2021 since the data has been tracked beginning in the year 2000 (Bureau of Labor Statistics, 2021). To further complicate the current work environment, increased numbers of employees are working remotely, challenging leaders to create social presence and transparent communication without having the luxury of being face to face. Analytics forecast that 25-30% of employees will work from home after COVID compared to 3% before the pandemic (Global Workplace Analytics, 2022). Leaders must carefully maneuver the resulting change occurring in company culture.

In the past, some leaders may have been hesitant to acknowledge that part of the role is to seek awareness of and to support employees' needs that originate outside of work; rather, they maintained focus on accountability, business strategy and organizational outcomes. Even those who have leaned toward servant leadership may sense an impetus to change. The amplified overlap of personal and work life cannot be denied as the pandemic's end is not yet secure. A leader's primary goal is to maximize and sustain long-term organizational performance through the performance of others. With the unprecedented environment, what is the best way to achieve this while also providing an intrinsic meaningful role for the leader?

BACKGROUND

Leadership Styles and Performance

There are many different types of leadership styles, for example, transformational leadership, transactional leadership, charismatic leadership, laissez-faire leadership, situational leadership, and servant leadership. According to Saragih et al. (2021), leadership style has a positive effect on both employee motivation and performance. In addition, Saliu et al. (2018) found a relationship between leadership style and motivation, which is a contributing factor to employee performance and organizational performance. Each leadership style has benefits, and some may propose limitations.

To begin with, transformational leadership embodies uniting employees around the organization's vision and mission (Kuntz, et al., 2019). Through communication and creating passion, transformational leaders can inspire their teams and improve morale (Futcher, 2019). Transformational leaders are especially effective in an environment encompassing change. The transformational leader can engage employees in becoming change agents and fostering principles and values to implement the change. Results of transformational leadership include positive behavior and performance outcomes (Kuntz, et al., 2019). This type of leader is more likely than others to involve the team in decision-making and to support employees' development and growth (Bass & Riggio, 2006). Critics of this leadership style purport that credibility and trust may be more important than magnetism. If transformational leaders are not perceived as credible and competent, they will not motivate and engage their employees (Williams, et al., 2018). Holding employees accountable may supersede the attraction characterized by transformational leaders. Transformational leaders may have a reputation for rallying team members around a cause, but they may lack the follow through that is essential to produce results. Ultimately, the leader is responsible for the end outcome. Furthermore, transformational leaders may communicate by implementing committees. This can backfire when employees feel uncomfortable with challenging the leader, in instances of conflict arising, or lack of participation (Futcher, 2019).

Transactional leadership is commonly characterized by rewards for performance or a transaction. Domingues et al. (2017) found a positive relationship between transactional leadership and employee performance. Their research discovered that employees under transactional leaders may be motivated by clear roles, expectations, and well-defined direction. Transactional leaders practice inclusivity with their employees in the goal setting process. Through collaboration, they involve their employees in identifying and establishing organizational and individual objectives. They provide their employees with the tools and resources to achieve these goals, follow up on employee contributions, and offer feedback to motivate employees. They are proficient in showing the connection between progress and performance results (Kuntz, et al., 2019). Transactional leadership has been shown to impact organizational commitment due to a positive perception of organizational justice which is related to employee trust, satisfaction, and retention. Some might argue that transactional leaders focus heavily on reward and punishment from a top-down approach. Transactional leaders are historically more likely to see the leader-member exchange as critical to management: employee does "x" (or does not do "x"); manager does "y." Negative consequences may decrease the employee's level of involvement and thereby, be considered commanding (You-De, et al., 2013). Further, external consequences in transactional leadership may not produce the longer-lasting changes in performance when employees behave from an internal locus of control (Bass & Riggio, 2006).

Some political leaders have been described as charismatic leaders (Ansary, 2019). Charismatic leadership is characterized by captivating leaders that create a compelling future vision that employees desire to follow (Shea & Howell, 1999). Charismatic leaders typically have an alluring personality and stimulate employees' yearning to follow them (Ansary, 2019). Research supports the correlation between a charismatic leadership style and positive organizational performance. This style of leadership relies on the emotional connection of the employee to the future direction of the organization. Nevertheless, there is also support that a laissez-fair, a "hands off" or passive, leadership style can elicit the same results as an inspirational, motivational leader with the difference being the amount of follow-up given to the employee on their progress. Overall, the situation may play a significant role between the two (Shea & Howell, 1999).

Therefore, situational leadership may be a crucial element. Situational leadership is keen to the present environment or current goal of the organization. It allows the leader to be adaptable and change their approach based on the current situation (Walls, 2019). For example, a leader may adopt a transformational leadership style to aid employees in achievement of a long-term goal and a more transactional leadership style with a short-term tactical objective. This approach may require enhanced emotional intelligence to be fluid and flexible based on different situations in the environment. Situational leadership considers two employee factors: competence and commitment (motivation and confidence). The leader identifies the employee's level of each factor and takes a specific behavioral posture in response to the paired result: Telling, Selling, Participating, or Delegating. However, should leadership be solely focused on these two situational aspects more than the employee's best interest?

Accordingly, servant leadership or leadership with morality is focused on the employee and has been found to improve performance by 15-20% and result in an increase in group productivity by 20-50% (Bande, et al., 2016). Research has found this association between considerate leaders and employee performance through the reduction of stress and ambiguity (Shea & Howell, 1999). By serving their employees and prioritizing those needs, servant leaders develop self-motivated employees who are committed to the organization and their goals (Bande, et al., 2016). Through their actions, communications, and attitude with a high-level understanding of their moral responsibility, servant leaders can create a positive climate of elevated engagement and resulting performance levels. Servant leadership is an extension of ethical behavior, maintaining integrity at the forefront of the leadership style. It may even cause the leader's employees to exceed their own performance expectations (Peng & Chen, 2020).

Subsequently, with the challenges created from the pandemic, a diverse workforce will be comprised of diverse backgrounds, experience and education levels, personality and thinking styles. We propose a tailored style of leadership based on the employee's best interest as it interacts with the context; whether that be transformational, transactional, charismatic, laissez-faire, or situational, each style should include a facet of servant leadership. Specifically, the practice of placing the employees' best interests above the leader's self-interest. Servant leadership is genuinely concerned with the employee's growth and achievements. The servant leader is engaged with the employee and provides an atmosphere of development for them. Servant leaders are humble and transparent and offer a work environment of high moral principles (Bande, et al., 2016). According to Bande et al. (2016), supporting employees with positivity and instilling trust can enhance their ability to deal with change and adapt. Having empathy and understanding of their unique situation is more important post COVID-19.

Table 1. Summary of Leadership Styles

Leadership Style	Traits
Transformational Leadership	• Role model • Expects transformation from team • Motivates through optimism to achieve innovation • Effective for long-term goals
Transactional Leadership	• Positive and negative reinforcement • Clear direction • Accountability • Effective for short-term goals
Charismatic Leadership	• Inspirational personality • Passionate • Creates sense of action • Effective to raise morale
Laissez-Faire Leadership	• Provides autonomy • Not directly involved • Trusts the team • Effective with a skilled, experienced team
Situational Leadership	• Adaptive • Behavior based on situation • Monitors team readiness • Effective for reinvention
Servant Leadership	• Self-less • Leads with integrity • Bestows team with credit • Effective for creating a positive culture

(Blanken, 2019)

Retention

The pandemic resulted in labor supply shortage consequences due to increased early retirements and parents withdrawing from the labor force to support children with school and daycare closures. In addition, organizations reduced spending on efforts to increase market share, instead allocating resources on the issues that the pandemic created (Hernandez, 2022). Therefore, retention of an organization's workforce is of utmost importance to prevent further shortages.

A generous leader who focuses on growth and places their employees' interests above their own creates an environment filled with loyalty and commitment (Lapointe & Vandenberghe, 2018). This leads to retention and avoids voluntary turnover, which can significantly impact the bottom line. Elevated levels of voluntary resignation result in both tangible and intangible outcomes. For example, tangible outcomes include separation costs (exit interviews), replacement costs (advertising, recruitment, and selection), new-hire training, and other administrative costs (Dess & Shaw, 2001). In addition, intangible outcomes include loss of productivity, decreased morale, and increased voluntary turnover among other employees. Therefore, voluntary resignation can be a vicious cycle (Green & Puetzer, 2002). All the above-mentioned outcomes can directly or indirectly affect the profitability of organizations.

Research supports that servant leaders strengthen the bond employees have with the organization, therefore, increasing commitment levels through an emotional connection with the leaders. In response to the leader's self-less behavior, employees feel that they have a voice. Feeling heard also increases the level of commitment to the organization. By creating a positive work climate, both affective (emotional) and normative (obligatory) commitment levels are amplified (Lapointe & Vandenberghe, 2018).

Figure 1. Leadership Styles and Servant Leadership

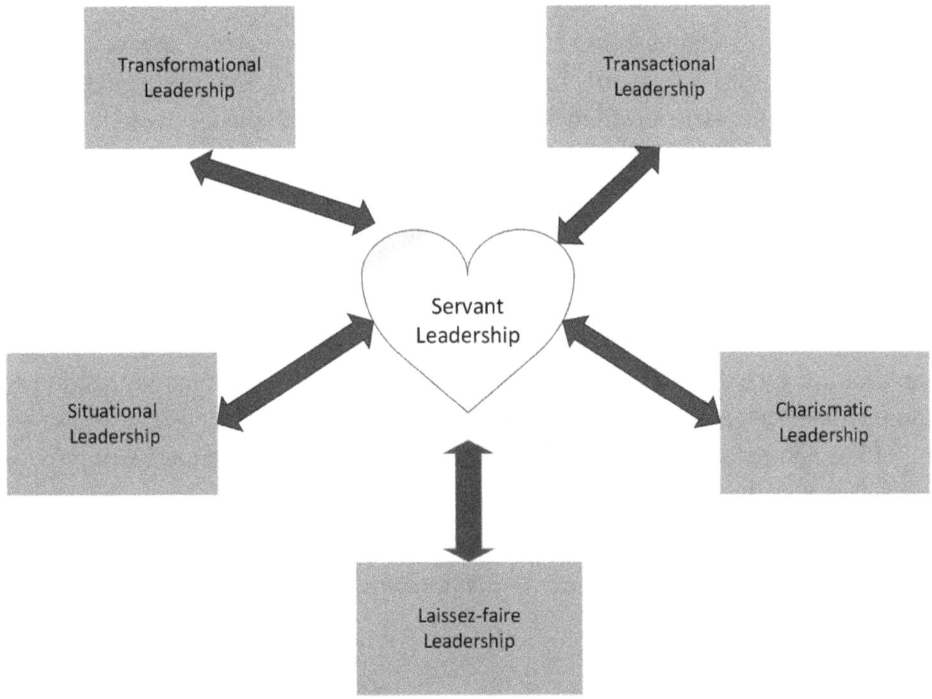

Therefore, tailored leadership is more critical than ever before. Enhancing the performance of leadership to impact associates' and company outcomes is the current call to action. There is little discussion in the literature of a method to integrate the best features of the varying leadership models in a way that leaders at all levels may find competence and the confidence to maneuver in the current environment. Figure 1 illustrates the existing types of leadership styles.

Ultimately, the organization's goal is to achieve its outcomes, and the leader's goal is to drive behavior that will achieve those results. At the same time, the leader cannot lose sight of leading human beings who have different aspirations, backgrounds, and challenges. Embracing that today's "best practice" leadership style is a journey to a new reality, the natural consequences of tailored leadership will be enhanced performance, increased retention, and loyalty, as well as greater collaboration toward a cohesive vision.

MAIN FOCUS OF THE CHAPTER

The purpose of this chapter is to review research on effective leadership post-COVID. Rather than establish a set of principles or guidelines—another leadership model—this chapter offers a user-friendly framework to remember critical considerations as a leader "triages" situations in a dynamic, unexpected and, at times, volatile environment. "Change, by its nature, leaves people and organizations feeling confused, vulnerable and fractured at a time when resilience, cohesion and collaboration are necessary to perform at the highest levels" (Chima & Guttman, 2020). While research suggests there may be upwards of fifty implications and issues that intersect with leading an organization post COVID-19, what is clear is that our leadership models were not built for this type of change. Leadership style and decision-making may need to be more multi-faceted in the current environment to be effective but also be accessible if leaders and managers are to utilize the approach. Last, the chapter will provide considerations for future research.

Figure 2. The 5 Facets to Consider in Leadership Approach

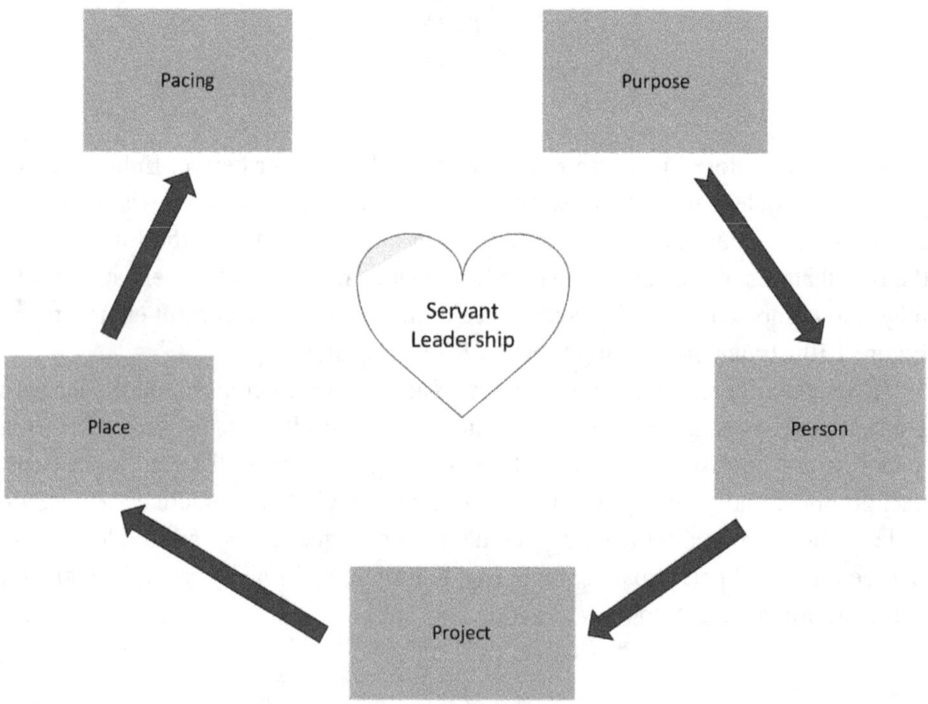

General Introduction to the 5 P's

An esteemed mentor once said, "A client may have a "Problem," but you cannot forget the "Person" or the "Place." The appropriate action or decision will be different, depending on all three." Indeed, experience has proven that the best approach to an individual or team challenge has been in consideration of the interrelationship between the three. Whether pondering three employees who do not reach a particular performance goal and identifying the next steps; or assisting three students who fail a final assessment and determining plans to aid learning, one dimension of leadership cannot be isolated. It could be argued that the joy and reward of leadership is successful outcomes, not despite the variations but through seeing the opportunity in them. The three Ps from the mentor above, with the additional considerations of "Place" and "Pacing," were investigated to suggest a framework of tailored leadership with servant leadership. See Figure 2 for a graphic representation of the 5 Ps focused through the lens of servant leadership.

The First P Is for Purpose (the Why)

The choices that company leaders and managers make are the X factor in helping employees fulfill their purpose at work and upwards of 60% of employees want more of "that". (Dhingra, et al., 2020, p.1)

Why start with purpose? Daniel Goleman, one of the primary researchers in Emotional Intelligence, suggests that a shared sense of purpose or shared vision was the strongest predictor of organizational-leader effectiveness, engagement, organizational citizenship, and product innovation in pre-pandemic times (Goleman, et al., 2001). Purposeful employees make a greater effort and have an increased likelihood of being innovative. This has continued to be the case post COVID-19. Take for example, USAA, a North American insurance company that gleans nearly 10,000 ideas from employees each year. Despite COVID, they reported that 900 novel ideas were awarded patents, including 25 by one of their security guards (Dhingra, et al., 2022). Organizational success toward a common mission continues to drive performance. How does the current project, task, or issue impact the company's purpose? The leader's purpose?

Since the pandemic, leaders have faced a situation in which company mission may have been trumped by a greater purpose—the health and safety of not only employees and stakeholders but also of all humanity. Leaders have worn the hat of company CEO or VP of HR (Human Resource) with expanded responsibility to be transparent about CDC guidelines, operational changes, and working from home. Through these messages, research indicates that a leaders' goal has been to reinforce safety and health, including psychological safety, and the idea that work is a place where employees can learn to be their best selves and to prosper despite the changes and pressure (Baker, 2021; Whitwell, 2021). This has been called "a moral higher calling" around which the teams unite, even an altruistic venture (McKinsey, 2020; Van Bavel, et al., 2020). It is well documented that groups that face a tragic or uncertain event tend to bond. Van Bavel et al. (2020) stipulate that the opportunity to fight together for a common good may increase cooperation, shared values and increase employees' sense of shared identity. Post COVID-19, each challenge may require a leader to consider global and community impacts in a novel way.

Next, organization leaders have sought to understand the intersection between company mission and the world's situation. Walmart CEO Doug McMillon shared this critical question: "What can we do to help other people through this crisis that does good for this company and society? (Chima & Guttman, 2020, np)." By establishing a corporate purpose that overlaps with society and with individuals' purpose, companies can provide employees with meaningful ways to increase their impact (Dhingra, et.al., 2022; Dressler, 2020). Having input and choice may increase individual contributors' sense of control, even hope, in challenging times. Thereafter, transitioning such conversation about a task or project's company purpose to the individual's strengths that they can employ, reflecting on their own sense of purpose and self-development, may build resilience (Whitwell, 2021). Nearly 2/3 of US based employees surveyed (n> 1000) said that COVID 19 caused them to reflect on their purpose in life, including what kind of work they did. Further, 70% of employees reported that their purpose was defined by their work (Dhingra, et al, 2022). To that end, sense of purpose and perceived meaningful work impacted health, resilience, retention, and employee engagement (Ghadi, et al., 2013). Finally, Rob Cross of MIT suggests that focusing on purpose as a company, department or work group may generate a greater sense of collaboration and teamwork (MacDonald, 2021). Senior leaders may consider emphasizing a "transition journey" with amplified communication, specific steps, and a clear destination to which the leadership and employees are travelling together post COVID-19 (Jadoul, et al., 2020). In this way, it may be that moving forward at work supports some sense of moving toward "normal." Jadoul et al. (2020) suggests that through conversation of individual and team strengths, effective leaders take hold of the opportunity to boost the level of engagement as the team rallies around learning new skills, a new model, or merely being in concert with one another.

Purpose--Tailored Through the Lens of a Servant Leader

Beyond information, employees expect that leaders will provide hope, guidance, and comfort, especially in times of disruption and confusion (Lagowska, et al., 2020).

Consider Bob Mulholland, Senior Vice President, and head of client relations at Merrill Lynch at Two World Financial Center on September 11, 2001. It is said that all his people escaped and thrived because "...he helped himself and his people find meaning in the face of chaos and madness" (Goleman, et al., 2001, p.41).

Questions on Purpose for Self-reflection

- How must leaders continue to message concern for employee health and safety, cascading from top leadership to frontline managers?
- In what ways does corporate purpose support humanity? What opportunities to increase corporate social responsibility are present?
- What cultural values and employee strengths may be magnified to generate a sense of team, resilience, and growth? Through what action steps?

The Second P Is for Person (the Who)

'Take care of your people, take care of your future, and take care of your stakeholders" (Pasmore & Mallis, 2022, p. 1).

Past models have looked to competence and commitment (motivation, confidence) as the employee factors that may affect the appropriate choice of leadership style (Patterson, et al., 2021). Research has recognized that direct reports have differing personalities, divergent thinking styles, and varying approaches to conflict resolution (Mayer, 2000; van Thiel, 2021). Post COVID-19, an effective leader needs to be consistently attuned to individual employees' attributes and needs during tactical and strategic decision-making. Beyond ability, motivation and confidence which are addressed in situational leadership models, factors to consider related to an individual direct report may include the following aspects: level of trust in the existing member-leader relationship; emotional intelligence (EI); and presence of mental health challenges such as stress or depression.

In Gallup's *State of the American Manager* (2015) report of 27 million employees surveyed, more than 70% of the variance in employee engagement scores across all business units was related to leader decisions and actions. In the short run, connecting with employees requires an investment of time and effort. Helping employees set goals, asking about progress, and reinforcing their strengths increases engagement. In such a time as COVID, leaders with these established employee relationships may have a significant advantage. To that end, Ahern and Low (2020) purport that leaders consider the level of trust in the employee relationship as a criterion in each employee situation. Trust is defined as an individual's expectation or belief that the actions or motives of another person are honest, fair, and based on integrity. Although impacted by past actions, it is dynamic and can have important impacts on resilience, perceived credibility, and success (Van Bavel, et al, 2020). In times of uncertainty and fear, employees may have trouble trusting even those people who are normally viewed as trustworthy; an effective leader will consistently connect with warmth and strength, increasing employee engagement and openness (Bauer, 2021; Dimitriadis & Psychogios, 2016; Glaser, 2014). The relationships built translate into social capital that a leader can "spend" to glean ideas, gain followership, and bridge challenging conversations (Podsakoff, et.al, 1990; Zhu, et.al, 2013).

Second, post-COVID, effective leaders, are those who consistently seek to understand and manage the emotions of others. Specifically, these leaders attend to both home and work considerations of the employees and strive to know the interests, needs and "usual" behaviors and attitudes of each person on the team (Lawton-Misra & Pretorius, 2021). In this way, a leader with Emotional Intelligence is more able to perceive when an employee may be showing signs of stress through changes in performance, behavior, or attitude. This effective leader is more able to display cognitive and emotional empathy, impact employee perspective, and provide support (Druskat & Wolfe, 2018; Goleman, 2013). Does the employee or team involved have low or high emotional intelligence? Are they managing emotions? What does the leader think they are feeling? How can the leader respond while also showing their self-reflection as a solid example for others as they navigate obstacles (Kraemer, 2015)?

Consider resilience, one specific outcome of high emotional intelligence on the part of an individual or team. As humans, we are amazed and perplexed at how some of the holocaust survivors have created positive and meaningful lives. How does a person fall one time after another, but persist until they become an Olympic ice-skater? How does a Kansas father whose wife and four children are whisked out of the car and killed in a flood, become a pastor who ministers to others? An aspect of their rebounding is resilience; Clark (2016) defines it as …"the ability to endure and overcome adversity based on tenacious inner motivation" (p. 110). Some individuals possess heightened levels of resilience enhancing their survival mechanisms. They have the ability and foresight to reach out to others for aid and care (mayoclinic.org). In the face of the pandemic, some employees encountered added challenges with extended lockdown(s), homeschooling children, job loss, and physical challenges. Employees who lack resilience may experience more psychological distress and become overwhelmed at work due to these trials (Very Well Mind, 2022).

To that end, individualizing a leader's approach to an employee, post pandemic, must include a cursory awareness of potential mental health challenges. Factors of consideration include a direct report's novel stressors such as the following elements:

changes in time constraints due to teaching children at home; loss of loved ones; loss of abilities; personal beliefs; food and supply insecurity; family loss of jobs; and more (Parks, 2021). It is not that the leader has the responsibility to manage others' well-being, nor should they avoid crucial feedback. Instead, logically contemplate if mental health is an impact on the person's or team's performance. What does the leader know, or should have known, about the physical or mental status of the employee? In this instance, does the leader himself have the competence and confidence to approach this delicate situation to offer support?

In addition, research suggests that senior leaders may provide guidance to middle managers, helping them personalize the day-to-day experience of direct reports. As a tailored leader takes care to examine the "person" aspect in each scenario, these writers expect that the leader's leader is doing the same. Outside the scope of this chapter but worthy of mention is to recognize that employee factors such as race and gender have interplay. These and other aspects of diversity are related to effective individualized consideration by a leader, especially in times of conflict and crisis (Thomas, et al., 2009). Therefore, a critical step in today's workplace is to equip managers with the skills to engage in impactful and diversity-sensitive employee conversations and to provide information about available support and services (Parks, 2020).

Person--Tailored Through the Lens of a Servant Leader

Dimitriadis and Psychogios (2016) remind us that

"...Our brain...needs the right sources and input of information for it to create great output in the form of ideas, emotions and actions...a leader understands the importance of human connections and networks and works tirelessly to build bridges that matter." (p.179)

For most individuals, "distancing" is directly opposed to the human instinct to connect and can decrease ability to cope with stress or to bounce back in a crisis (VanBavel, et al., 2020). What a tailored leader considers and applies in this "Person" area will result in dividends on a personal and business level.

Questions on Person for Self-reflection

- How do leaders continue to adjust approach and tone appropriately when engaging the workforce?

- What unique circumstances is an employee (or team) facing that may not be fully understood? How can a leader increase their knowledge of diversity, equity, and inclusion?
- In what ways is the company equipping leaders and managers with the resources they need to support the overall well-being of their teams (and of themselves) (Deloitte, 2022)?

The Third P Is for Project (the What)

"You can't choose to die on every hill, but this is one hill I'm willing to die on" (Riles, 1975, np).

For purposes of this chapter, the possibilities of which Project a leader is considering are infinite. This facet relates to the work itself, the task at hand. A senior leader may need to restructure reporting relationships in the organization; a Vice President of Human Resources may be forced to consider layoffs; or a front-line manager may be thinking about how to handle an employee who has continued to miss too much work. Rather, it is suggested that a few universal factors related to decision-making that taken with the other four P's (purpose, person, place, and pacing) make a notable difference post-COVID-19. These aspects are as follows: alignment with strategy/mission, actual or forecasted impact, and frequency of the issue.

The quotation from Wilson Riles which begins this section is meant to remind tailored leaders that a primary consideration about the Project or the "what" of work assignments is the critical nature (or the lack thereof) of that concern. Does the challenge or issue relate to the organization's strategy? Phrases such as "Choose your battles carefully! ", "Don't fight a battle you can't win!" or even "Is it mission critical?" apply here (Brobst, 2012, p. 1). COVID-19 is clearly indicated as a difficult or "strong" situation, suggesting that intensity of emotional reactions, helplessness and lack of control is possible (Meyer, et al., 2010). Especially in times of crisis and stress, even the most successful leader may be "emotionally hijacked" by passion for a project, or if fatigued, more "irked" by an associate's behavior (Gohm, et al., 2005). An action step or a change in thinking that aligns with current strategic goals is more likely a "fight" in which to be engaged. The tailored leader's question to himself may then be, "How can this step be taken with purpose, with consideration of those involved, communicated in the best venue, and appropriately paced?"

Beyond alignment with strategy, the conclusion to move ahead with a project or to address an employee challenge may be related to the actual or forecasted impact it has. The current state and projected outcomes are consistent factors in decision-making in standard business situations. However, post-COVID, it is critical that an effective leader not only shares the business case with transparency to stakeholders but welcomes questions, even employee challenges, to generate a greater level of understanding among the team. In this way, employees, other leaders, and shareholders may readily accept the change and see its relationship to individual and organizational success.

Last, an effective leader considers whether the present issue is a first-time occurrence or a repeating pattern. In other words, is the Project (challenge, issue, behavior) an anomaly or is there a trend that needs addressing? Under extreme stress such as that created by the pandemic, an employee or vendor may buckle for a moment, and the leader sees traits or behaviors that are unexpected, including absenteeism or turnover (Gohm, et al, 2005; Gupta & Beehr, 1979). In this unique time in history, is tackling this project, task, or issue necessary?

Project-Tailored Through the Lens of a Servant Leader

The "what" of many workplaces has shifted dramatically over the past three years. Whether accepting virtual tasks rather than in-person or covering someone's absence due to COVID illness, duties may have changed. A tailored leader is aware of the fatigue that can come from changes in tasks, teams, and mindset; they actively align roles and responsibilities for the best person-job fit (Rudroff, et al., 2020). Which strategic goals may energize the workforce as well as strengthen company performance? Which projects may serve our customers or clients to offer hope and enhance resilience?

Questions on Project for Self-reflection

- How has the company changed the norms for collaboration to design work assignments effectively in this unique situation?
- To encourage repeat behavior, what can leaders do to recognize individuals and teams who have delivered high performance during the pandemic (Deloitte, 2022)?
- What projects could include community partners as the company moves forward through the challenge of the pandemic?

The Fourth P Is for Place (the Where)

"This is not leadership as usual; this is leadership on the edge." -Dr. Carolos Pellegrini (Ball, 2020, p. 2)

The facet Place nudges a leader to remember that the contextual environment surrounding a person and problem may influence the best decision or outcome. Research suggests several factors to be considered in this realm, beyond this chapter's discussion of the overriding context of COVID-19 and the ongoing pandemic. These aspects include the state of the nation, e-leadership and presence, and the preponderance of mental health issues.

The state of the nation has been described as gloomy, tense, and uncertain (Gallup, 2021; Jones, 2022). Declines in national unity are attributed, but not limited to, the following concerns: racial unrest, vaccination status, political polarization, wealth and the economy, terrorism, crime, and women's rights (Jones, 2022; MacDonald, 2022; VanBavel, et al., 2020). Diversity, Equity, and Inclusion ranks as one of top ten trends in corporate HR according to the Society for Industrial and Organizational Psychology's (SIOP) survey (Stark, 2021). Certainly, these aspects cannot be overlooked as a tailored leader serves their team. In the best case, if embraced as part of the environment in which a company lays, leaders are being provided an opportunity to speak about differences and reinforce organizational commitment to respect all people. In a company in which silos exist or "in and out groups," a leaders' modeling of selflessness, kindness, and generosity to all may foster a sense of company unity, championing differences rather than tightening group boundaries (Van Bavel, et al, 2020). Organizational unity may counterbalance some of the unrest employees are experiencing outside company walls. Further, leadership with the collaboration of Human Resources may find this an appropriate time to establish (or re-establish) cultural norms, to train on harassment prevention, and to deepen DEI awareness, simultaneously preventing discrimination (Rowell & Torres, 2022).

Regardless of the industry, in this time, research suggests that leaders be omni-present, investing time and effort to be visible to employees in both usual and unexpected ways and places (Ball, 2020). Visibility and responsibility show accountability to and shared risk with the followers, providing opportunity to model personal vulnerability, resilience, and dependability (Ahern & Low, 2020). What may seem like overcommunication reassures employees and informs them about changes in world health guidelines and the company's responses. Among the greatest challenges in being omni-present has been bridging the move to telework. How does a leader "manage by walking around" in a virtual environment? In what ways can the same rich communication that occurs face to face happen in Zoom or Teams? Research suggests that the e-environment tests the following aspects of effective leadership: establishing virtual presence; creating and maintaining trusting relationships; generating a positive, motivated work atmosphere that spurs productivity; and leader support in problem-solving (Contreras, et al., 2020; Lilian, 2014; Roman, et al., 2019). Campbell (2021) reminds us:

"The leadership challenge in this is that we tend to use what we are the most comfortable with when it comes to technology and not necessarily what is the most effective or impactful for our audience. This can be a big problem when your team is highly distributed" (p. 3).

If a leader is open to learning from what other online institutions have been practicing for years, creating presence online is within reach. Six practices recommended by Poll (2014) for online instructors may apply in e-leadership for organizations with dispersed workers: 1) build an online community; 2) clarify expectations and objectives; 3) utilize the best tools for online communication; 4) promote consistent exchange of communication; 5) provide timely and actionable feedback; and 6) create an employee-centered environment. [*note: in Poll's research the final suggestion was a student-centered environment.] These six actions build trust, transparency and motivation for productivity and engagement.

Last, as discussed in prior sections, the pandemic has impacted the mental health of many employees through loss of loved ones, plant closings or layoffs, and general uncertainty surrounding public health (Parks, 2020). In consideration of Place, it is critical to recognize this context and affirm leaders' dedication to praise and be positive in public, and as often as possible, to handle correction or negativity in private. In other words, leaders' modeling the caution to choose where communication occurs is of utmost importance. In what size meeting is information best shared? In what venue? In addition to traditional conversations leaders have held in private, Ahern and Low (2020) reinforce that an effective leader upholds confidentiality of vaccination status and COVID-related disabilities. Further, empathetic leaders may ask privately for permission to share a direct report's tragic circumstances with teammates, so the team can provide operational and emotional support.

Place--Tailored Through the Lens of a Servant Leader

Have you ever looked for your favorite space in the grocery store parking lot? Or noticed that students tend to sit in the same seats, even if not assigned in the classroom? Do you have a lucky sweater for your favorite football team's games? These simple questions are reminders that context, familiarity, and patterns carry some weight. A setting of national uncertainty, leader presence, and the locations in which a leader interacts with employees may have critical impact. Acting with awareness of physical aspects facing employees will improve an effective leader's responses and propel successful outcomes.

Questions on Place for Self-reflection

- How are company leaders continuing to heighten visibility and instill trust in teams (Deloitte, 2022)?
- In what ways is the company leadership enhancing reputation and employment brand, related to their response to COVID-19?
- If the journey through the pandemic could be plotted, where is your company in its travels? In what ways can you reinforce the distance covered successfully while establishing the destination is not yet in sight? How does this journey align with the one planned pre-pandemic?

The 5th P Is for Pacing (the When)

"Waiting for the right pitch is the most important thing for a batter." Ted Williams, legendary baseball player (Clark, 2016, p. 137)

Clark (2016) states, "What is true in baseball is true in life and leadership: Brilliant moves are based on timing…the problems you do not solve offensively you will eventually have to solve defensively" (p. 137). Organizations that get the timing right, balancing between longevity of the organization, the needs of society and supporting well-being of employees, will have differentiated themselves (Pasmore & Mallis, 2022). Yet, employees are experiencing mental health issues, uncertainty, and decreased trust post COVID-19; rapid change can be perceived as threatening or stress-inducing (Schecter & Gould, 2020). This juxtaposition must bring the importance of pacing into the leadership discussion. Mid-level leaders who are tailoring their approach to both business and employees' needs may act as a conduit for information. Factors to consider regarding pace include the following: pace of work tasks, pace of intervention, and pace of feedback.

As effective leaders, a paced approach leads to better outcomes and may improve employee experience (Collins, 2011). Jim Collins (2011) in *Great by Choice* suggests three reasons why a consistent pace leads to success: pace builds confidence to perform in adverse circumstances; the discipline required to maintain a consistent pace reduces the likelihood of failure when there are disruptions; the pace and discipline themselves lend to maintaining control in an out-of-control situation. Taken individually, pacesetting can be seen universally in several scenarios. Automation imposes pace on assembly lines as workers pack four items to a box from an automated conveyor belt. Project Management software helps leaders set stretch goals for team timelines to create and track staff accomplishments on a new university course. Progress bars on the computer indicate how far into a cross-country flight the passengers have travelled, how much longer they must wait for a customer service representative (CSR), and how long to complete software updates (Conrad, et al., 2011). Knowing that progress is taking place builds anticipation and motivation which may also lead to a sense of satisfaction with oneself or, in our examples, with the pilot, software or company CSR (Latham, 2004). When adversity arises, the momentum to continue can be sustained by the motivation or even the mere existence of the goal that was set and the employees' commitment to it (Latham & Locke, 2007). Self-efficacy has been found to relate to what goals a person may choose and may increase as the employee sees the impact their efforts and pace have on heightened performance.

For purposes of this article, brief review of self-control—the underlying process driving discipline—will be offered. Tangney, et al. (2016) defined self-control as "...the capacity to override, modify or suppress undesirable behavior when pursuing long-term goals..." (p. 275). According to a recent metanalysis of more than 120 management papers, Yam et al. (2017) suggest that leaders may increase employee self-control in three important ways: 1) be mindful about the importance of sleep for restoration; long work hours can impact behavior and well-being; 2) train service-oriented employees (who must put on a smile to the customer) to have empathy in a genuine way; 3) promote an ethical culture with a clear code of ethics. Self-control is not limitless but instead may be seen as a well from which one draws throughout the day that needs replenished. Further, being practiced in self-control, or discipline as Collins calls it, may contribute to resilience discussed in prior sections. In the overlap between pace and discipline, the ability to gain control when situations are out of control may create habits that help to propel us and may give us comfort. In other words, the practice of pace and discipline may make some reactions automatic. As Rudyard Kipling (1943) might put it, "...If you can keep your head when all about you are losing theirs..." then a leader can lend guidance to others, and the others can help as well.

Second, when considering Pace, a tailored leader may find it useful to create a pause, a time during which decisions are not made and people intentionally "marinate" on a challenge (Schecter & Gould, 2020). While it can be difficult to remain undecided, insightful questions and possibilities can result. Research has shown that when the leader delays their response to solve an issue, an employee or team will continue to brainstorm to resolve it themselves (Kokemuller, 2022). Further, employees may accept greater responsibility for the outcome of the decision having been involved. Post COVID-19, pacing leader interventions and involvement is impactful.

Last, appropriate timing of feedback to a manager or employee is critical. Compartmentalizing the facet of Pace only, according to B.F. Skinner, renowned psychologist, reinforcement theory suggests that feedback is best provided as closely to the execution of a good or unruly behavior (Gordan, 2014). In this way, the recipient of the feedback has a clear "line of sight" between action (the behavior) and response (the feedback). Also, the details of the occurrence are fresh in the minds of those involved. Especially post COVID-19, a sympathetic leader may question if delaying the conversation is best, not wanting to add stress to an already tense situation. However, it may be true that more immediate responses are preferable since that timing more closely mirrors what would occur in a "normal" time. Combined with the other factors discussed in this article, what is best for the individual? What is best for the company? What is ethical? What timing protects the organization legally?

Pacing--Tailored Through the Lens of a Servant Leader

In the book *24 Hour Rule*, author Charles Fred (2019) suggests:

Pausing resets our cognitive power, gives us more control and provides crucial perspective. A conscious pause, especially by people in positions of authority, can fundamentally change a course of action, improve and strengthen relationships, reduce stress, and save lives" (p. 28).

The words "work/life balance" may seem overused. However, employees and managers (and leaders) must have time and space in which to regenerate, refresh and rest. Otherwise, even with positive intentions, our best work cannot be performed, and burnout may ensue (Antle & Silver, 2016). Through the servant leadership lens, a tailored leader aids employees and teams by carefully managing the pacing and timing of work assignment, joint decision-making, and giving feedback.

Questions on Pacing for Self-reflection

- What is the appropriate cadence for ongoing communication about the pandemic and the company's response/plans, to keep employees engaged and informed but not overwhelmed (Deloitte, 2022)?
- In what ways does senior leadership need to engage differently post COVID-19 to make timely decisions about related operational or human resources issues?
- How can leaders pursue employee feedback to identify opportunities for improvement in pacing and timing of work assignments and company "temperature checks"?

Table 2. Summary of Facets of the Five P Framework and Considerations

Facet of Leadership Style	Considerations
Purpose	•relationship to vision and mission •impact on humanity •intersection between mission and humanity
Person	•level of trust in the member-leader relationship •emotional intelligence and resilience •mental health issues
Place	•context of global setting: post COVID-19 •e-leadership and leader presence •where to provide feedback
Project	•relationship to strategy •actual or forecasted impact •new issue or repeated pattern
Pacing	•pace of task assignment •pace of intervention •timing of feedback

FUTURE RESEARCH DIRECTIONS

There are multiple opportunities for future research on tailored leadership's advantages in the workplace. Based on previous research, there is a correlation between leadership style and individual performance and resulting organizational performance (Saliu, et al., 2018). It would be of value for scholars to conduct empirical research surrounding the above framework of tailored leadership including the facet of servant leadership. Recommended topics to explore include the relationship among tailored leadership and performance, organizational commitment, and retention particularly in a post-COVID environment. Of interest may be to compare two organizations: one in which leaders utilize the framework provided here and another that does not. How does employee engagement and satisfaction differ? It may be interesting to measure the employee's engagement pre-introduction of this framework as a baseline.

Next, there may be aspects of the leaders themselves that affect the use of this framework. Is there a generational difference in acceptance of the facets and considerations suggested? Do leaders of a certain generation hesitate to apply a particular facet? Similarly, research could investigate whether a leader's gender relates to effectiveness of this framework post COVID-19 and acceptance of the tailored style. What is the outcome if both variables, generation, and gender, are measured?

Future research may also consider what role the company's culture plays in which leadership style is most effective. Specifically, does an organization with an authoritarian culture find this framework useful, or is it more successful with an open, creative culture? If a company is a learning organization, are the leaders more effective in incorporating this tailored style? How does the industry in which the organization operates relate?

These writers recognize that this chapter leaves open-ended questions. It may be that the next step in research is to create a valid and reliable self-evaluation that allows a leader to assess their own performance and tendencies in a series of situations. The scenarios would vary in detail such as the purpose, people involved, pace/timing, project/issue, and place. The results would suggest in which areas the leader naturally ponders certain considerations and other areas to which giving more attention may be beneficial. Next, post self-evaluation, the instrument may give specific suggestions which an interested leader could follow. It may be challenging to identify exemplars amid the variables, but the results could make for interesting learning as part of a leadership training program. Judgement and decision-making skills could be said to be a saw that can always become sharper.

SUMMARY

In summary, we are in unknown territory with the COVID-19 pandemic including the aftermath and probable workplace consequences. As leaders navigate through this unparalleled environment, this writing suggests a unique leadership approach while serving the employee during this challenging time keeping in mind the final objectives of employee and organizational performance as well as increased retention and market share.

This article summarizes six different types of leadership styles (transformational, transactional, charismatic, laissez-faire, situational, and servant leadership) including their valuable contributions as well as potential limitations. The authors have provided a suggested framework entitled "tailored leadership" which encompasses the consideration of five aspects (the "five P's") all through the lens of the servant leader. The authors illustrate the value of supporting the individual employee with care and trust to enhance growth and instill loyalty and optimal individual and organizational performance levels. The authors go a step further and provide guidelines to achieve tailored leadership during this dynamic and unexpected environment. Overall, the framework presented highlights how the leader can tailor their leadership style based on not only the person (most importantly), but a five-step framework that also takes into consideration the purpose, project, place, and pace of the current circumstance. These five factors (the "Five P's) are a framework aid in developing the leader's "tailored leadership" style to benefit the workforce and the organization. All these items are seen through the lens of the servant leader. This is what makes "tailored leadership" special; never losing sight that the leader is in the role to serve the individual and meanwhile steer the organization with heart, support, and integrity.

This framework is a set of principles to reach the end outcome of gaining market share in the best interests of the workforce and the organization.

This framework brings insights that are deserving of consideration. It is not meant to be prescriptive, guiding one to audit every simple decision against this list of facets. Instead, to understand that as the world at large and the world of work continue to be more agile and complex, our approach to exceptional leadership must be as well.

REFERENCES

Ahern, S., & Loh, E. (2020). *Leadership during the COVID-19 pandemic: building and sustaining trust in times of uncertainty.* doi:10.1136/leader-2020-000271

Ansary, A. S. (2019). *Pros and Cons of Charismatic Leadership.* Retrieved on 2/7/22 from http://careercliff/com/pros-and-cons-of-charismatic-leadership

Antle, M., & Silver, R. (2016). Circadian Insights into Motivated Behavior. In Behavioral Neuroscience of Motivation. Springer International Publishing, AG Switzerland.

Baker, M. (2021). *9 Future of Work Trends Post-COVID-19.* Retrieved on 2.5.22 from https://www.gartner.com/smarterwithgartner/9-future-of-work-trends-post-covid-19

Ball, C. (2020). Leadership during the COVID crisis and beyond. *Canadian Journal of Surgery*, *63*(4), E370–E371. doi:10.1503/cjs.016020 PMID:32813482

Bande, B., Fernandez-Ferrin, P., Varela-Neira, C., & Otero-Neira, C. (2016). Exploring the relationship among servant leadership, intrinsic motivation, and performance in an industrial sales setting. *Journal of Business and Industrial Marketing*, *31*(2), 219–231. doi:10.1108/JBIM-03-2014-0046

Bass, B., & Riggio, R. (2006). *Transformational Leadership* (2nd ed.). Lawrence Erlbaum Associates, Publisher. doi:10.4324/9781410617095

Bauer, T., Erdogan, B., Cauglin, D., & Truxillo, D. (2021). *Fundamentals of Human Resource Management: People, Data and Analytics.* Sage Publications.

Blanken, R. (2019). *Eight Leadership Styles and When to Use Them.* The Center for Association Leadership. Retrieved on 2/14/22 at https://www.asaecenter.org/resources/articles/an_plus/2019/december/eight-leadership-styles-and-when-to-use-them

Brobst, D. (2012). *Choose Your Battles Wisely.* https://www.donbrobst.com/2012/05/choosing-your-battles-wisely/

Bureau of Labor Statistics. (n.d.). *Number of quits at all-time high November 2021*. Retrieved on 2/5/22 from https://www.bls.gov/opub/ted/2022/number-of-quits-at-all-time-high-in-november-2021.htm

Campbell, T. (2022). The Future of Leadership in a Remote-Work World. *Journal of AHIMA*. https://journal.ahima.org/the-future-of-leadership-in-a-remote-work-world

Chima, A., & Gutman, R. What it Takes to Lead Through an Era of Exponential Change. *Harvard Business Review*.

Clark, T. (2016). *Leading with Character & Competence*. Berrett-Koehler Publishers, Inc.

Collins, J., & Hansen, M. (2011). *Great by Choice: Uncertainty, Chaos and Luck—Why Some Thrive Despite Them All*. HarperCollins Publishers.

Conrad, F., Couper, M., Tourangeau, R., & Peytchev, A. (2010, September 1). The Impact or Progress Indicators on Task Completion. *Interacting with Computers, 22*(5), 417–427. doi:10.1016/j.intcom.2010.03.001 PMID:20676386

Contreras, F., Baykal, E., & Abid, G. (2020). E-Leadership and Teleworking in Times of COVID-19 and Beyond: What we Know and Where Do We Go. *Frontiers in Psychology, 1*, 1–10. doi:10.3389/fpsyg.2020.590271 PMID:33362656

Covey, S. R. (2004). The 7 habits of highly effective people: Powerful lessons in personal change (25th ann. ed.). New York: Simon & Schuster.

Deloitte. (2022). *Workforce Strategies for a Post-COVID-19 recovery: Workbook*. Retrieved on 1.12.22 from https://www2.deloitte.com/content/dam/Deloitte/uk/Documents/human-capital/deloitte-uk-workforce-strategies-for-post-covid-recovery-workbook.pdf

Dess, G., & Shaw, J. (2001). Voluntary turnover, social capital, and organizational performance. *Academy of Management Review, 26*(3), 446-456.

Dhingra, N., Samo, A., Schaninger, B., & Schrimper, M. (n.d.). *Help Your Employees Find Purpose—or Watch Them Leave*. McKinsey & Company, Retrieved on 1/22/22 from https://www.mckinsey.com/business-functions/people-and-organizational-performance/our-insights/help-your-employees-find-purpose-or-watch-them-leave

Dimitrias, N., & Psychogios, A. (2016). *Neuroscience for Leaders: A brain-adaptive leadership approach*. Kogan Page Limited.

Domingues, J., Vieira, V. A., & Agnihotri, R. (2017). The interactive effects of goal orientation and leadership style on sales performance. *Marketing Letters, 28*(4), 637–649. doi:10.100711002-017-9436-3

Dressler, G. (2020). *Human Resource Management* (16th ed.). Pearson.

Druskat, V., & Wolff, S. (2001). Building the Emotional Intelligence of Groups. In D. Goleman, A. McKee, & S. Achor (Eds.), *Everyday Emotional Intelligence* (pp. 73–93). Harvard Business School Publishing Corporation.

Flynn, J. (2020). *Smart Strategies for Managing a Dispersed Workforce*. Retrieved on 2.12.22 from https://www.insight.com/en_US/content-and-resources/tech-journal/fall-2020/smart-strategies-for-managing-a-dispersed-workforce.html

Fred, C. (2019). *The 24-Hour Rule: Leading in a Frenetic World*. Magnusson-Skor Publishing, LLC.

Futcher, C. (2019). *Transformational Leadership Advantages and Disadvantages Explored*. Retrieved on 2/7/22 from https://info.cavendishwood.com/blog/transformational-leadership-advantages-and-disadvantages-explored

Gallup. (2015). *State of the American Manager*. Retrieved on 2.24.22 from https://www.gallup.com/services/182138/state-american-manager.aspx

Gallup. (2021). *Americans Offer Gloomy State of the Nation Report*. Retrieved on 1/22/22 from https://news.gallup.com/poll/389309/americans-offer-gloomy-state-nation-report

Ghadi, M. Y., Fernando, M., & Caputi, P. (2013). Transformational leadership and work engagement; the mediating effect of meaning in work. *Leadership and Organization Development Journal, 34*(6), 532–550. doi:10.1108/LODJ-10-2011-0110

Glaser, J. (2014). *Conversational Intelligence: How Great Leaders Build Trust and Get Extraordinary Results*. Bibliomotion, Inc.

Global Workplace Analytics. (n.d.). *Work-at-Home After COVID-19 – Our Forecast*. Retrieved on 2/5/22 from https://globalworkplaceanalytics.com/work-at-home-after-covid-19-our-forecast

Gohm, C., Corser, G., & Dalsky, D. (2005). Emotional Intelligence under Stress: Useful, unnecessary, or irrelevant? *Personality and Individual Differences, 39*(6), 1017–1028. doi:10.1016/j.paid.2005.03.018

Goleman, D. (2013). What is Empathy? In D. Goleman, A. McKee, & S. Achor (Eds.), *Everyday Emotional Intelligence* (pp. 293–297). Harvard Business School Publishing Corporation.

Goleman, D., Boyatzis, R., & McKee, A. (2001). Primal Leadership: The Hidden Driver of Great Performance. In D. Goleman, A. McKee, & S. Achor (Eds.), *Everyday Emotional Intelligence* (pp. 25–44). Harvard Business School Publishing Corporation.

Gordan, M. (2021). A Review of B. F. Skinner's 'Reinforcement Theory of Motivation.'. *International Journal of Research in Education Methodology.*, *5*(3), 680–688. doi:10.24297/ijrem.v5i3.3892

Green, M. T., & Puetzer, M. (2002). The value of mentoring: A strategic approach to retention and recruitment. *Journal of Nursing Care Quality*, *17*(1), 67–74. doi:10.1097/00001786-200210000-00008 PMID:12369749

Gupta, N., & Beehr, T. (1979). Job Stress and Employee Behaviors. *Organizational Behavior and Human Performance*, *23*(3), 373–387. doi:10.1016/0030-5073(79)90004-7 PMID:10242304

Hernandez, R. (2022). Will the COVID-19 pandemic reduce potential output? *Monthly Labor Review.* Retrieved on 2/5/22 from https://www.bls.gov/opub/mlr/2022/beyond-bls/pdf/will-the-covid-19-pandemic-reduce-potential-output

Jadoul, Q., Nascimento, A., Salo, O., & Willi, R. (2020). *Agility in the Time of COVID-19: Changing Your Operating Model in an Age of Turbulence.* McKinsey & Company. https://www.mckinsey.com/business-functions/people-and-organizational-performance/our-insights/agility-in-the-time-of-covid-19-changing-your-operating-model-in-an-age-of-turbulence

Jones, J. (2022). *Americans Offer Gloomy State of the Nation Report.* Retrieved on 2.2.22 from https://news.gallup.com/poll/389309/americans-offer-gloomy-state-nation-report.aspx?version=print

Kahneman, D., & Tversky, A. (1979). Prospect Theory: An Analysis of Decisions Under Risk. *Econometrica*, *47*(2), 263–291. doi:10.2307/1914185

Kipling, R. (1943). *If.* Retrieved on 2.12.22 from https://www.poetryfoundation.org/poems/46473/if--

Kokemuller, N. (2022). The Importance of Timing in Leadership. *Small Business Chronicle*. Retrieved on 1/12/22 from https://smallbusiness.chron.com/importance-timing-leadership-51987.html

Kraemer, H. (2015). *Becoming the Best: Build a World-Class Organization through Values-Based Leadership*. John Wiley & Sons, Inc.

Kuntz, J., Davies, B., & Naswall, K. (2019). From transactional to transformational: Exploring the role of leadership style on CEO succession outcomes. *Leadership and Organization Development Journal, 40*(7), 815–827. doi:10.1108/LODJ-01-2019-0012

Lagowska, U., Sobral, F., & Furtado, L. (2020). Leadership under Crises: A Research Agenda for the Post-COVID-19 Era. *Brazilian Administration Review, 17*(2), 1–5. doi:10.1590/1807-7692bar2020200062

Lapointe, E., & Vandenberghe, C. (2018). Examination of the Relationships Between Servant Leadership, Organizational Commitment, and Voice and Antisocial Behaviors. *Journal of Business Ethics, 148*(1), 99–115. doi:10.100710551-015-3002-9

Latham, G. (2004). The Motivational Benefits of Goal setting. *The Academy of Management Executive, 18*(4), 126-129.

Latham, G., & Locke, E. (2007). New Development in and Directions for Goal-Setting Research. *European Psychologist, 12*(4), 290–300. doi:10.1027/1016-9040.12.4.290

Lawton-Misra, N., & Pretorius, T. (2021). Leading with heart: Academic leadership during the Covid-19 crisis. *South African Journal of Psychology. Suid-Afrikaanse Tydskrif vir Sielkunde, 51*(2), 205–214. doi:10.1177/0081246321992979

Lilian, S. (2014). Virtual Teams: Opportunities and challenges for e-leaders. *Procedia: Social and Behavioral Sciences, 110*, 1251–1261. doi:10.1016/j.sbspro.2013.12.972

Lumen Learning. (n.d.). *Listening: Three A's of Active Listening*. Retrieved on 2/15/22 from https://courses.lumenlearning.com/ivytech-comm101-master/chapter/chapter-4-three-as-of-active-listening/

MacDonald, M. (2022). *Six Ways Leaders Can Adapt to the Workplace of 2022*. Retrieved on 2.4.22 from https://sloanreview.mit.edu/article/six-ways-leaders-can-adapt-to-the-workplace-of-2022/

Mayer, C.-H., & Louw, L. (2012). Managing cross-cultural conflict in organizations. *International Journal of Cross Cultural Management, 12*(1), 3–8. doi:10.1177/1470595811413104

Mayo Clinic. (n.d.a). *COVID-19 (coronavirus): Long-term effects*. Retrieved on November 5, 2021, from https://www.mayoclinic.org/diseases-conditions/coronavirus/in-depth/coronavirus-long-term-effects/art-20490351

Mayo Clinic. (n.d.b). *Resilience Training*. Retrieved 2/12/22 from Mayoclinic.org/tests-procedures/resilience-training/in-depth/resilience/art-20046311

McKinsey. (2020). *Psychological safety, emotional intelligence, and leadership in a time of flux*. July 2, 2020. Retrieved on 1/21/22 from https://www.mckinsey.com/featured-insights/leadership/psychological-safety-emotional-intelligence-and-leadership-in-a-time-of-flux

Meyer, R. K., Dalal, R. S., & Hermida, R. (2010). A review and synthesis of situational strength in the organizational sciences. *Journal of Management*, *36*(1), 121–140. doi:10.1177/0149206309349309

Parks, R. (2021). *COVID-19 Mental Health Crises: Holistic Understanding and Solutions*. Parks Press.

Pasmore, W., & Mallis, E. (2022). *Re-set Your Organization for the Post Covid Future with Creative Leadership*. Center for Creative Leadership. Retrieved on 1.12.22 from https://www.ccl.org/articles/white-papers/reset-your-organization-for-post-covid-future/

Patterson, K., Grenny, J., McMillan, R., & Switzler, A. (2004). *Crucial Confrontations: Tools for Resolving Broken Promises, Violated Expectations, and Bad Behavior*. McGraw-Hill.

Peng, J. C., & Chen, S. W. (2021). Servant Leadership and Service Performance: A Multilevel Mediation Model. *Psychological Reports*, *124*(4), 1738–1760. doi:10.1177/0033294120950302 PMID:32838663

Podsakoff, P. M., MacKenzie, S. B., Moorman, R. H., & Fetter, R. (1990). Transformational leader behaviors, and their effects on followers' trust in leader, satisfaction, and organizational citizenship behaviors. *The Leadership Quarterly*, *1*(2), 107–142. doi:10.1016/1048-9843(90)90009-7

Poll, K., Widen, J., & Weller, S. (2014). Six Instructional Best Practices for Online Engagement and Retention. *Journal of Online Doctoral Education*, *1*(1), 56–72.

Radraf, A., Ferreira, M., Sosa, J., & Filip, I. (2021). Emergent Crisis of COVID-19 Pandemic: Mental Health Challenges and Opportunities. *Frontiers in Psychology*, *12*, 631008. doi:10.3389/fpsyt.2021.631008 PMID:34349675

Riles, W. (1975). ECE in California Passes Its First Tests. *Phi Delta Kappan*, *57*(1), 3–7.

Roman, A., Van Wart, M., Wang, X., Liu, C., Kim, S., & McCarthy, A. (2019). Defining e-leadership as competence in ICT-mediated communications: An exploratory assessment. *Public Administration Review, 79*(6), 853–866. doi:10.1111/puar.12980

Rowell, S., & Torres, B. (2022). *Strategic Insights for 2022: Linking DEI and Harassment Prevention Initiatives.* Retrieved on 2.1.12 from https://www.shrm.org/learningandcareer/learning/webcasts/pages/0122rowell.aspx

Rudroff, T., Fietsam, A., Deters, J., Bruant, A., & Kamholz, J. (2020, December 19). Post-COVID-19 Fatigue: Potential Contributing Factors. *Brain Sciences, 10*(12), 1012. doi:10.3390/brainsci10121012 PMID:33352638

Saliu, U. A., Wankasi, A. J., Eromosele, G. O., & Olukade, A. O. (2018). Leadership Styles and Motivation on Job Performance of Library Personnel in Public University Libraries in North Central Nigeria. *Library and Philosophy Practice.* https://link.gale.com/apps/doc/A580773835/AONE?=anon~585617ba&sid=googleScholar&xid=7ceece62

Saragih, R., Pradana, M., Wijaksana, T., Arwiyah, Y., & Fadhillah, N. (2021). Antecedents of Work Performance: The Effect of Leadership Style, Work Motivation and Discipline. *Journal of Management Information and Decision Sciences, 24*(6), 1–8.

Schecter, T., & Gould, M. (2020). Lead from Your Heart: The Art of Relationship-Based Leadership. HTI Institute.

Shea, C., & Howell, J. (1999). Charismatic Leadership and Task Feedback: A Laboratory Study of their Effects on Self-Efficacy and Task Performance. *The Leadership Quarterly, 10*(3), 375–396. doi:10.1016/S1048-9843(99)00020-X

Stark, A. (2021). Top 10 Work Trends for 2021. *SIOP Member News.* Retrieved on 1.22.22 from https://www.siop.org/Research-Publications/Items-of-Interest/ArtMID/19366/ArticleID/4914/Top-10-Work-Trends-for-2021

Tangney, J., Baumeister, R., & Boone, A. (2004). High self-control predicts good adjustment, less pathology, better grades, and interpersonal success. *Journal of Personality, 72*(2), 271–324. doi:10.1111/j.0022-3506.2004.00263.x PMID:15016066

Thomas, K. W., Thomas, G. F., & Schaubhut, N. (2008). Conflict styles of men and women at six organization levels. *International Journal of Conflict Management, 14*(2), 1–38. doi:10.1108/10444060810856085

Van Thiel, E. (2021). *Big 5 Personality Test*. Retrieved on 1/25/22 at https://www.123test.com/personality-test/

VanBavel, J., Baicker, K., Boggio, P., Capraro, V., Cichocka, A., Cikara, M., Crockett, M., Crum, A., Douglas, K., Druckman, J., & Drury, J. (2020). Using Social and Behavioural Science to Support Covid-19 Pandemic Response. *Nature Human Behaviour*, *4*(5), 460–471. doi:10.103841562-020-0884-z PMID:32355299

Very Well Mind. (n.d.). *What is resilience?* Retrieved 2/12/22 from Verywellmind.com/what-is-resilience-2795059

Walls, E. (2019). The Value of Situational Leadership. *Community Practitioner*, 31–33.

Whitwell, G. (2021). *Resetting the Leadership Agenda Post-COVID-19*. AACSB. Retrieved on 1/12/22 from https://www.aacsb.edu/insights/articles/2021/01/resetting-the-leadership-agenda-post-covid-19

Williams, R. Jr, Raffo, D. M., & Clark, L. A. (2018). Charisma as an attribute of transformational leaders: What about credibility? *Journal of Management Development*, *37*(6), 512–524. doi:10.1108/JMD-03-2018-0088

Yam, K., Lian, H., Ferris, D., & Brown, D. (2017). Leadership Takes Self-Control. Here's What We Know About It. *Harvard Business Review*. Retrieved on 2.11.22 from https://hbr.org/2017/06/leadership-takes-self-control-heres-what-we-know-about-it

You-De, D., You-Yu, D., Kuan-Yang, C., & Hui-Chun, W. (2013). Transformational vs. transactional leadership: Which is better? A study on employees of international tourist hotels in Taipei City. *International Journal of Contemporary Hospitality Management*, *25*(5), 760–778. doi:10.1108/IJCHM-Dec-2011-0223

Zhu, W., Avolio, B. J., & Walumbwa, F. O. (2009). Moderating role of follower characteristics with transformational leadership and follower work engagement. *Group & Organization Management*, *34*(5), 590–619. doi:10.1177/1059601108331242

Zhu, W., Newman, A., Miao, Q., & Hooke, A. (2013). Revisiting the mediating role of trust in transformational leadership effects: Do different types of trust make a difference? *The Leadership Quarterly*, *24*(1), 94–105. doi:10.1016/j.leaqua.2012.08.004

ADDITIONAL READING

Crask, E. (2021). *The Role of Pace and Consistency in Employee Experience Strategy*. SHL Blog. June 17, 2021. Retrieved on 1.12.22 from https://www.shl.com/resources/by-type-blog/2020/the-role-of-pace-and-consistency-in-employee-experience-strategy/

Detert, J., & Burris, E. (2007). Leadership Behavior and Employee Voice: Is the Door Really Open? *Academy of Management Journal*, *50*(4), 869–884. doi:10.5465/amj.2007.26279183

Donthu, N., & Gustafsson, A. (2020, September). Effects of COVID-19 on Business and research. *Journal of Business Research*, *17*, 284–289. doi:10.1016/j.jbusres.2020.06.008 PMID:32536736

Durning, S. J., Cervero, R. M., & Roberts, L. W. (2022). The Need for Listening Leaders. *Journal. Association of American Medical Colleges*, *97*(2), 165–166. doi:10.1097/ACM.0000000000004520 PMID:35084389

Flynn, J. (2020). *Smart Strategies for Managing a Dispersed Workforce*. https://www.insight.com/en_US/content-and-resources/tech-journal/fall-2020/smart-strategies-for-managing-a-dispersed-workforce.html

Khan, H. (2021). *Leadership Reinvented*. Rockridge Press.

Schneider, S. K., & George, W. M. (2010). Servant leadership versus transformational leadership in voluntary service organizations. *Leadership and Organization Development Journal*, *32*(1), 60–77. doi:10.1108/01437731111099283

Situ, Q., Li, J., & Dou, K. (2016). Reexamining the linear and U-shaped relationships between self-control and emotional and behavioral problems. *Asian Journal of Psychology*, *19*(2), 177–185. doi:10.1111/ajsp.12118

KEY TERMS AND DEFINITIONS

COVID-19: Coronavirus (SARS-CoV-2).

E-Leadership: Guiding, motivating, and leading employees, some of whom are working from multiple locations (possibly their homes), requiring effective use of technology.

Emotional Intelligence: The ability to be aware of and manage one's own emotions and the emotions of others.

Endowment Effect: A cognitive bias by which a person over-values something that they "own" or believe is theirs.

Individual Performance: The quality and quantity of an employee's performance.

Organizational Performance: Maintaining or gaining market share through the combination of individual and team performance.

Presence: The actual or perceived availability and visibility of a leader that may bring hope and comfort to employees in uncertain times.

Retention: An organization's ability to prevent turnover and sustain their workforce.

Resilience: An individual's capacity to cope with adversity.

Tailored Leadership: Shaping the leadership style based on the employee's needs including the lens of servant leadership and taking into consideration the facets of purpose (why), project (what), place (where), and pacing (when/timing).

Chapter 7
Learn, Unlearn, and Relearn: A Step Towards Bringing Resilience in Business Organizations in the Post-COVID-19 Regime

Nabanita Ghosh
Christ University, India

Anubha Srivastava
https://orcid.org/0000-0003-2289-3105
Christ University, India

Sunita Kumar
https://orcid.org/0000-0002-0628-1873
Christ University, India

Daniel Gnanaraj
Christ University, India

ABSTRACT

The world on a whole has undergone a paradigm shift in its journey with the onset of the COVID-19 threat. Business houses of any stature tremendously suffered towards their consistent and competitive survival. Means and measures are few to implement and very challenging in reaping the benefits. Sheer fall in GDP, dropping in the rate of industrialization and productivity are no unknown facts in the present-day scenario. Indians are unfortunate to embrace the pandemic at the juncture of planning and forecasting of India becoming a five trillion dollar economy by 2025.

DOI: 10.4018/978-1-6684-4358-3.ch007

INTRODUCTION

The world under the clutches of covid 19 had succumbed to implementing enormous changes in every sphere of its performance in order to sustain itself. India has moved into the post covid dispensation where the country will prioritize economic expansion and sustainability for secured growth and influence on the world stage. Continued disruption is observed in sectors all around. Sectors like automobile, food and catering, tourism, education, healthcare, banking, and insurance are battered though not beaten by the continencies of the pandemic. The survival of the companies called for transformation in the mode of operation with the hope that the performance will be gradually seamless in the new normal scenario.

The term technology is an ever-green concept for newer inclusions in the light of change management. The corporates across dimensions are forced to face the challenges of the change management which was in literature before the onset of Covid 19. Now due to the cascading effects in various performance metrics, the word change has gained momentum and accelerated the need for the inclusions and exclusions principle. In the pre covid scenario, the management attempted to continue their current style of operation as resistance to change is a familiar topic among the employees and demands a mammoth task and now due to the severity of the virus the turnaround strategy is realized in the truest spirit and perform or perish became the mantra for companies, consequently resorted to change their flow of work by majorly nurturing the leadership style, culture, and technology. Some of the strategies for cost management, cost curtailment, and cost monitoring which were in the stage of conception for many industries became an obvious concern, and therefore with a blessing in disguise the industries were forced to unearth their hidden talents in shaping their functions and deliberations in the virus contained regime. The commonality in this transformation is the urgency to introduce such changes with unequal vibes in responses among the different generations due to constraints and traditional blocks. The use of technology may demand the involvement of less task force and more machines, still, empowerment of humans with the requisite skills is a must for an organization's digital transformation journey to come true. In this reformatory period, the success of the introduction of the newer methodologies depended greatly on the business leaders' ability to respond to uncertainties in productivity and growth jolts. The strategists, decision-makers and the policymakers all together were victimized in their approach as the paws of the Covid repeatedly brought questions about the art and architecture of management of organizational function with resource constraints.

The labour shortage, lack of supply of raw material, and less availability of finance multiplied with the need for emphasis on technology made the situation tough and competitive.

The shift in organizational performance became evident in its redefined and restructured situation-centric leadership style, culture, and technology. At this backdrop, the paper attempts to analyse the following changes:

1. The shift in leadership style in organizations in the post covid scenario.
2. The shift in the cultural climate in organizations in the post covid scenario.
3. The shift towards the technology from the traditional work style in organizations in the post covid scenario.
4. Compliance procedure followed in the organization in the reformatory period.

BACKGROUND

The study primarily rests on the aspect of the promotion of culture, technology and leadership revolution in the organizations in the covid saturated regime for sustainable survival among the tough competitors. The pre covid scenario during the period of 2018-19 indicated the need for a shift in the work, life and culture. The disruption was initiated but the thunderbolts of the pandemic kept that in abeyance and after the resumption of the new normal the organizations faced the challenges of the changes (Forbes, 2020). The signal of transformation since pre covid is a mini-crisis in the post covid era and the compounded mini-crises are the disasters of today. The companies tried to introduce the change in the working style, culture and emphasize the digitization but the fear of make-over delayed the process and with the advent of covid 19 the world is forced to respond to the moves by compulsorily taking out from the comfortable fundamentals. Taking a look at the pulse of the remote work before and after covid 19, it is observed that there is a significant gap of the seasoned remote workers and new remote workers. Major segments excluding the core IT, a majority of the nation's economy underwent the learning phase of different work options. Despite tremendous uncertainties, bounded literacies, and difficulties in providing uniform facilities, the organizations by and large took a bold step to bring the change in leadership, culture, and technology for an obvious purpose in the post covid period (pipefy, 2022). The paper is divided into three levels to substantiate the transformation in leadership, culture, and technology in the organizations in the post covid scenario.

TRANSFORMATION IN LEADERSHIP

Leadership is one of the important functions of management. Leadership is an integral part of our life and it's important for every organization. The common leadership styles vehemently followed in the business organizations need to be remodelled by keeping in mind that the leadership has to adapt the modified level of operations to maximize the output without fostering burnout. Wherever there are organized groups of people working towards a common goal, leadership becomes essential. During a time of crisis, the leader is the one who builds up confidence and zeal among people to move forward. COVID-19 has given us many opportunities to learn and relearn leadership. COVID-19 brought many complex issues not only related to the health care system but also various other issues like social-economic, emotional, technological, and environmental to name a few.

These dynamic issues have shaken the consistency of our life and also given us the opportunity to relook at our wisdom and progress and also redefine the leadership style. This pandemic also allowed us to look at a holistic approach to leadership. COVID-19 has touched the lives of everyone directly or indirectly. It has impacted every aspect of life ranging from personal to professional including health and finances (Antonakis,2020; Mckibbin and Fernando,2020). It has shown impacts on the physical as well as the mental state of people working across the organization. The degree of uncertainties has increased tremendously across the boards (Charoensukmongkol and Phungsoonthorn, 2020). The style of leadership has evolved to adopt the new changes in the workforce, surroundings and business prospects (both sides where opportunities are changing in either direction). The response is more agile and human-centric because of unseen situations (Dirani et al., 2020).

Effective leadership can foster teamwork, and cultivate a sense of belonging which motivates and inspires people to channelise their energy for the betterment of the organization and business. As Steve Jobs rightly said, "*Management is about persuading people to do things they do not want to do, while leadership is about inspiring people to do things, they never thought they could.*" So, when we see the current scenario it's all about the culmination of every aspect of life that a leader needs to deal with. Whether it is workforce motivation, culture, or technology the leaders need to understand the basic principles to lead the people in a difficult situation. Difficult situations could be anything either technology-related or working with diverse groups of people who belong to different cultures. Here we need to understand that culture and technology also have its own rules and regulations which leaders need to follow. Covid 19 has made leaders understand the situation through a different lens.

Evaluating Leadership During Crises

When the world was under lockdown the majority of employees and managers implemented lockdowns due to the COVID-19 pandemic during spring 2020 (Vargas,

2020). The importance of effective interest in the effect of working from home (WFH) on employees, their productivity and their well-being during the pandemic has highlighted the importance to understand how managers should manage and lead across distances to support their employees' mental health and performance (Adenle, 2020; Giurge & Bohns, 2020; World Health Organisation, 2020). If we dive deep into the issues related to the pandemic, we can analyse through a few questions and evaluate whole leadership effectiveness in dealing with the situation (Allen, E. 2020).

- The first question that we can ask is about the holistic approach. Whether the leader has taken a balanced approach to respond to the situation.
- How effectively data has been used to combat the situation and take a relevant decision?
- Does the leader honestly and efficiently cooperate to solve the problem?
- How well do they create a friendly environment to encourage and motivate people to build friendly surroundings?

The leader has to wear various lenses to deal with this situation. Some of the leaders they have done that. Henry Fayol suggested a few qualities which proved to be helpful in any situation and these include physical, mental, moral, educational, technical, and experience. This is relevant in today's context also concerning crises, especially during covid. There are various leadership theories on leadership behaviour and style. Different authors hold different views on the qualities that are considered essential for effective leadership. Some emphasize the personal qualities and traits of leadership. Others emphasize the actual behaviour and action of the leader. Some people emphasize situations in which leadership is to be exercised. More specifically, how does a leader ensure that their group/team is greater than the sum of its parts? Given the current coronavirus disease 2019 (COVID-19) pandemic and the incredible practice-changing shifts that have occurred as a result, therefore came the need for exceptional leadership within (Surg, CJ.2020).

Trait Theory

According to the Trait theory, there are certain personal qualities and traits which are essential to being a successful leader. The advocates of this theory believe that persons who are leaders are psychologically better adjusted to display better judgment and be engaged in social activities. They seek more information, give more information and take the lead in interpreting social activities in interpreting or summing up a situation.

The Behavioural Theory

Leadership involves an interpersonal relationship between leaders and subordinates in which the behaviour of the leader toward the subordinate constitutes the most critical element.

The good behaviour of the leader raises the morale, and builds up confidence and spirit among the team members. Social exchange theory is one of the most important theories (Balu, 1964).

Subordinates or the people who find their leader as a role model are likely to feel responsible towards their leader and show a better attitude towards the work (Liborius, 2014). During covid 19 some companies showed how leaders motivated and boosted the morale of their employees.

The Situational Theory

The situational theories emphasize not the personal qualities or traits of a leader, but the situation in which they operate. During COVID various leaders of various countries showed the power and traits of how to handle the situation and take their people along with them and this was one of the key elements which we observed during this pandemic.

The Follower Theory

The essence of leadership is followership and it is the willingness of people to follow that makes a person a leader. The members of a group tend to follow only those whom they recognize as providing means for achieving their desires, wants, and needs.

But in today's context, if we look at leadership, it is all about fostering collaboration, and teamwork across the team is an important aspect of leadership. So, one needs to share information and be transparent in distributing the authority and responsibility. During a crisis like situation, a leader might feel like control information and only provide based on the need but doing this might demotivate the team, so for a successful function one needs to encourage their team and be transparent and this is the biggest lesson learned by the company during the pandemic (D'Auria, G. and De Smet, A.2020).

Table 1. Leadership style changes during and after COVID

SN	Leadership Style	During and After COVID
1	Democratic	It is one of the major types of leadership that continued 'as is' during the covid time. To make it more effective, micromanagement should be removed.
2	Authoritarian	If worked before then it must have worked during covid too. Although, a leader must consider both mental and physical states during the covid time because both were impacted.
3	Delegative	The delegation of tasks without micromanagement worked wonders during covid.
4	Visionary	The vision communication medium has changed from physical to digital, which has its challenges. If vision is communicated and received properly then no change in leadership style is necessary.
5	Coaching	The coaching communication medium has changed from physical to digital, which has its challenges. If coaching is communicated and received properly then no change in leadership style is necessary.
6	Affiliative	The mental and physical state has been fluctuating during the covid time and hence impacting the relationship between leader and team members. In the new situation, how effectively the new relationship is taken care, of decides the effectiveness of Affiliative leadership.
7	Pace-setting	All team members are working in different situations, hence setting achievable expectations is key to the success of Pacesetting leadership.
8	Commanding	The command's effectiveness depends on the mental and physical state of the leaders. Hence, commanding leadership must consider these new states.
9	Transformational	Once both leaders and team members acknowledge the new situations (covid) then they help each other to advance a higher level of morale and motivation.

During the covid, it emerged that participative leadership is the most effective and important for the success of the organization. In participative leadership, individuals are involved in an independent decision-making process. Always active involvement in the process of decision-making encourages employees to believe and trust their leader. When leaders have faith and trust in their employees, they always try to take on additional responsibilities and perform better this also helps the organizations to do better. (Lu et al.,2015). Participative leaders motivate their employees by involving them to be a part of leadership (Kahai et al.,1997; somech,2003) this method makes their employees feel that leaders value their idea and suggestions (Sagnak,2016).

TRANSFORMATION IN CULTURE

Culture is not something you fix like a car or roaster. Leaders must have a disciplined approach to assessing and measuring the culture at defined periods. This process should inform their strategic plans and supporting priorities. (Tim Kuppler, Director of Culture and Organization Development at Human Synergistics International)

The COVID-19 pandemic has changed the world in more ways than one. The unprecedented economic and social challenges that COVID-19 posed to the world have caused much uncertainty and anxiety among the public. The sudden shift from office work to the working from home culture was a disruptive force. Organizational culture in the last two years has undergone a sea-change. Organizational culture is largely shaped by the "collective norms of behavior exhibited by the individuals within an organization". Thus, by being a collective norm of behaviour, organizational culture is something that does not change very much or very quickly. In fact, it is something that changes over a long period of time. Change in organizational culture could be brought by many external factors that cannot be controlled. Economic challenges in the past have brought changes to the organizational culture in the past (Organizational Culture and COVID-19). Cultural change can be very difficult for an organization to go through. Nevertheless, this long-standing belief regarding organizational culture changed within a matter of days when the pandemic struck. Many organizations have shifted their priority from exploration to safety and resilience. Organizational leaders in the post-COVID world have adapted themselves to the new challenges. One of the first changes that people noticed in organizational culture is that of the immediate and wide deployment and reliance on digital technologies. Many digital technologies, both free and premium versions, have enabled organizations to shift their work home.

The digital tools available for collaborative work today such as "videoconferencing, digital shared file storage, simultaneous multi-authoring of documents, digital whiteboards, screen-sharing, smartphone chat groups" (How the pandemic can change workplace culture for the better by James Thomas) have enabled people to make this shift quickly. As people shifted from a common office workplace to their homes, people in leadership positions are wondering about ways to build an organizational culture in such a scenario.

This transition has also led to the abandoning of the belief that work can be done better only when we come together face-to-face. While remote working may seem cheap, it can come with hidden costs to the organizational culture. Also, apart from the challenges that an organization may face, individuals faced several challenges working from home. As people work from home, they may spend more time in front of their computers and other digital devices. Even though they may clock long hours of work than they normally do in the office, their productivity would certainly take a hit.

The term "zoom fatigue" was coined in the pandemic years to refer to the tiredness and disorientation that employees may experience due to long hours of online collaborative work. Both experienced and novice workers faced several challenges. The newcomers to many organizations found it challenging to undergo the onboarding process and learn the organizational culture of their new organizations. Researchers have begun asking pertinent questions about the organizational culture by asking if it would be possible to shift the organizational culture from the physical workspaces to a virtual one. The organizational culture may also see a rise or fall in the empty symbolic rituals of corporate life.

The companies tried to roll out a rotating remote work culture to fight out the monotony of working alone and also to maintain a healthy relationship with the team members in this covid crisis. Pairing up with similar role-based employees will effectuate the trust-building, enable alternate go-to communications, and also would result in upskilling (hrdirector.com).

The four types of culture commonly known in the organizational setup till the entry of dragon covid are Clan culture, Adhocracy culture, Market culture, and Hierarchy culture in today's dynamic environment though market culture seems relevant, taking the Amazon as an example, still, the sole profit-making objective in a competitive framework need to be replaced with strong collaborative culture for an enduring existence and so over a period, the companies namely Google, Zappos, Tom's of Maine adopted clan culture, a family-oriented bonding which elevates the working mood and enriches the departmental cohesion. The modern days employees are disinterested in an adhocracy work culture and since privatization in the country is gradually mushrooming, therefore it is no wonder that the future PPP (Public-private-partnership) model will not embrace even the Hierarchy culture (personio.com).

The covid has compelled the organizations to induce a spectacular cultural shift from regular face-to-face bonding to digital bonding keeping in mind the need to bridge between the significance of the smart achievement of goals and relational gravity.

TRANSFORMATION IN TECHNOLOGY

Technology has swept away the shackles of centuries-old traditions and reinforced the implementation of a modern version of living to tide over the covid crisis (Bernard Marr). From the top urbanites to the innermost rural belt the need to respond to the technological changes is felt

Each industry has adopted the most compatible form of technology to use and the glimpses of such transformation in various sectors are narrated below.

Automobile Industry

The industry faced downturns though not a complete shutdown. The pre covid changed in terms of revolution in driverless cars, automated factories, ride-sharing, and smart booking for in and out of station services. The covid pushed up the demand for electric vehicles and due to continuous lockdowns, the CO2 emissions got reduced and a reflection on the desirability of traditional vehicles is observed from the environmental perspective. People initiated the learning process of accepting the usage of electric cars over the gasoline and diesel-powered vehicles though the Government initiatives for the promotion is the ultimate, yet revolution in terms of technology is evident (World Economic Forum). Manual intervention substituted with automation is the need of the hour to protect human lives from the deadly virus against which the management of the livelihood became a paramount concern among the entrepreneurs and decision-makers and so the rampant technological shift stood at its maxim in all the countries with special mentioning for the developing and underdeveloped countries where the zeal to know the know-how needs conversion to learn to use the know-how. The change in technology on account of covid could therefore shift the automobile industry into a new world of on-screen vehicular control and touch-less gas station payments and maintenance and integrated digitization slowly and steadily would pave the way for unavoidable universal acceptance to change (www.accenture.com, April 2020).

Food and Catering Industry

The foodservice landscape witnessed the tremors of the pandemic and reeled under the shock waves for the last two years and is currently showing signs of recovery with the customers' intention to return to in-person dining. Canada's foodservice segment was a $95 Billion industry catering to the job requirement of 1.2 million faced loss in revenues, increasing operational expenses, acute shortage of labour, and debt burden. yet the industry kick-started keeping in abeyance the repercussions of the pandemic and moved towards the transition of revival from survival with Governmental support. The industry learned to deal with the limited manpower and other necessary resources, strengthened the working of the innovative ideas to add newer attributes to the services making it more lucrative to the customers, focussing on the removal of redundant activities, creating a competitive edge on the deliverables and on a whole generating a formula for maximizing the customers' retention index. Adoption of new technology became more evident in terms of rendering value-added services to customers without quality compromise and quantity reduction and filling up the gaps in services provided so far keeping the cost-effectiveness in mind (Restaurants Canada, The voice of Foodservice, 2021)

Tourism Industry

The tourism enterprise is a composition of products involving transport, accommodation, catering, natural resources, entertainment, and other facilities and services, such as shops and banks, and other tour operators. (Sinclair & Stabler, 1997). The tourism industry is one of the potential industries driving not only the national economy but also the global economy. A tourism entrepreneur is a person or group of persons who are managing and producing tourism products (Khanka, 1999). The tourism industry started in the 17th century in Western Europe and became a giant business center in the latter half of the 20th century with the progressive deregulation and decoupling of air travel from the flag carriers. The industry over some time has reconfigured to fit itself in the frame of tourism entrepreneurship as an entrepreneurial philosophy that reduces unemployment rates, promotes economic growth, and changes the lifestyle of the tourists and tour operators. The tourism industry was brought into the light in 1945 when Sir John Sargent Committee was set up for the promotion and development of the said industry post which the Indian Government undertook to develop the tourism industry in a planned manner coinciding with the 2nd five-year plan in 1956. Momentum is observed in the 1980s and in 1982 a national policy in tourism was announced, later in 1988, a national committee on tourism formulated sustainable growth plans. In 1992 the national action plan was made and in 1996, the national strategy for the promotion of tourism was drafted. In 1997 the new tourism policy recognized the role of central and state government, public and private -sector undertakings in its phase-wise development. From the late 1990s till 2019 the industry undeniably contributed to the country's economic growth at a faster pace as the country unearthed its potential in tourism resources and also experienced the transformation in lifestyle from a luxurious approach to a need-based approach. The rampant growth rate in the tourism industry motivated the entrepreneurs to extend the tourism business and take the industry to the next level.

The total contribution of travel and tourism to the Indian GDP for the period 2013 to 2019 constantly exhibited an accelerating trend, the figure went from 172.91 billion to 268.29 billion in US $. The foreign exchange earnings in the year 2000 were at US $ 3.46 billion and reached up to 29.96 billion in 2019. The number of foreign tourists arriving in the year 2000 was 2.65 million and came up to 10.93 million in the year 2019. The occupancy rate in the hotels with annual fluctuations continually from 2001 to 2019 jumped to 65.7% from 57.2%. The data reflects that the tourism industry reported positive signs of growth rate and the development of various modes of accommodation increased in the above-mentioned timeline and the tour operations also stayed upward. The country ranked 34th in the overall global tourism index in 2019, a jump of 6 places from 40th in the year 2018. The lesson from the global shock made the underpinning statement pertinent that the entrepreneurs who operated frugally will be perished by it covid or other disaster and those who can rethink to strategize and are not resistant to change will survive. The key to surviving in the entrepreneurial plunge is to have astounding but rational ways outs in the arsenal. The data concerning foreign tourists' arrival, revenue generation, and occupancy rates in the hotel got substantially reduced due to the threat of covid 19 and called for the implementation of strategic measures to survive. Tourism like other business sectors needs to upgrade in the line of digital reformatory moves (Newman, 2018). The operators have started implementing various solutions to make themselves in the digital platform and offer a basket of innovative digital experiences. The travel cycle of any customer has become tremendously digitized which showed its signs of presence in the pre covid days but with the onset of the global shocks, travel, and tourism can be brought back to normalcy by assisting the customers with the digital assets. The possible technological changes in the post covid times are:

1. Use of mobile applications: The ease and comfort of the customer can be best realized when they make use of their mobile applications right from the travel plan till the end of the tour. Booking of tickets, booking the accommodation at a reasonable discounted rate without physical intervention, placing the order of the meals and also the e- arrangements for the planned visits if executed without wastage of time, the journey turns out to be an assured and insured pleasure trip and to overcome the inconsistencies in the flow of revenue on account of covid -19, the technological outburst in the tourism industry can help to Spotify the lost glory.
2. Artificial intelligence and use of Chatbots: The benefit of the application of these measures in the tourism industry is lying in its availability at any time, any place, and in any desired language.
3. Internet of things: It helps to grow in terms of connection with items, services, and people. Mining the Internet of things enables tour operators to promote their tourism products and services and also maximizes the satisfaction of the tourists due to the ready accessibility of services across the globe and 24/7.

Education

With the onset of the Pandemic, the right to education remained as a sole right in the pages of literature, and reality revealed catastrophic disruption of education, affecting close to 1.6 billion learners in almost 190 countries and all continents.

The pandemic impacted 94% of the world's student population, more prominently to the extent of 99% in low- and mid-income countries. Looking at the world of education before Covid 19, there was already a challenge to meet the target of Sustainable development goal 4: quality education in low- and middle-income countries and with the deepening crisis of the pandemic, historical dealignment in education from the pre-primary to secondary schools, technical and vocational education and training institutes, universities, adult learning centres, and skill development establishments became noticeable. India also experienced a digital divide in its traditional education system (IAS Working Paper, 2021). The change was warranted in the education system of our country but the shockwaves of the pandemic upscaled the need for change for an improved educated India got translated into a platform of forcible technology-enabled education. Different media of communication are being adopted by the educational authorities to replace traditional learning with techno education models but the reception of these methods was found to be mixed. Urban learners became enthused in taking up the online mode of education due to the uninterrupted availability of resources at their end on one hand but on the other hand, the rural learners found it tough to materialize their dreams for education (Mukesh Rawal, 2021). The structural changes to revamp the education sector are the need of the hour. The ambience of traditional learning has become now the present day's ambiguity by compulsory adoption of online learning platforms. In August 2021, the 328th Indian Parliamentary standing committee report unveiled a scathing indictment of the learning loss due to continued school closures. The severity of this learning gap became distinguished for the vulnerable sections of the society like students emanating from the downtrodden backgrounds, rural and tribal areas, and also the marginalized segments of the society and girls (supportscroll.in).

The rapidity of transformation to digitized education also impacted the higher education segments as the closures of colleges, universities, and other higher educational institutions over this two-year lockdown period but these institutions at least reap the benefits of utilizing the opportunity of implementing the various types of pedagogy across the higher education ecosystem. Strategies like blended learning, and Online Program Management Partnership, were considered for enhancement of teaching and learning with asynchronous (Canvas, Blackboard, D2L) and synchronous (Zoom, Google meet) modes and it is believed that it would yield benefit to the learner's community if these techniques can be layered into face-to-face instructions (Joshua Kim, Inside Higher Ed, 2020). To transform the education into the digital mode calls for not only the support of private entities to supply the digital platforms at an economical rate but at the same time, the effort must be given to facilitate the use of no-cost use, reuse, repurposing, and adaption. Public education can be defined and controlled by the contents and methods within the framework of pedagogical design and healthy relationship between teacher and students. Investment for education needs to be revised by changing the allocation of public expenditure on education and education-related resources else the lives of the underprivileged students who are severely impacted due to pandemic-induced disruption would be antithetical to the basic tenets of justice and human rights (countercurrents.org, 2020). The relevant transformation in the education system became more demanding in the primary and pre-primary level in comparison with higher education levels, else the unethical priorities in the child education would put the girl child into forceful domestic duties and premature marriages. So, the Indian Government across the nation should promote unilateral opportunities for education to reduce the regional disparity, gender bias, and increased dropouts irrespective of the threats of the pandemic.

Healthcare

The shift in the healthcare industry has started much before covid-19 still the pandemic has thrown multiple challenges as the health care services are most sought after in the world. Telehealth which was in its natal stage became over mature service in the days of covid-19. John Kravitz of Geisinger Health observed a 500% increase in telehealth visits during the first few weeks of the coronavirus outbreak. The Government has taken measures to reimburse even these telehealth visits by treating them at par with clinical footfalls, needless to mention that patient's walk-in is directly proportional to the acuity needs. The strategy of patient monitoring has also transformed from traditional systems to remote patient monitoring and home-based monitoring which helped to manage and create a care continuum. Internet of medical things gained substantial ground during the pandemic allowing the providers to deliver proactive services at a distance and the uptake of connected devices and digital health technologies went better (Peek, N., Sujan, M., & Scott, P, 2020). New cyber security concerns increase cloud adoption in healthcare. Changes in care delivery models also have implications for associated IT infrastructure. Infrastructure being very porous, its security is important and elusive. Not only is the technology change paramount in the healthcare industry concerning medical education also some major changes are seen. As a result of technological interventions, more grown-up telemedicine visits with patients, e-meetings with the colleagues, online interviews and therefore the trainees are also becoming techno-omnivores who are comfortable with using the technology more than the classroom learning or laboratory experience (Paddy Padmanabhan, CIO, 2020). With the launch of the national portal for the digital exchange of covid-19 patient data in the Netherlands, in a week major health care institutions along with the Government created an online portal that allows Dutch hospitals to share seamlessly the covid-19 patient details. The pandemic has also changed the entire healthcare supply chain landscape. The steep fall in the demand for health care products and services has dampened not only streams of cash flows in this sector but also sparked a positive thrive which made the companies be more analytical and calculated to deal with the unprecedented fluctuations of demand. To reduce the supply chain instability, the focus is given to hiking the onboarding of potential suppliers and the adoption of newer regulations. The need for home testing and other testing protocols might change and a mass sharing of patient-related information across the stakeholders of health care systems is also to be executed with appropriate adoption of data protection measures and cyber security mechanisms. To mitigate the use of hospital real estate for electives, the industry can create a new dimension of the avenue by exploring the possibility of serving the community with ambulatory service centres.

Banking

The industry has divergent issues and faced challenges for decades. The industry faced a screeching halt in the growth rate of its revenue, services, and transactions. Recollection if done concerning financial distress from 1930 to 2020, would be revealing that financial institutions faced the most due to the unprecedented tremors of the world. A historic perspective of the covid crash highlights the intensity of distress varied across 1930's Great depression, 1940's Hitler's invasion of France, 1987's Black Monday, 2001's dot-com bubble, 2008's subprime crisis to 2020's covid-19 pandemic (Alice, Nripen, Mandar) The outcome has shattered the planned operations which became prominent in piling Non-performing Assets, reduced demands for loans, disruptions in the payments, and delayed receivables. On top of the financial dents, the forbearance measures are taken by the banks at the request of the regulatory bodies to extend moratorium facilities and new loans to the MSMEs for the rejuvenation of their illiquid industries. To cope with the crisis the multiple bank branches were completely shut down on account of their inability to run the show.

As a cost-cutting and cost control measure the banks have merged and leveraged them. This situational move forced the customers and account holders to shift to the online banking transactions, social distancing norms succumbed the customers to opt for the mobile app to execute their transactions (Citizens' Voice, The Wilkes-Barre Insurance Newsnet). The realization of the bankers came through late that the chatting option to connect with customers is the need for survival and not an innovative feature. The deployment of funds towards the implementation of digitization in the banking industry is stiff and competitive but unavoidable as the significance of the use of Artificial Intelligence, and Blockchain technology is immense in the context of operations and functions, and consequently, the banking industry diverts their budget to reap the benefits of the applications. At this juncture, it is wise to consider the Dwight D Eisenhower matrix adopted in the banking sector for prioritizing the task about the intensity of the urgency by reducing chaos (Vasiliy Ivanov, Forbes Business Council). Below given is the Eisenhower matrix adopted in the banks.

Figure 1. Eisenhower Matrix in the Banking Industry

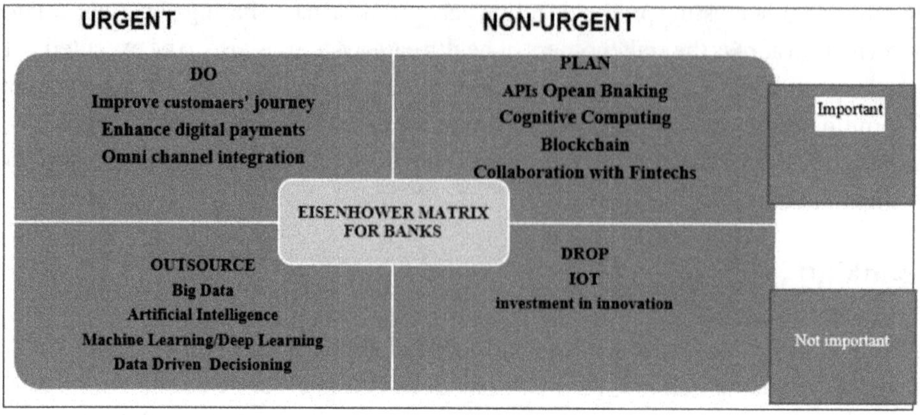

The matrix focuses on the aspects which are of prime importance and forgo those which are redundant and less valuable and so the above pictorial presentation depicts the important and unimportant areas to consider and ignore respectively. At present, the major drive is that the countries are moving towards self-reliance and there is a talk of strategic autonomy. Some of the modifications observed in the banking industry are: (businesstoday.in/banking)

1. Architectural transformation: As the banking industry is moving away from the clumpy rigidity of monolithic setup, the future banking sector would be more fluid and responsive with lighter but flexible microservices architecture.
2. With the advent of better cloud security, the banks are looking for a public cloud eco-system as a viable long-term option.
3. Banks are inclined to the usage of the blockchain model for the cost-effective setup and also to ensure the veracity of information and probability of detection of frauds.
4. Banks are open for API banking which will reduce go-to-market cycles, improve the decision making, and increase responsiveness as the customers are free to the bank as per their ease and convenience and this would eventually lead to data democratization and socio-economic upliftment.
5. With the ease of doing business gaining prominence the frauds are on the rise and the banks are making tighter KYC norms and rigorous background checks. The repository of credit risk data in a present real-time rule-based system would benefit the banks to impose stricter control measures before making them disproportional.

6. Banks are replacing the manual methods of model validation with faster and more efficient automated validation for better accuracy and transparency in the operation.

Insurance

The impact of covid 19 has also impacted the insurance business in India and across the globe. Concerning cybersecurity, the sudden change in technology and the work style of the employees make them vulnerable to new cyber risks. The nature of cyber threats is changing as the attackers are exploiting the uncertainties, the unprecedented situation, and rapid IT and organizational change (PWC, 2020). The immediate impacts on the insurance organization due to the covid 19 crisis are as follows:

1. Absence of workforce and declined efficiency.
2. Psychological stress and panic.
3. Over-dependence on technology due to rapid shift in remote working.
4. Over-reliance on digital platforms for sale and renewal of policies.

Apart from the above, there is an exponential rise in cyber-attacks as the cybercriminals are using the comparatively newly registered covid related domains to run social engineering campaigns such as phishing to distribute malware, trojans, and ransomware. Under this situation, the insurance organizations have adopted some measures to mitigate the risks.

The measures are: Secure newly implemented remote working practices, ensure the continuity of critical security functions, and counter opportunistic threats to take the advantage of the situation. Insurance companies offering cyber insurance should also take ultra-care by introducing some proactive measures, post-claim measures, and fresh considerations for underwriting, and by these measures, the insurance companies would be able to run most cost-effectively which was almost non-existent in earlier times. The work-from-home options reduced significantly the travel and rental costs and there came fungibility of skills available across the industry. The change in the compliance of rules also needs to be examined in a post-crisis period in the organization.

Analysis of Survey Reports as Evidence of Changes in Covid

The covid-19 situation has not just brought about a change in the day-to-day life of the people but has also transformed drastically the way the companies do business. If the various reports and surveys are to be believed there has been a digitalization in the customer and supply chain interactions. There is a huge shift in the acceleration of business in the past two years compared to what the situation was earlier.

When the pandemic had hit the world, most of the businesses and the economy of the country went down creating a loss situation in the market. There were new practices adopted by the companies. The e-commerce business transformed drastically and the consumers moved dramatically towards online channels.

Below is the survey conducted by Mc Kinsey which confirms the rapid shift in consumers interacting through online channels. It also shows that rates of adoption are years ahead of where they were when previous surveys were conducted—and even more in developed Asia than in other regions.

Figure 2. Accelerated Digitized Customers Interactions in Covid
Source: Mckinsey & Company Report

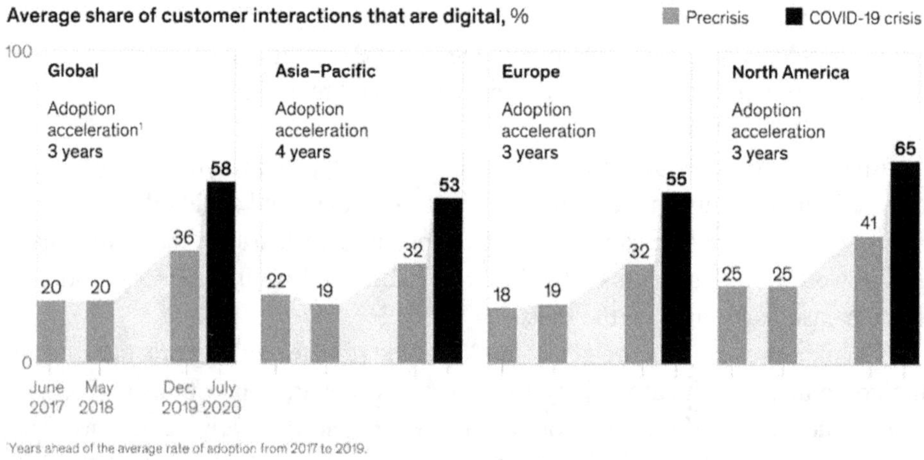

There was one more survey conducted which again showed a leap in terms of digitally enhanced offerings. The results suggest that there was a seven-year increase, on average, at the rate the companies have been developing in terms of products and services.

The findings of the survey conducted by McKinsey suggest that during the crisis, companies have probably refocused their offerings rather than made huge leaps in product development in the span of a few months.

Figure 3. Digitization leap in the share of offerings across business areas
Source: Mckinsey & Company Report

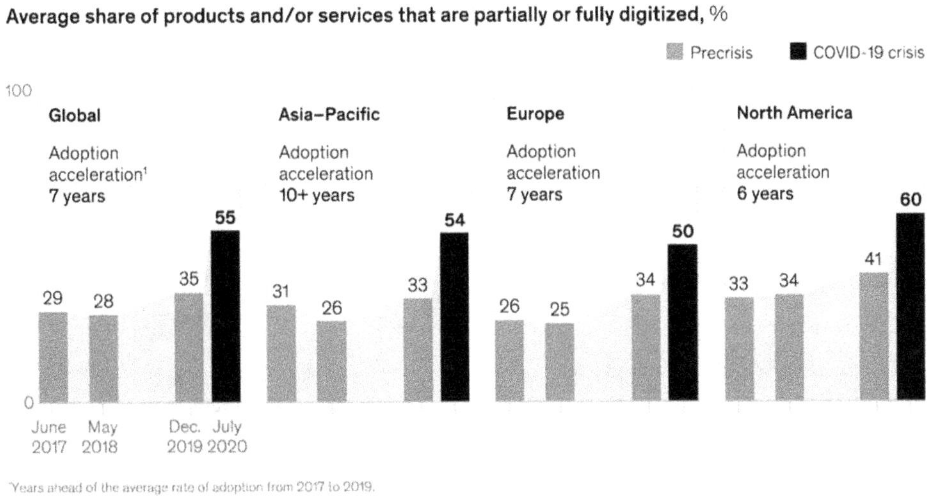

There has been an overall change whether it be businesses, entrepreneurs, doctors, consumers, or education. There was a need for digitalization in every sector post-covid 19 situations. These remarkable changes are shown in the image below which is again a survey done by McKinsey and the report submitted by them.

Figure 4. Quick change in the responses of the company in covid crisis
Source: Mckinsey & Company Report

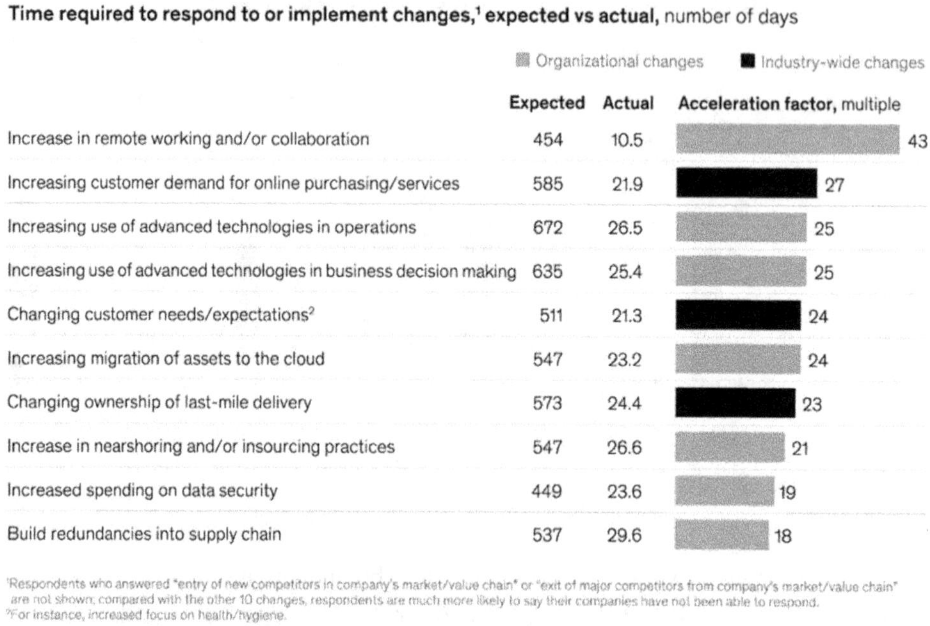

The images above clearly show how the companies have benefited with the help of digital transformation. Factually, all of us have seen that the companies which were small or could not afford the cost which was involved in the transformation to the digital mode could not survive. And the companies which were able to spend that amount of expenditure for the transformation are still leading in the market. We homo-sapiens have now grappled with the phenomenon known as the Internet. Like we say every coin has two sides, and the same applies to technology also. With the advancement of technology, and digitalization in every sphere, there have been issues pertaining to cyberspace as well. With the thrust to go digital, new legal challenges are faced. Unfortunately, even though there are stringent legislations available for the internet, technology, cyberspace, companies, and consumers but still the common man is relatively uninformed about the legal implications emanating from these technologies, both positive and negative. Major areas of law in which technology has been applied and issues are reported are technology and crime, technology and IPR, technology and contracts, technology and evidence when now we talk specifically about digital evidence, taxation issues in the digital economy, and the biggest one and the most challenging one-cyberspace and internet, as they share with our universe the attribute of inflation, inhering accelerating pace of change (Nappinai, N.S., 2009) There is an interplay of technology vis-à-vis these areas. Along with the advent of new technologies, new avenues of transacting business have emerged, making to a large extent paper transactions a thing of the past. E-commerce and e-banking are the order of the day. Laws that have been passed in India, viz., Information technology act, amendments which are made in the Evidence Act, Indian penal Code, Code of Criminal Procedure, and other related laws are equally complex and not yet tested in the courts. Higher courts themselves find it difficult to grapple with the complex issues which emerge for interpretation since they come along with unfamiliar technical and legal jargon. Law in the world of the internet is a constantly changing and evolving beast. It neither has parental guidance nor memory genes to guide its path to the lighter horizon. It had to rely on the existing logic and purposive interpretation in its formative years. Laws and their application in this new domain have relied on usage, trial and error for their development, and it has not hesitated to watch, learn and imitate its neighbors and then evolve beyond its learning. The first and most important learning was to standardize the laws across domains to ensure enforcement and adherence under multiple jurisdictions (Everard J., 2000).

Uniform model laws were chartered to create standardization. This however did not result in uniform model laws applied by all the countries. Laws on the internet therefore cannot be read in isolation and cannot be pigeon hooded in a tight slot.

Laws that regulate the development of technology are also driven by it, and most of them have evolved from the development of the technology, broadly classified as ‘Technology Laws’.

Technology vs. Cyber Laws

Technology laws lay scattered throughout the larger fabric of law and the justice delivery system. The world of cyber now reaches much farther than just the internet in the electronic domain. Cyber impacts almost every domain of human activity including the political, economic, social, legal, or environmental including a person’s social environment. Today, cyber is almost interchangeable with technology. The technological advances have thrown a strong challenge at existing norms and beliefs. They question the existing laws, concepts, and interpretations and encourage governments, legislators, and enforcement mechanisms to push boundaries to evolve laws to meet technology’s challenges. Technology has further unsettled “settled laws and principles” and emboldened the judiciary to become more adventurous by venturing into uncharted territories (Eichensehr, K.E., 2014). It has erased boundaries not only with the reach of technology but also concerning enforcement. The challenge to the territory-the jurisdictional issues is another challenge faced. With the shrinking borders, international laws have evolved, especially in the private international law sector. Countries across the globe proactively involved themselves in the evolving process of harmonizing laws across boundaries and through multiple jurisdictions. Nation-states engaged in formulating soft laws which work for trade and commerce (Bhuiyan, A., 2014)

Technology encompasses every walk of life, and each such application of technology consequently is subject to regulation. The debates for example on net neutrality highlighted the issue of regulating the internet. These debates highlighted the fact that the internet is neither free nor regulated/unregulated. Another example can be of creating a website, for which the very first step would be to obtain a domain name, which leads to contractual obligations between the user and the registry handling it. Contract laws and IPR laws come into play in such situations for developing the website and hosting it. Apart from this fundamental rights also have a role to play concerning the use of data collected online including personal data. Constitutional rights, therefore, take primacy. Finally, the content on the website may give rise to criminal prosecutions. It's just an example of how a new paradigm in law may be created in such a scenario.

Technology law is still in its fledgling state and testing the water with baby steps. Every day brings some new developments in technology. Rigid drafting of laws, therefore, is neither suitable nor advisable in such a scenario.

Also, another drawback is that the internet grew with bare minimum legal requirements and restrictions. Hence, we can say that technology laws bring with them complexities, dangers, obstacles, and challenges at the same time promise to be exciting and adventurous.

CONCLUSION

Business organizations, in post covid scenario, reoriented their production and rendering of services in the VUCA (Volatility, Uncertainty, Complexity, and Ambiguity) world considering the changes in the behaviour of the stakeholders in a state of redefined culture, transitioned technology and leadership style. Incessant depression-induced loss of disposable income in the hands of consumers, loss of lives, and livelihood which triggered the organizations to re-flourish in a dual-friendly business model for the mutual benefit of the customers and producers. The mostly observed three types of behaviours are sustained behaviour, transformed behaviour, and collapsed behaviour. Sustained behaviour refers to the likeliness of returning to the age-old practices once the crisis is over. Transformed behaviour is the shift in the trend with the advent of the crisis. Collapsed behaviour is the one that is either to perish or will have a rebirth. The organizations for a sustainable green survival revamped themselves with a cost-effective performance strategy by eliminating the redundant activities and strengthening the value-added productions subject to no sacrifices to the qualities and attributes of the products and services by making optimum utilization of manpower, material, and financial resources in a participative leadership approach.

Indians are unfortunate to embrace the pandemic at the juncture of planning and forecasting India to become a 5 trillion-dollar economy by 2025. Therefore, efforts are made to make this dream come true if not in the proposed period but at least within some reasonable time horizon.

In a covid sustained scenario, the outlook is reoriented in the name of learning the new, unlearning the outdated techniques, and relearning for a sustainable future as literacy of the revolutionized world does not confine within the limits of mere reading and writing but also simultaneously to learn, unlearn and also to relearn. (Alvin Toffler)

REFERENCES

Antonakis, J. (2020). Leadership to defeat COVID-19. *Group Processes & Intergroup Relations*, *24*, 210–215. doi:10.1177/1368430220981418

Banking on technology: 5 tech trends every bank should be prepared for in 2021. (n.d.). Retrieved Jan 13, 2022, from https://www.businesstoday.in/opinion/columns/story/five-technology-trends-to-shape-indian-banking-and-financial-services-in-2021-285594-2021-01-27

Barrios, L. M., Prowse, A., & Vargas, V. R. (2020). Sustainable development and women's leadership: A participatory exploration of capabilities in Colombian Caribbean Fisher Communities. *Journal of Cleaner Production*, *264*, 121277. https://doi.org/10.1016/j.jclepro.2020.121277

Bhuiyan, A. (2014). *Internet Governance and Global South: Demand for a New Framework*. Palgrave Macmillian.

Blau, P. M. (1964). Justice in social exchange. *Sociological Inquiry*, *34*, 193–206. doi:10.1111/j.1475-682X.1964.tb0058

Børge Brende, P. (n.d.). *India's roadmap for Success and prosperity in the post-COVID world*. World Economic Forum. Retrieved Dec 16, 2021, from https://www.weforum.org/agenda/2021/11/india-s-future-and-role-in-the-post-covid-19-world/

Carsten Linz, M. E. C. (n.d.). *How to transform your business model for a post-COVID future*. World Economic Forum. Retrieved Jan 14, 2022, from https://www.weforum.org/agenda/2020/11/transform-business-model-post-covid-future/

Cavozas, E. A., & Morin, G. (1994). *Cyberspace and the tha Law: Your rights and duties in the Online World*. MIIT press.

Charoensukmongkol, P., & Phungsoonthorn, T. (2020). The effectiveness of supervisor support in lessening perceived uncertainties and emotional exhaustion of university employees during the COVID-19 crisis: The constraining role of organizational intransigence. *The Journal of General Psychology*, 1–20. doi:10.1080/00221309.2020.1800582

Covid-19 and the automotive industry: Changes, challenges and the future. (n.d.). *Dartmouth Undergraduate Journal of Science*. Retrieved Jan 5, 2022, from https://sites.dartmouth.edu/dujs/2020/11/09/covid-19-and-the-automotive-industry-changes-challenges-and-the-future/

Covid-19: How the crisis impacts high tech. (n.d.). Retrieved Dec 16, 2021, from https://www.accenture.com/_acnmedia/PDF-121/Accenture-High-Tech-COVID-19-Impacts.pdf

Covid-19 impact on the Indian insurance industry - pwc.in. (n.d.). Retrieved Jan 3, 2022, from https://www.pwc.in/assets/pdfs/services/crisis-management/covid-19/covid-19-impact-on-the-indian-insurance-industry.pdf

D'Auria, G., & De Smet, A. (2020). *Leadership in a Crises: Responding to the coronavirus outbreak and future challenges*. McKinsey & Company.

Dirani, K. M., Abadi, M., Alizadeh, A., Barhate, B., Garza, R. C., Gunasekara, N., & ... . (2020). Leadership competencies and the essential role of human resource development in times of crisis: A response to Covid-19 pandemic. *Human Resource Development International, 23*, 380–394. doi:10.1080/13678868.2020.1780078

Education in a post-COVID world-nine ideas for public action. (n.d.). Retrieved Jan 16, 2022, from https://en.unesco.org/sites/default/files/education_in_a_post-covid_world-nine_ideas_for_public_action.pdf

Eichensehr, K. E. (2014). Cyber law of Nations. *The Georgetown Law Journal, 103*, 317.

Everard, J. (2000). *VirtualmStates: The Internet and the boundaries of the Nation State*. Psycology Press.

Four types of company culture: Which one is yours? (2021, November 26). Retrieved Jan 25, 2022, from https://www.personio.com/hr-lexicon/company-culture-types/

Giurge, L., & Bohns, V. (2020). 3 Tips to Avoid WFH Burnout. *Harvard Business Review.*

Heinegg, V., & Heintschel, W. (2012, June). *Legal Implications of Territorial Soverignty in Cyberspace*. Academic Press.

Impact of covid-19 on Technology. (2021, July 13). Retrieved Jan 16, 2022, from https://bernardmarr.com/impact-of-covid-19-on-technology/

ISAS.NUS.EDU.SG. (n.d.). Retrieved Dec 6, 2021, from https://www.isas.nus.edu.sg/papers/education-during-the-covid-19-pandemic-in-india/

January 2021: Educational Resurgence Journal: DYPCOED, Pimpri. (n.d.). Retrieved Dec 14, 2021, from https://coed.dypvp.edu.in/educational-resurgence-journal/jan-2021.aspx

Kahai, S. S., Sosik, J. J., & Avolio, B. J. (1997). Effects of leadership style and problem structure on workgroup process and outcomes in an electronic meeting system environment. *Pers. Psychol., 50*, 121–146. doi: .1997.tb00903.x doi:10.1111/j.1744-6570

Liborius, P. (2014). Who is worthy of being followed? The impact of leaders' character and the moderating role of followers' personality. *The Journal of Psychology, 148*, 347–385. doi:10.1080/00223980.2013.801335

Limited, I. (n.d.). *Navigating the next for financial services.* Infosys. Retrieved Jan 11, 2022, from https://www.infosys.com/industries/financial-services.html

Lu, J., Jiang, X., Yu, H., & Li, D. (2015). Building collaborative structures for teachers' autonomy and self-efficacy: The mediating role of participative management and learning culture. *School Effectiveness and School Improvement, 26*, 240–257. doi: 10.1080/09243453.2014.888086

McKibbin, W. J., & Fernando, R. (2020). *The global macroeconomic impacts of COVID-19: seven scenarios*. doi:10.2139srn.3547729

McKinsey & Company. (2022, April 13). *Covid-19: Implications for business in 2021*. McKinsey & Company. Retrieved April16, 2022, from https://www.mckinsey.com/business-functions/risk-and-resilience/our-insights/covid-19-implications-for-business-2021

Mukherjee, M. (2022, January 30). *The lost generation: How unending school closures have crippled the education of India's children*. Retrieved Feb 6, 2022, from https://scroll.in/article/1016178/the-lost-generation-how-unending-school-closures-have-crippled-the-education-of-india-s-children

Nappinai, N. S. (2009). Cybercrime Law in India. Has law kept pace with emerging trends? An Emperical Study. *Journal of International Commercial Law and Technology*, *5*(1), 22–28.

No right to education in the pandemic? (2020, December 7). Retrieved Jan 1, 2022, from https://countercurrents.org/2020/12/no-right-to-education-in-the-pandemic/

Paddy Padmanabhan, C. (2020, March 27). *How the COVID-19 pandemic is reshaping healthcare with technology*. CIO. Retrieved Dec4, 2021, from https://www.cio.com/article/193198/how-the-covid-19-pandemic-is-reshaping-healthcare-with-technology.html

Peek, N., Sujan, M., & Scott, P. (2020, June 1). Digital Health and care in pandemic times: Impact of covid-19. *BMJ Health & Care Informatics*. Retrieved Nov 16, 2021, from https://informatics.bmj.com/content/27/1/e100166

Pulse of Remote Work: Before & After covid-19. (2022, March 8). Retrieved May 6, 2022, from https://www.pipefy.com/blog/pulse-of-remote-work-before-after-covid-19

Sagnak, M. (2016). Participative leadership and change-oriented organizational citizenship: The mediating effect of intrinsic motivation. *Eurasian J. Educ. Res.*, *62*, 181–194.

Saura, J. R., Reyes-Menendez, A., & Palos-Sanchez, P. R. (1970, January 1). *The Digital Tourism Business: A systematic review of Essential Digital Marketing Strategies and trends*. IGI Global. Retrieved Dec 17, 2021, from https://www.igi-global.com/chapter/the-digital-tourism-business

Schools in the pandemic: Seeking Education Equity for India's Marginalized Kids. (2021, April 8). Retrieved Jan 2, 2022, from https://news.microsoft.com/en-in/features/schools-in-the-pandemic-seeking-education-equity-for-indias-marginalized-kids/

Sinclair and stabler 1997 have defined the tourism enterprise as a composition: Course hero. (n.d.). Retrieved Nov 6, 2021, from https://www.coursehero.com/file/p244ji8g/Sinclair-and-Stabler-1997

Teaching and learning after covid-19. (n.d.). Retrieved Jan 4, 2022, from https://www.insidehighered.com/digital-learning/blogs/learning-innovation/teaching-and-learning-after-covid-19

The role of entrepreneurship in tourism industry: An overview. (n.d.). Retrieved Feb 4, 2022, from https://www.researchgate.net/publication/316240958_The_Role_of_Entrepreneurship_in_Tourism_Industry_An_Overview

Voice, C. (2021, March 28). *Covid-19 pandemic has changed the Banking Industry*. InsuranceNewsNet. Retrieved Dec 24, 2021, from https://insurancenewsnet.com/oarticle/covid-19-pandemic-has-changed

Weston, M. (2020, June 12). *How covid-19 has impacted organisational culture*. theHRDIRECTOR. Retrieved Dec 26, 2021, from https://www.thehrdirector.com/how-covid-19-has-impacted-organisational-culture/

Williams, M. (2020, August 21). Council post: How to scale your business operations fast without burning your money. *Forbes*. Retrieved Jan 16, 2022, from https://www.forbes.com/sites/forbesbusinesscouncil/2020/08/24/how-to-scale-your-business-operations-fast-without-burning-your-money/

Yahoo. (n.d.). *Recovery and rebirth for restaurants in a post-pandemic world*. Yahoo! Finance. Retrieved Nov 5, 2021, from https://ca.finance.yahoo.com/news/recovery-rebirth-restaurants-post-pandemic-171100354.html

Chapter 8
In Times of Uncertainty, Organizational Cultural Shifts Appear Seamless With Emotionally Intelligent Transformational Leaders at the Helm

Lezlie Banks
Western Governors University, USA

ABSTRACT

Dealing with confusing times requires drastic measures. COVID-19, one of the most significant events in 2019, created an immediate frenzy. January 6th, 2021 showcased how voting results can sway people to make far-reaching choices. February 2022, the war between Russia and Ukraine generated trepidation for others' well-being. The anxiety and emotional turmoil generated from circumstances discussed above will not go away easily. Yet, the aftermath of these conditions will lessen with emotionally strong leaders at the forefront. Using emotions to help create relationships reduces anxiety. This chapter focused on understanding the origins of emotions, recognizing the benefits of emotional intelligence and transformational leadership styles, drawing attention to how these behaviors create learning organizations, and proposing the best ways to move forward in turbulent times.

DOI: 10.4018/978-1-6684-4358-3.ch008

INTRODUCTION

The world changed in 2019. COVID-19 forced people to reevaluate views on safety and health (Elbarazi, 2022). So far, 979,725 deaths ensued in the United States due to the coronavirus (Cable News Network.com). January 6th, 2021 brought additional panic as people witnessed a mob of 2,000-2,500 supporters of Donald Trump storm the White House in Washington, D.C. (Cable News Network.com) Additional civil unrest happened in February, 2022 when Russia invaded Ukraine (Cable News Network.com). These instances created fear and panic (Elbarazi). This new found anxiety carried over into the workforce. Employers began determining alternate plans for work arrangements while also addressing employees' wellbeing (Supramanian, Shahruddin, & Sekar, 2022). When faced with turbulent times, maintaining mental health remains a top priority (Lawrence, 2021). As a result, businesses quickly realized the importance of providing outlets for employees who needed additional emotional support.

The ability to maintain emotions begins in the brain because the brain facilitates emotional advancement. Edelman & Van Knippenberg (2018) assert leaders must react appropriately to emotions in order to effectively work with and relate to others. Responding accurately throughout times of doubt requires a combination of skills (Karimi, Leggat, Bartram, Afshari, Sarkeshik, & Verulava, 2021). Today, organizations remain under immense pressure to respond to an ever-changing workforce (Karimi, et. al, 2021). These changes instigated an increased focus on hiring and retaining leaders with emotional intelligence abilities because these qualities help employees embrace change (Görgens-Ekermans & Roux 2021). In the workplace, a person's capacity to manage their own emotions is crucial. Psychologists affirm this skill is known as emotional intelligence (Supramanian, Shahruddin, & Sekar, 2022). In today's post-pandemic areas, EI skills outweigh intelligence because emotionally intelligent leaders relate better to followers (Supramanian, et. al.).

To connect with followers, leaders must acknowledge their emotions by putting themselves in someone else's shoes and continuing to monitor their own emotional state during difficult situations (Mateiaş 2018). In some cases, gaining emotional intelligence skills persists as a work in progress since problem-solving skills also require expertise and patience (Bayraktar, Yalçınkaya, Eyitmiş, 2021). Once mastered, possessing aptitude and talent proves necessary because emotions, attitudes and values assist leaders with creating a sound problem solving process (Bayraktar, et. al). As a result, relying on the tenets of emotional intelligence will help lead to effective problem-solving and creative solutions (Bayraktar, et. al). As leaders witnessed first-hand, innovative solutions help begin the process of getting things back to normal.

When the world began to change, organizations experienced a climate shift due to social, cultural and political differences (Karasneh, & Momani 2020). These recent outcomes illustrate the importance of emotional intelligence (Karasneh & Al-Momani). In order to balance retention and performance, leaders must develop swift changes to current processes. Adjusting systemic opportunities requires brainpower combined with strategy. These necessary updates would not have been possible without also exercising the principles of emotional intelligence. Leaders with high emotional intelligence enhance job satisfaction, increase job performance and improve organizational situations (Lai; Tang; Lu; Lee; & Lin, (2020) & Raman, Peng, & Chi (2021). Based on these findings, emotional intelligence and effective leadership remain an integral part in shaping and creating a positive and functional organizational atmosphere.

Additionally, the necessity to create effective organizations begins with the leader and employee engaging in authentic relationships and positive interactions (Karimi, Leggat, Bartram, Afshari, Sarkeshik, & Verulava, 2021). To rebuild departments and teams, leaders must create new strategies to encourage happiness and security which will lead to job satisfaction (Edelman & Van Knippenberg). Even though the aftermath of the pandemic taught leaders the importance of embracing changes during times of crisis, employees still need reassurance that leaders have a plan in place (Karimi, et. al, 2021). Preparing a plan will ensure seamless cultural transitions during times of ambiguity (Supramanian, Shahruddin, & Sekar, 2022).

This chapter will provide an overview of how the brain guides emotions and how traumatic events affect the brain's recovery system; discuss the relationship between emotional development and empathy; reveal the importance of transformational leadership; share similarities between emotional intelligence and transformational leadership and explain how learning organizations allow employees to move forward as high performing teams.

THE BRAIN

Bradberry and Greaves (2009) believed EI begins in the brain; specifically, near the spinal cord and insisted the design of the brain enables people to express emotions. Researchers discovered that the brain responds to emotional incentives (Balconi & Pozzoli, 2009). The frontal brain responds to emotional signals (Balconi, & Pozzoli). In particular, the frontal brain also responds to facial expressions. Empirical data confirms the right and frontal lobes of the brain address emotional activities (Balconi, & Pozzoli; Goleman, 2011). In order for rational thoughts to occur, the right and left brain must work collectively (Goleman,). Emotional competence equals exemplary work performance because people in control of their emotions react well under pressure and understand others' emotions (Goleman,). Specifically, the left frontal lobe determines motivation and the right frontal lobe assists with positive emotions (Balconi &Pozzoli). Dolcos, Alexandru & Dolcos (2011) posit emotions affect cognition and behavior because feelings originate in the brain.

Fox and Shackman (2017) state the amygdala (part of the brain that stores positive and negative emotions) operates in concert with the limbic system (part of the brain that contains knowledge and emotional development) and facilitates cognitive and behavioral effects of emotions. The amygdala lives in the emotional region of the brain (Lawrence, 2021). After information circulates through the amygdala, it travels to the prefrontal cortex which exists as the problem-solving area (Lawrence). The cerebral cortex recognizes limitations that are related to separate cognitive functions (Jiang, Qiao, & Li. 2021). Further, the prefrontal cortex (cognitive center) supports rational thought; so, often this section is also known as the executive center of the brain (Jiang, Qiao, & Li). However, once stress sets in, the prefrontal cortex becomes taken over by the amygdala and the limbic system (Lawrence). All things considered the amygdala helps ensure survival (Lawrence). Balancing emotions between the prefrontal cortex, and the amygdala will aid in improving emotional intelligence skills (Lawrence). In fact, emotional intelligence is similar to social cognition; predominantly linked with the amygdala (Jiang, Qiao, & Li).

Emotional memories, located in the posterior neocortices, control behavior and provide feedback to the amygdala (Ross, 2022). The amygdala has the ability to increase or decrease memories stored in the neocortex through a direct link to the hippocampal pattern (Ross). Based on prior research, it would be reasonable to assume emotions are generated by the amygdala (Saarimäki, Glereanb, Smirnov, Mynttinenb, Jääskeläinen, Sams, & Nummenmaa, 2022). Primary emotions (reactions) are located on the right side of the brain (Ross). Interestingly, secondary emotions (connections) are located on the left side (Ross,).

In essence, emotions occur in both the right and left hemispheres (Ross, 2022). Emotional behaviors (anxiety or happiness) labeled as qualities with specific classification, often receive information directly from the brain (Ross). Social emotions are based on individual beliefs and how these beliefs occur while interacting with others. Ross (2022) affirms, even though the human brain is naturally equipped to learn and understand social emotions, this does not always apply to unrecognized behaviors.

Researchers state the frontal lobe has a positive effect on intelligence (Jiang & Li, 2021). Yet, interpreting how the human brain works remains a work in progress for neuroscientists (Jiang, Qiao, & Li). Surprisingly, emotional intelligence traits appear debatable on its construct as to a specific portion of the brain (Jiang Numkena, Qiao, & Li) because both sides of the brain house emotions. In today's climate, focusing on positive emotions will help employees move forward during ambiguous and extraneous times. Leaders with emotional intelligence skills can assist employees who experienced traumatic events during COVID-19 (Jiang, Qiao, & Li).

Traumatic events influence people. Though some people cope well with stress and anxiety, others may struggle from long-term symptoms (Goërtz, Van Herck Delbressine Vaes, Meys Machado, Houben-Wilke, Burtin, Posthuma, Franssen, van Loon, Hajian Spies, Vijlbrief, van 't Hul, Janssen and Spruit, 2020). These long-term struggles often include: distressing memories, nightmares, mental exhaustion, or physical inactivity (Goërtz, et. al,). Unfortunately, trauma creates longstanding effects on a person's well-being (Goërtz, et. al). The impact on mental health, commonly described by people suffering from a traumatic event, ranges from anxiety and depression to post-traumatic stress disorder (Jiang, Qiao, & Li, 2021). Further, people also often report prolonged cognitive setbacks commonly known as "brain fog." (Burton, Aughterson, Fancourt, D & Philip, 2022).

In addition to the cerebral areas, brain alterations affect the amygdala, which as referenced earlier, contains emotions, and memory (Avalew & Avenew). Research indicates emotions and memory are a part of the social brain (Lawrence, 2021). What's more, the social brain helps people interact in social situations (Lawrence). When interacting with employees' leaders accept a person's mental state; then, based on the initial exchange, they make predictions about future behavior (Avalew & Avenew, 2022). Since behaviors are contagious, leaders must gain personal emotional control and also assist employees with balancing their emotions (Valenzano, Scarinci, Monda, Sessa, Messina, Monda, Precenzano, Mollica, Carotenuto, Messina, & Cibelli, 2020). The ability to maintain emotional control will enable leaders to flawlessly provide an optimistic outlook for employees (Valenzano, et. al). Leading with a positive attitude helps with transforming the organization and improving employees' emotional wellbeing (Avalew & Avenew).

EMOTIONS

During the 17th century, Hall (1604) described the word "emotion" which emerged from the French word *émotion*, as physical disorder. In the 18th century, the definition of emotion expanded to mean mental experiences (Dixon, 2012). Centuries later, Freud (1921) advised emotions allowed people to express feelings. Lange (1922) affirmed emotions exist as the most important facets in life. In 1990, Salovey & Mayer (1990) declared emotions dictate everyday responses through signals received from the brain. Lazarus (1991) confirmed, "Emotions play a central role in the significant events of our lives" (p. 4). Clore (1992) affirmed emotions create positive cognitive responses.

Emotions determine a persons' outlook on life and emotions stem from different situations and scenarios (Batson, Shaw, & Oleson, 1992). Often, these signals help build emotional relationships (Batson, et. al.). Hatfield, Cacioppo, & Rapson (1994) affirmed emotions are contagious. According to Goleman (1995), cognitive growth enables self-perception and the ability to read and relate to other people. Clearly, managing emotions leads to better life experiences (Goleman, Boyatzis & McKee, 2002).

Years later, the definitions of emotion expanded to include different groups. Izard (2007) discussed two emotions: basic and complex. Basic emotions (interest, joy, sadness, and anger) develop during childhood and complex emotions occur during teenage and adulthood years (Izard, 2007). The ability to control emotions, emotional regulation, encourages individuals to monitor and recognize emotions (Giuliani & Gross, 2009). According to Dolcos, Alexandru, & Dolcos, (2011), two additional categories of emotions developed: bottom-up emotions and top-down emotions. Bottom-up emotions begin in the amygdala (Dolcos, et.al,). Top-down emotions rely on information from the amygdala; however, these emotions receive data from the dorsomedial pre-frontal cortex (Dolcos, et.al, 2011). Thus, emotions affect cognition and behavior because feelings originate in the brain (Dolcos, Alexandru, & Dolcos, 2011).

Individual beliefs depend on a person's current emotional state (Adophs, 2017). According to Lubbadeh, (2020) science proves emotions include thoughts. Identifying emotional signals remains vital in social interactions, because these signals help decipher an individuals' behavioral reactions (Ng, Lai, Brown, Järvinen, Halgren, Bellugi, & Trauner, 2018). Some analyses assert emotions represent vital features necessary for high performing leaders versus strictly relying on technical competencies and logical intelligence (Karimi, Leggat, Bartram, Afshari, Sarkeshik, & Verulava, 2021). According to Saarimäki, Glereanb, Smirnov, Mynttinenb, Jääskeläinen, Sams, & Nummenmaa, (2022), our emotions occur in several classifications. For that reason, effective leaders must maintain emotional control while assisting followers with adapting to change (Ng, et. al.). The best way to approach this is for the leader to acknowledge and appreciate others' feelings (Karimi, et. al.). In fact, leaders who utilize empathy during discussions tend to have better relationships with direct reports and peers (Lubbadeh, 2020).

EMPATHY

Empathy derives from a neuroscientific model (Rolls, 2005). This model suggests that when a person observes varying levels of emotions in others, he or she instantly reacts according to his or her emotional state. Empathy, an emotion exhibited by emotionally intelligent leaders, encourages followers to openly share ideas (Polychroniou, 2008). Later, Singer and Lamm (2009) stated empathy means, "The ability to react to the emotions exuded from someone else by putting yourself into their current emotional state" (p. 1).

In 2011 Goleman insisted that empathy occurs in three different forms: cognitive, emotional and compassion. Managers and supervisors who utilize cognitive empathy can easily relate to their followers because the leader relays information in basic and simple terms (Goleman, 2011). Emotional empathy helps leaders relate to followers using feelings and understanding (Goleman). Leaders who use emotional empathy often create an immediate personal connection with followers (Goleman). For this reason, successful leaders usually possess emotional empathy characteristics (Goleman). Compassion empathy encourages leaders to readily provide input and solutions when asked. Empathic leaders know when to offer feedback and when to remain quiet (Goleman). Still, one particular emotion, empathy, continues to surface in discussions regarding emotional cognition (McRae, Misra, Prasad, Pereira, & Gross, 2011).

A combination of empathy and emotional discipline contribute to high levels of emotional intelligence ((Lubbadeh, 2020). Edelman & Van Knippenberg (2018) posit emotional balance between leaders and followers creates productive environments. Positively administering empathy during interactions helps motivate followers (Edelman & Van Knippenberg). Experts agree empathy remains the most important leadership trait in today's organizational climate (Edelman & Van Knippenberg). Previously, leaders believed empathy only existed as a "soft skill" and this talent was not crucial to success; however, a renewed appreciation for this trait formed after the pandemic (Votaw, 2022). Leading with empathy helps support people even when they are not performing at high levels (Votaw). Still, leaders must use a combination of intellect and empathy to become trustworthy (Votaw).

INTELLIGENCE

Studies on intelligence started in 1920. Based on his research, Thorndike (1920) created three subsets of intelligence known as abstract intelligence; mechanical intelligence, and social intelligence. Thorndike (1920) defined social intelligence as, "The ability to understand men and women, boys and girls—to act wisely in human relations" (p. 18). Moreover, Thorndike stated the basic premise of effective relationships rests with socially competent behaviors and actions. Abstract intelligence is focused on concepts; mechanical intelligence refers to comprehending machinery and instruments; and social intelligence includes concern for others' emotions as the focal point (Thorndike, 1920). Spearman (1927) defined intelligence as individual abilities and skills. Conversely, Thurstone (1938) disagreed and stated intelligence relies on different coping capabilities (e.g., communication skills and technical skills). Piaget (1950) described intelligence in three phases: assimilation, accommodation, and equilibration (Piaget, 1950). Assimilation, a cognitive ability, supports prior thoughts and memories combined with new experiences; accommodation encourages current and previous experiences to interact and equilibration occurs when assimilation and accommodation merge (Piaget, 1950). Similar to Piaget, Gardner (1983) believed intelligence originated in two ways: interpersonal and intrapersonal. Interpersonal intelligence is used to determine other people's moods and enable cooperative working relationships; while intrapersonal intelligence is concentrated on one's own emotions (Gardner, 1983). Gardner (1983) believed intelligence originates at birth and people enter the world with levels of intelligence that alter over time.

Carroll (1993) created a model of intelligence that relied on cognitive abilities. This model, known as the three-stratum model, is focused on general intelligence. General intelligence includes fluid intelligence (ability to apply abstract thinking), crystallized intelligence (knowledge gained throughout life), broad memory (non-specific memories), broad speediness (how fast information travels), and broad visual analysis (Carroll, 1993). In 1997, Weschler stated the concepts of emotional intelligence emerged from social intelligence.

Goleman (2014) indicated leaders with social intelligence aid in helping employees deal with crises situations because this group exercises self-awareness; applies self-management techniques; remains socially aware; and manages relationships well. Each of these behaviors contribute to high levels of emotional intelligence. In 2019, García-Bullé advised social intelligence includes the capacity to build relationships by using a formula of empathy and confidence. All in all, social intelligence includes implementing emotional control in varying situations (García-Bullé). Throughout conversations, socially intelligent people empathize with others; listen well; read emotions well; pay close attention to their peers and subordinates' behavior and adjust their behavior appropriately; feel comfortable with different personalities; and handle their image by remaining authentic during interactions (García-Bullé). Distinctly, social intelligence requires a balance of emotional control and self-awareness to enhance connections (Garcia-Bulle,).

When leaders use skill sets of emotional Intelligence (EI) and social intelligence (SI), known as ESI, the organization manifests into a cohesive unit (García-Bullé, 2019). These behaviors coexist through a focus on self-awareness; self-management; and relationship management (Mateiaş 2018). Leaders who use ESI exist as top performers because followers relate well to this group (Mateiaş). Though similar, one difference separates these two areas: social intelligence focuses on the future (Mateiaş). Exercising these tenets helps people plan for the unknown (García-Bullé). On the other hand, emotional intelligence concentrates on the present because emotions transpire in real-time (García-Bullé). Previously, cognitive intelligence was the most crucial characteristic of a good leader (Mateiaş, 2018). Present-day research indicates people with emotional intelligence relate to others through a shared sense of emotional awareness (García-Bullé).

EMOTIONAL INTELLIGENCE

In 1960, the term emotional intelligence (EI) focused specifically on psychology. The pioneers of EI research, Mayer and Salovey (1990) defined emotional intelligence as, "the ability to monitor one's own and others' feelings and emotions, to discriminate among them and to use this information to guide one's thinking and actions" (p. 2). Another researcher who established EI as a noteworthy topic defined EI as, "the ability to handle ourselves well in personal and professional relationships" (Goleman, 1995, p. 5). Goleman's book, *Emotional Intelligence*, helped business leaders, psychologists, and higher education organizations understand the importance of self-awareness, self-management, social awareness, and relationship management. Established from extensive psychological research, having EI helps employees achieve success in the workplace (Goleman, 1995). Goleman (1995) assured leaders and followers that the ability to show emotions creates a positive organizational culture.

To comprehend why emotional intelligence helps leaders become effective, one must understand the concepts of emotion and intelligence (Mayer & Salovey, 1995). Bar-On (1997) believed EI included non-cognitive capabilities and competencies that influence the ability to adapt to different surroundings and stated that organizations rely on key emotional skills and abilities when selecting employees. Bar-On (1997), further stated that the data received from 300 companies proves that employers look for emotional competence more so than educational intelligence. Later, after conducting research in the field, Bar-On (1997) expanded the concept of emotionally competent behavior by establishing the Bar-On Emotional Quotient Inventory. This inventory measures the following skills: interpersonal, intrapersonal, stress management, adaptability, and general mood scale (Bar-On, 1997).

Emotionally intelligent people experience psychological wellbeing (Boyatzis, Goleman, & McKee, 2002). In fact, people with EI easily maintain a positive outlook by recognizing, understanding, generating, regulating and promoting their emotions (Goleman, Boyatzis & McKee, 2002). To prove this theory, Schutte, Malouff, Simunek, McKenley, & Hollander (2002), studied the relationship between high levels of emotional intelligence, self-esteem and positive moods. The researchers stated people with high levels of EI do not suffer from low-self-esteem, suicidal thoughts, lack of self-confidence, aggressive behavior, poor impulse control, drug and/or alcohol addiction, promiscuity, and personality disorders. On the other hand, individuals with low levels of EI remain void of emotion. To overcome these stumbling blocks, leaders must study and welcome the use of emotional intelligence skills (Karimi, et. al).

Leaders who successfully manage emotions and changes within the organizational environment are considered as emotionally intelligent (Kumar & Iyer. 2012). Petrides (2017) defined EI as characteristics in which people observe, understand, and manage their emotional state and the emotions of other people. Contemporary research insists EI is expressed as an awareness and perception of others feelings and relying on this information helps leaders engage respectfully in social situations (Di Fabio & Saklofske, 2018, Cole, 2019). In reality, individuals with high levels of EI employ and use adaptive strategies when interacting with others (O'Connor, Hill, Kaya, & Martin, 2019). Yet, emotional challenges linger, and leaders must find a way to proceed and connect with employees (Contreras, Baykal, & Abid, 2020). Up to date neuroscience research expanded our understanding of emotional intelligence (Lawrence, 2021). Today, EI focuses on interconnected social and emotional competencies that unite to determine the success in which people recognize and expresses individual feelings and understand how to relate and appreciate the feelings of others (Karimi, Leggat, Bartram, Afshari, Sarkeshik, & Verulava, 2021).

Goleman (2022) posits emotional intelligence attributes are learned over time. For example, when leaders receive training in social and emotional competencies, they will increase the ability to lead with emotional intelligence (Goleman). Once the training ends, leaders can appraise emotional intelligence competencies through the following three different models: the ability model; the trait model; and the mixed-methods model (Lubbadeh, 2020). These models determine different levels of EI as it relates to specific competencies (Goleman).

EMOTIONAL INTELLIGENCE MODELS

Salovey and Mayer (1990) created the ability model based on tactics used to decode information received from the sender. Petrides and Furman (2001) generated the trait model because this approach relies on specific personality traits to determine levels of EI. Confidence and inspiration, according to Boyatzis and Goleman (2007) explain a variance in emotional behaviors known as the mixed model.

Ability Model

Mayer, Caruso and Salovey (2002) created the four-branch ability model, known as the Mayer Salovey Caruso Emotional Intelligence Test (MSCEIT) to address four specific areas relevant to psychological research. The four-branch model includes observing emotion, using emotion to enable thoughts, understanding emotions, and managing emotions. According to Mayer, Caruso, and Salovey (2002), the sequence of the four branches are arranged in the order of importance. Branch 1 includes the ability to read facial expressions and non-verbal communication; Branch 2 includes thinking; Branch 3 includes the ability to analyze emotions; and Branch 4 includes the ability to maintain emotional thoughts and feelings (Mayer, Caruso, & Salovey, 2002). The ability model contains specific correlated abilities. The principle of this model is that emotions communicate different meanings (Goleman). Moreover, the ability model suggests people can acquire emotional intelligence skills and these skills increase with age (Supramanian, Shahruddin, & Sekar, 2022; Goleman, 2022). As a result, this model aligns with mental abilities (intelligence) and emotions (Supramanian, Shahruddin, & Sekar, 2022).

Trait Model

In this model, participants self-report behavioral competencies based on perception, in contrast to the ability model in which performance-based measures are encouraged (Petrides & Furman, 2001). The trait model includes 153 items measuring aspects related to emotional intelligence, including wellbeing, self-control, emotionality, and social ability. The Trait model analyzes emotional self-efficacies and experiences (Jacobs, Wollny, Seidler, & Wochatz, 2021). Essentially, trait models of EI examine self-perceptions that originate from emotional abilities (Supramanian, Shahruddin, & Sekar, 2022)

Mixed Model

The Emotional and Social Competence Inventory (ESCI), created by Goleman and Boyatzis (2007), measures four levels of EI: self-awareness, self-management, social awareness, and relationship management. The first component, self-awareness, measures the following areas: emotional self-awareness and self-confidence The second component self-management, measures self-control, conscientiousness, trustworthiness, adaptability, and initiative. The third component, social awareness, measures empathy, organizational awareness, and service orientation. Relationship management, the fourth component, measures influence, communication skills, team building, creating a vision, and conflict management (Goleman, McKee, Boyatzis, 2002).

The ESCI examines behaviors of EI and how these traits influence working with and relating to others (Goleman & Boyatzis, 2007). The mixed models of EI states mental abilities and personality traits such as hopefulness and wellbeing exist as important features in moving organizations forward (Supramanian, Shahruddin, &

Sekar, 2022). Though differences exist between each model, the validity remains unquestioned because each one evaluates emotional intelligence skills (O'Connor, Hill, Kaya, & Martin, 2019). Still, improving these behaviors requires leaders to take an introspective look. Undergoing constant change necessitates the need for self-improvement (O'Connor, et. al). When leaders develop emotional intelligence traits, they make genuine decisions and formulate empathic responses (Bayraktar, Yalçınkaya, Eyitmiş, 2021). When leaders use these two behaviors, they easily address problems that originated from the pandemic (Bayraktar, et. al).

The connection between emotional intelligence and leadership performance remains a widely researched topic (Kim & Kim, 2017). Researchers often study leadership success as it relates to motivating followers (Edelman & Van Knippenberg, 2018). Senior management teams constantly seek effective leaders who maintain positive emotional traits ((Edelman & Van Knippenberg, 2018). Leaders with high levels of emotional intelligence) use an engaging leadership style to achieve positive results from their followers (Goleman, 2022). Goleman advised organizations want leaders who acknowledge their emotional state and recognize their subordinates' emotional characteristics. These skills, indicative of EI, exist in four key areas: self-awareness; self-management; social awareness; and relationship management (Goleman)

Self-awareness is the ability of leaders to understand and manage emotions (Goleman, 2022). To remain successful, leaders must rely on this behavior to navigate through current pandemic challenges because self-awareness enables leaders to always remain present during interactions (Moroña & Biolik-Moroń, 2021). When leaders manage their internal state (i.e., impulses), they become instrumental in leading effective change in organizations as well as communicating with employees to help overcome issues created by the coronavirus (Baba, 2021). *Self-management* highlights maintaining self-control; using transparency; remaining adaptable; helping others achieve; taking initiative and embracing optimism (Goleman). Once leaders master the art of self-awareness and self-management, they must also focus on relationship management and social awareness (Moroña & Biolik-Moroń, 2021.

Relationship management places importance on influence; motivation, collaboration and change (Goleman). Similarly, *Social-awareness* areas contain an emphasis on empathy; awareness; and service (Goleman). Effectively managing relationships in social situations helps create positive social interactions; hence, relationship management and social awareness go hand in hand (Baba). Using a combination of EI techniques remains crucial to leading and implementing transformational changes in organizations that suffered from the aftermath of Covid-19 (Moroña & Biolik-Moroń, 2021). Moreover, these skills ensure organizational leaders make swift cultural changes and reset the mindset of employees (Baba, 2021; Moroña & Biolik-Moroń, 2021).

TRANSFORMATIONAL LEADERSHIP

In reality, transformational leaders unite followers by influencing their views (Burns, 1978). Bass (1990) expanded research on transactional and transformational leadership and advised that transformational leadership included the following behaviors: charisma, inspiration, intellectual stimulation, and individual consideration; while transactional leaders rely on specific tasks. Still, Bass (1990) asserted that transactional and transformational leaders can work together because both styles contribute to organizational effectiveness. Followers remain inspired because these leaders motivate by using rewards and incentives.

Empathy, a behavior of transformational leaders, enables workers to share personal experiences and feelings with managers (Burns, 1978). Burns (1978) shared teams identify with and relate to leaders who appear considerate of their feelings. Inspirational leaders inspire followers by telling a compelling story and providing rewards and motivational tactics (Bass, 1990). Leaders who use intellectual stimulation, according to Bass (1990), incorporate team work and create organizations through problem-solving and innovative methods. Leaders who believe in individual consideration pay close attention to their employees' emotional state. Transformational leaders work well under pressure, blossom in learning organizations, and work well in team environments (Senge, 1992).

Numerous studies proved that EI and transformational leadership share similar traits (Groves & LaRocca; Judge & Piccolo, 2004). Previous theories realized that incorporating ethical standards into decision making often provided a better outcome (Humphrey, Kellett, & Sleeth, 2006). The current transformational model utilizes five behavioral qualities: idealized attributes, idealized behaviors, inspirational motivation, intellectual stimulation, and individualized consideration (Avolio & Bass, 2007). Managers who employ idealized attributes and idealized behaviors quickly gain respect and trust from followers because this leadership style encourages a team-centered approach to problem solving (Avolio & Bass, 2007).

In fact, transformational leaders employ EI when making decisions and reacting to organizational opportunities (Groves & LaRocca; Nye, 2008). Inspirational motivation encourages forward thinking (Avolio & Bass, 2007; Groves & LaRocca, 2011). Intellectual stimulation encourages innovative and creative thinking, and individualized consideration supports effective relationships between leaders and followers (Groves & LaRocca, 2011). According to Bhatti, Maitlo, Shakiah, Hashmi, and Shakiah (2012), leaders recognize prospects, consider the big picture, serve as mentors, adapt to change quickly, take risks, communicate well, and listen effectively. Based on prior and current research, transformational leaders, who incorporate emotion and empathy surface as emotionally intelligent leaders.

Current research in leadership theories facilitated a reintroduced awareness of the similarities between emotional intelligence and transformational leadership (Cribb, 2018). Transformational leaders empower others, enhance self-efficacy, and experience personal growth (Cribb,). Even though leaders must utilize various styles, transformational leadership improves relationships through authentic and transparent communication (Cribb). It is safe to say that prior research corroborated the connection between emotional intelligence and transformational leadership traits (Görgens-Ekermans & Roux 2021). Today, the premise of transformational leadership, according to Görgens-Ekermans, and Roux focuses on how these leaders have an innate ability to inspire and guide followers to work collaboratively with peers and other leaders. Essentially, leaders who follow a transformational style create a team environment and these leaders also increase productivity and morale (Du Plessis & Waglay, 2020). When leaders use a combination of influence, enthusiasm and intellectual encouragement, employees feel empowered to succeed (Waglay, et. al, 2020). Osman (2020) posits, this group of leaders remain cognizant of others' emotional states and perceptions while remaining interested and energetic to help others. Through conclusive communication, transformational leaders grant individualized attention and help ensure positive outcomes (Osman).

Transformational leadership uses decisive behaviors that alter the way employees interpret the world around them (Morońa & Biolik-Moroń, 2021). Emotional intelligence and transformational leadership share the following similarities: inspirational motivation, idealized consideration and intellectual stimulation (Brown & Nwagbara, 2021). Theories of emotional intelligence and transformational leadership are rooted in the concept of social skills; hence, the two areas appear similar (Rinfret, Laplante, Lagacé, Deschamps & Privé, 2020). Sivanathan & Fekken (2020) posit employees believe leaders with emotional intelligence are effective, and influential and these behaviors translate into transformational leadership styles. Transformational leaders, who appropriately manage relationships with empathy and who are emotionally intelligent provide personal consideration to employees (Sivanathan, & Fekken, 2020). This is what employees need to move on from crises situations (Sivanathan, & Fekken).

EMOTIONALLY INTELLIGENT LEADERS

To successfully lead during uncertainty, emotionally intelligent leaders should incorporate Goleman's Emotional Intelligence Model consisting of Inner Focus, Other Focus, and Outer Focus (Goleman, 2014). Based on Goleman's research, Waterford (2019) added to the literature and advised leaders must concentrate on these three areas of focus: self (inner), people (other), and system (outer) awareness.

When leaders use an Inner focus, they depend on emotions and insights, which helps them make sound decision (Waterford, 2019). In fact, Inner Focus aligns well with emotional intelligence abilities. Leaders with high levels of self-awareness and high self-management follow an internal motivation path (Waterford). This means these leaders display self-confidence and a concrete acknowledgment of his or her competences and weaknesses. Similarly, leaders who maintain emotional control and remain flexible in uncertain situations also retain high levels of self-management (Waterford, 2019; Goleman, 2014).

Other Focus leaders (also representative of self-awareness and self-management) build lasting connections with peers, supervisors, and subordinates (Waterford,). Thus, leaders who utilize empathy during conversations and appear relatable to followers support an Other Focus (Waterford, 2019; Goleman, 2014). These leaders genuinely care about how others feel. This emphasis on people helps these leaders create collaborative team relationships (Waterford). Ultimately, leaders who exhibit strong people skills obtain higher than expected performance from their underlings (Waterford). This group easily builds connections within the organization as well as outside of the organization (Waterford, Goleman).

An Outer Focus leader, also needed during uncertainty, considers how outside forces contribute to the overall climate (Waterford). Effectively using an Outer Focus helps leaders move through the world. This style, closely related to social awareness and relationship management, rests on the idea of using compassion to relate to followers (Waterford). Outer Focus behaviors place an emphasis on external factors (i.e., societal and cultural) and how these elements affect the entire organization (Waterford). Further, Outer Focus leaders use increased empathy by relating to employees. Therefore, incorporating Goleman's 3 Focus model ensures leaders use a variety of skills to build relationships and see the big picture (Waterford).

Using a holistic approach to leadership addresses employee's needs (McKinsey, 2020). Still, leaders should also consider other areas to help people navigate through unprecedented times. Leading with emotional intelligence exists as a somewhat new model in modern leadership (Mateiaş, 2018). The concept of emotional awareness indicates individuals possess the ability to recognize, appreciate, and communicate, emotional responses (Smith; Bajaj; Dailey; Alkozei; Smith; Sanova, Williams, & Kilgore (2018). In order to adjust to the post-pandemic environment of today's organizations, leaders must adapt new ways to connect with followers (McKinsey). When leaders apply emotional intelligence and transformational leadership, employee's wellbeing will increase and the work atmosphere will turn into a productive environment (McKinsey). When emotions remain positive, employees and leaders will achieve organizational goals. Thus, to recover from the impact of COVID-19, leaders must display emotional insight (McKinsey).

Emotionally intelligent leaders remain composed; transparent; utilize self-management techniques; and maintain a team-centered approach when interrelating with direct reports (Osman, 2020). Therefore, leaders need to use emotional intelligence to remain effective (Raman, Peng, & Chi, C.I. 2021). Underlings gain inspiration from leaders through interactions and behaviors (Lai; Tang; Lu; Lee; & Lin, 2020). Subsequently, leaders who occupy emotional intelligence traits will easily obtain admiration from direct reports as well as increase productivity and performance (Osman, 2020).

To further guide employees through insecurity, Folstein (2021), recommends leaders add these 5 habits to their toolbox. Use selective vulnerability- this process builds trust and increases performance. Always emotionally proofread writing-put yourself in the readers' shoes. This will ensure the intended message is clear. Play chess not checkers-leaders must acknowledge the complexity of the work and be prepared for the unexpected. Understand your emotional expression tendency-always tread lightly with employees' emotions. Last, frame feedback as bridging the gap-consistently choose words carefully. Adding these routines to exchanges will enhance communication and clarity in everyday interactions and help ensure structural climates turn into learning organizations (Folstein).

The goal of nurturing your emotional intelligence rests with understanding your triggers (Lawrence, 2021). Before and after the pandemic, self-awareness is crucial for leaders. Mastering this concept helps leaders understand how their behavior affects employees' perceptions (Morońa & Biolik-Moroń, 2021). Leaders must engage in self-management. Managing internal conditions, resources and instincts helps lead effective change in organizations (Baba, 2021). Hesitation and doubt still exist due to the coronavirus. To overcome uneasy feelings, the relationship between leaders and followers should follow an emotion based transformative leadership style (Morońa & Biolik-Moroń). This style enables leaders to view the organization as a whole, or rather, as an intricate system of several different areas (Morońa & Biolik-Moroń).

LEARNING ORGANIZATIONS

In 1938, organizations were defined as a system of consciously coordinated activities (Owens, 2004). Around 1945, Herbert A. Simon concluded organizations were grouped together through bounded rationality (Owens). Almost 35 years later, Clay (1982) contributed to the research by declaring organizations exist as active transformers consisting of inputs and outputs. Later, in 1992 Senge stated, to function as a cohesive unit, continuous learning must exist (Senge, p.3). As a result, Senge coined the term, "learning organization." Senge (1992) defined learning organizations as: "a place where people continually expand their capacity to create results, they truly desire, where new and expansive patterns of thinking are nurtured, where collective aspiration is set free, and where people are continually learning how to learn together." (p. 3). Additionally, Senge theorized learning assists organizations with change management and change acceptance. A learning approach to problem solving appears in leaders who exhibit high levels of EI (Senge, 1992).

Senge's five disciplines appear beneficial for today's organizations: shared vision; mental models; team learning; personal mastery and systems thinking. When organizations create a shared vision, everyone accepts the future of the company. Mental models enable leaders to take a reflective look (Stockard. 2006). In fact, organizations will not move forward into learning communities without successful teams. In ambiguous times, team work appears crucial in completing daily and monthly tasks. Personal mastery includes individual learning. Systems thinking, according to Senge (1992) exemplifies an effective organization because every system within an organization must join forces. Moreover, Senge believed learning organizations build a united vision and support individual contributions to the organization. He also argued that in order for organizations to remain effective, the organization must incorporate a learning approach; and employ emotionally intelligent transformational leaders who possess the ability to create learning organizations (Senge).

These companies will serve as a protected place because stakeholders are dedicated to functioning collaboratively by sharing the vision of the organization; incorporating mental models to prevent biased decisions; use team learning as a way to collect diverse ideas and suggestions; help everyone personally master various processes; and utilize systems thinking in order to encourage innovative problem solving (Senge). This consideration allows time to ponder how the individual leader uses specific data to make decisions (Stockard, 2006). Shared principles exist between learning organizations and management science because both theories focus on human relations, effective management, organizational behavior, and systems thinking (Hasan, Ansari, & Masoudi, 2011). Since leaders must adapt to change quickly, a transformative leadership style appears to be the most beneficial in terms of reorganizing systems (Hasan, Ansari, & Masoudi, 2011; Purhaghshenas & Esmatnia, 2012). Additionally, leaders must create shared insight, motivate staff, actively question and listen to employees, and provide time and resources when needed (Maden, 2012; Purhaghshenas & Esmatnia, 2012). Accordingly, transformational emotionally intelligent leaders ensure continuous organizational learning.

Leaders in learning organizations remain influential by using charisma and treating employees as people (Robinson, 2020). Using charisma and influence, representative of transformational leaders appears necessary to gain buy-in from employees (Robinson, 2020). The intricacy of organizational systems requires leaders to examine levels of performance and how it relates to individual behavior (McGee & Crowley-Koch, 2021). Considering individual and team performance appears necessary to ensure everything runs smoothly post pandemic times (McGee & Crowley-Koch). Team learning focuses on the need for effective teams and team leaders (Robinson, 2020). Considering organizations as systems enables evaluations of workers and processes (Andiola, Downey & Westermann, 2021). Presently, leaders must consider how these components became impacted by the crisis. Empathizing with employees, as it relates to changes within the system which occurred because of the pandemic's barriers, must happen to establish new models for achievement and transformation within the organization (Andiola, et. al.). In addition to reorganizing systems, emotionally intelligent leaders also improve the organizational culture to facilitate employee safety and emotional wellbeing (McGee & Crowley-Koch).

Organizational Culture

Culture presents value and context that controls how employees consciously and subconsciously think and behave (Schein 1985). Schein defines culture as (1985, p. 9) ''a pattern of basic assumptions invented, discovered or developed by a given group as it learns to cope with its problems of external adaptation and internal integrations that has worked well enough to be considered valid, and therefore, to be taught to new members as the correct way to perceive, think, and feel in relation to those problems.'' Chatman and O'Reilly (2016) posit organizational culture is grounded in tradition, enables order, provides a collective identity, provides uniqueness to an organization, and, more importantly, is communicated and accepted by new employees.

Culture, a component of the organization's system, exists as a noteworthy part of performance (Binder, 2016). Research focusing on culture is important because it deals directly with helping employees within the organization (Van Tiem & Burns, 2020). Subsequently, grasping what employees value and listening when employees share experiences and expectations appears useful when making unified cultural adjustments to organizations (Van Tiem & Burns).

The leader's definition of performance, as it relates to an employee's expectations, remains a consideration for forming effective and ever-changing cultures (Andiola, Downey, & Westermann, 2020). The catalyst for cultural change rests with how the leader communicates with employees (Westermann). Working through a crisis and the implications of change heightens the complexity ensuring seamless transitions (Andiola, Downey, & Westermann, 2020). Occurrences of crises, as well as leadership decisions during catastrophes, alter and update cultures (Andiola, Downey, & Westermann, 2020).

A leader's aptitude to encourage employees to accept changes in the organization requires transformational leadership and emotional intelligence skills (Andiola, Downey, & Westermann, 2020). To facilitate a cultural shift, leaders must address challenges upfront versus waiting for problems to happen (Andiola, et. al.) Effective leaders lead through adversity and uncertainty during unfamiliar times (Van Tiem, & Burns, 2020). To appropriately deal with the aftermath of COVID-19, leaders in organizations must meet the needs of employees by focusing on mental health and emotional wellbeing (Van Tiem, & Burns).

Internal and external setbacks, due to the aftermath of COVID-19, affect the way employees and leaders' function at work (Sharpcloud, 2021). The level of insecurity during the pandemic experienced by employees; particularly leaders, is unique (Lubbadeh, 2020). As people adjust to new organizational environments, some situations require flexibility (Contreras, Baykal, & Abid, 2020). Though leaders plan for the unexpected, the pandemic created additional needs to review current processes; but most importantly, place a focus on helping employees cope and move forward (Sharpcloud, 2021). In order to swiftly implement changes in culture, leaders must remember to rely on their emotions when interacting with employees (Andiola, Downey, & Westermann, 2020). Providing empathic reactions and using positive language will help employees overcome anxiety (Van Tiem & Burns, 2020).

Due to the uncertainty and ambiguity of the pandemic, the relationship between leaders and followers became strained (Van Tiem & Burns, 2020). Reshaping the connection requires a leader who relies on emotion, uses empathy, transforms processes, and considers the entire organizational system before making a decision (Trimble, 2020). Consequently, the pandemic created a shift from autocratic leadership styles to democratic and transformational styles. These distinct styles allow leaders to show emotion (McKinsey, 2020. According to McKinsey (2020) failure to use these methods will likely result in depressed and unmotivated employees. Leaders with EI foster relationships; achieve success; incorporate analytical and strategic thinking skills, and incorporate forward-thinking (Ayalew & Ayenew, 2022).

Drigas & Papoutsi, 2018; Gilar-Corbi, Pozo-Rico, Sánchez, & Castejón, 2019; & Sefela, 2017) hypothesize existing research highlights leaders' levels of emotional intelligence generates management effectiveness within organizations. For example, maintaining emotional levels aids leaders in making informed decisions (Anoye & Kouamé, 2018). Accordingly, avoiding incorporating emotional intelligence capabilities into management decisions may prevent leaders from implementing a cultural shift within the organization (Sefela, 2017). Collin, Herranen, Paloniemi, Auvinen, Riivari, Sintonen, & Lemmertty (2018) affirmed leadership includes motivating and empowering individuals and teams to achieve success. Clearly, an EI deficiency limits progress and change (Collin, et. al). Equally, leaders with high EI help create collaborative environments which increase productivity (Lubbadeh, 2022).

Employing a blend of emotional intelligence and transformational leadership skills helps employees embrace change (Aguilar, 2021). To make a seamless cultural shift, strong leadership skills merged with emotional responses seem necessary (Aguilar).

Moreover, conversations must include empathic responses because empathy enables the leader to show vulnerability. (Lubbadeh. 2022). Companies must also introduce transformational leadership behaviors to facilitate a learning organization atmosphere (Lubbadeh). Also, leaders must continue recognizing and acknowledging employees need ongoing support to address and improve their mental health and emotional wellbeing (Ayalew & Ayenew, 2022).

MOVING FORWARD

Leaders must focus on improving emotional wellbeing for each and every employee. Interpreting how to approach this endeavor must start with knowing how the brain reacts to distressing situations (Elbarazi, 2022). Avalew and Avenew, (2022), theorize changes linked to COVID-19 and trauma affect the cerebral areas (limbic system). These changes occur in the form of stress, headaches, sleep deprivation and/or dizziness (Avalew & Avenew). The Institute for Health and Human Portal (2021), reports when emotions increase the following changes occur to the brain: cognitive abilities, social interactions, and decision-making (Avalew & Avenew). As a result, some employees may have a difficult time communicating experiences with leaders (Avalew & Avenew). Therefore, leaders must initiate programs to help improve employees' emotional wellbeing.

When leaders focus on improving employees' mental health, the organizational climate improves (Elbarazi, 2022). The best way to foster an environment of care comes from the leaders' ability to maintain emotional control (Avalew & Avenew, 2022). This extra effort will create a welcoming and relaxing atmosphere that employees really need in order to feel less stressed (Elbarazi, 2022). Assuredly, for employees to feel safe, leaders must integrate emotional responses during interactions (Elbarazi). Leaders who make decisions based on empathy gain trust easily from employees (Votaw, 2022). To help employees move into a new phase after the pandemic, leaders should incorporate empathic responses during one-on-one conversations and during team-based meetings (Votaw). Mastering the art of empathy while also preserving emotional control helps leaders progress intellectually (Elbarazi).

To move the organization into a unified status, leaders must hone in on their emotions and help transform the workplace into a safe haven for everyone involved (Sharpcloud, 2021). Leaders can accomplish this by perfecting emotional intelligence skills (Lubbadeh, 2020). Improving these behaviors: self-awareness; self-management; relationship management and social awareness enables leaders to consider the best approach to use when collaborating with employees (Goleman. 2022).

Sivanathan & Fekken (2020) admit followers perceive leaders with emotional intelligence as more effective, persuasive and transformative. Today, emotionally intelligent transformational leaders exist as the best group to rebuild teams after the pandemic (Mateias, 2018). This approach to leadership encourages empathic responses and drives motivation (Sivanathan & Fekken, 2020). These helpful behaviors are required to help leaders create lasting changes (Baba, 2021). In turn, rebuilding the workforce; moving into a positive cultural shift, and reducing anxiety requires transformational leadership in combination with emotional intelligence skills (Baba, 2021). This is the only approach that will ensure a smooth transition (Baba).

CONCLUSION

Coronavirus deaths still occur daily; the January 6th riot will remain etched in history as a traumatic event; and the conflict between Russia and Ukraine persists. To combat uncomfortable feelings, leaders must acknowledge how the brain affects decisions and how trauma changes the way the brain reacts to situations. To understand and communicate with the current workforce, leaders must also acknowledge their emotional state and use these emotions to authentically communicate with peers and direct reports. Leaders with high emotional intelligence and transformational leadership styles are necessary. Leaders who exhibit these traits provide a necessary calmness for employees. Interactions with emotionally intelligent transformational leaders help people deal with and cope with varying emotions. Within current and previous research, discussions centered around the history of emotions and empathy. Social intelligence also emerged as another facet involving balancing emotions during conversations. A deep investigation highlighted the benefits of emotional intelligence and transformational leadership. These qualities help workers increase productivity and build better relationships throughout the organization. To lead during unfamiliar and ambiguous instances, leaders must employ emotional intelligence behaviors (self-management; self-awareness; relationship management; and social awareness). Leaders who grasp these traits also exist as transformational leaders who use influence to transform organizations. Moreover, when leaders use a combination of emotionally intelligent behaviors and transformational leadership styles learning communities emerge. Clearly, for employees to feel a sense of normalcy in times of doubt, leaders must remain in sync with their emotions and others' emotions.

REFERENCES

Adolphs, R. (2021). Social Cognitive and Affective Neuroscience. doi:10.1093can/nsw153

Aguilar Yuste, M. (2021). Leadership and Emotional Intelligence during a Crisis. *Revista de Relaciones Internacionales. Estrategia y Seguridad, 16*(2), 47–60. doi:10.18359/ries.5619

Ahmad, R., Imran, H., & Mehmood, T. (2009). Emotional intelligence as a determinant of self-esteem and depression among adolescents. *Pakistan Journal of Psychology, 40*(1), 77–88.

Andiola, L. M., Downey, D. H., & Westermann, K. D. (2020). Examining Climate and Culture in Audit Firms: Insights, Practice Implications, and Future Research Directions. *Auditing, 39*(4), 1–29. doi:10.2308/AJPT-19-107

Anoye, A. B., & Kouamé, J. S. (2018). Leadership challenges in virtual team environment. *International Journal of Scientific & Technology Research, 7*(7).

Avolio, B. J., & Bass, B. M. (2007). *Multifactor Leadership Questionnaire, sample set manual, forms, and scoring key*. Mind Garden.

Ayalew, M., & Ayenew, Z. (2022). Do Paradoxical Virtual Leadership and Emotional Intelligence have Relationships? In Particular from Technology Dependence, Geographical Dispersion, and Human Capital Tensions. *International Journal of Organizational Leadership, 11*(1), 1–25. doi:10.33844/ijol.2022.60616

Baba, M. M. (2020). Navigating COVID-19 with emotional intelligence. *International Journal of Social Psychiatry, 66*(8), 810-820.

Balconi, M., & Pozzoli, U. (2009). Arousal affect on emotional face comprehension, frequency band changes, in different time intervals. *Physiology & Behavior, 97*, 460–462. doi:10.1016/j.physbeh.2009.03.023 PMID:19341748

Bar-on, R. (1997). *The Baron emotional quotient inventory (EQ-I) technical manual*. Multihealth Systems.

Bar-On, R., & Parker, D. A. (2000). *The handbook of emotional intelligence*. Jossey-Bass.

Bass, B. M. (1990). *Bass and Stogdill's handbook of leadership*. Free Press.

Batson, C. D., Shaw, L. L., & Oleson, K. C. (1992). Distinguishing affect, mood, and emotion: Toward functionally based conceptual distinctions. *Review of Personality and Social Psychology, 13*, 294.

Bayraktar, E., Yalçınkaya, M., & Eyitmiş, A. M. (2021). Examination of emotional intelligence and problem-solving skills of teachers working in secondary education institutions. *International Online Journal of Education & Teaching, 8*(4), 2228–2247.

Bhatti, N., Maitlo, G., Shaikh, N., Hashmi, M., & Shaikh, F. (2012). The impact of autocratic and democratic leadership style on job satisfaction. *International Business Research*, *5*(2), 192–193. doi:10.5539/ibr.v5n2p192

Boyatzis, R. E., Goleman, D., & Rhee, K. S. (2002). *Clustering competence in emotional intelligence insights from the emotional competency inventory*. Harvard Business Review.

Bradberry, T., & Greaves, J. (2009). *Emotional Intelligence 2.0*. Simon & Schuster.

Brown, C., & Nwagbara, U. (2021). Leading Change with the Heart: Exploring the Relationship between Emotional Intelligence and Transformational Leadership in the Era of Covid-19 Pandemic Challenges. *Economic Insights - Trends and Challenges*, *3*(3), 1–12. doi:10.51865/EITC.2021.03.01

Burton, A., Aughterson, H., Fancourt, D., & Philip, K. E. J. (2022). Factors shaping the mental health and well-being of people experiencing persistent COVID-19 symptoms or 'long COVID': Qualitative study. *BJPsych Open*, *8*(2), e72. Advance online publication. doi:10.1192/bjo.2022.38 PMID:35307048

Călinici, M. S., Călinici, T., Balazsi, R., & Miclea, M. (2020). Development and validation of a new ability-based measure of emotional intelligence: Cluj Emotional Intelligence Scale CEIS. *Cognition, Brain, Behavior: An Interdisciplinary Journal*, *24*(4), 335–363. doi:10.24193/cbb.2020.24.18

Carroll, J. B. (1993). *Human cognitive abilities: A survey of factor-analytic Studies*. Cambridge University Press. doi:10.1017/CBO9780511571312

Clay, N. (1982) *Paraprofessional utilization and reading achievement gains of third and fourth grade compensatory education students* [Published Dissertation]. University of Michigan, Ann Arbor, MI.

Clore, G. L. (1992). *Cognitive phenomenology: Feelings and the construction of judgment*. Erlbaum.

Collin, K., Herranen, S., Paloniemi, S., Auvinen, T., Riivari, E., Sintonen, T., & Lemmertty, S. (2018). Leadership as an enabler of professional agency and creativity: Case studies from the Finnish information technology sector. *International Journal of Training and Development*, *22*(3), 222–232. doi:10.1111/ijtd.12130

Contreras, F., Baykal, E., & Abid, G. (2020). E-leadership and teleworking in times of COVID-19 and beyond: What we know and where do we go. *Frontiers in Psychology*, *11*, 3484. doi:10.3389/fpsyg.2020.590271 PMID:33362656

Cribb, L. C. (2018). *Exploring effective virtual leadership and the use of computer-mediated communication technology on a virtual team* [Master's thesis]. University of Alberta.

Di Fabio, A., & Saklofske, D. H. (2018). The contributions of personality and emotional intelligence to resiliency. *Personality and Individual Differences*, *123*, 140–144. doi:10.1016/j.paid.2017.11.012

Dixon, T. (2012). *Emotion: The history of a key word in crisis.* doi:10.1177/1754073912445814

Dolcos, F., Alexandru, D. I., & Dolcos, S. (2011). Neural correlates of emotion-cognition interactions: a review of evidence from brain imaging investigations. *Journal of Cognitive Psychology*. Retrieved from: https://www.tandfonline.com/doi/full/10.1080/20445911.2011.594433

Drigas, A. S., & Papoutsi, C. (2018). A new layered model on emotional intelligence. *Behavioral Sciences (Basel, Switzerland)*, *8*(5), 45. doi:10.3390/bs8050045 PMID:29724021

Du Plessis, M., Waglay, A., & Becker, J. (2020). A leader member exchange perspective. *SA Journal of Industrial Psychology*, *46*. Advance online publication. doi:10.4102ajip.v46i0.1762

Edelman, P., & Van Knippenberg, D. (2018). Emotional intelligence, management of subordinate's emotions, and leadership effectiveness. *Leadership & Organization Development Journal. Bradford.*, *39*(5), 592–607. doi:10.1108/LODJ-04-2018-0154

Elbarazi, A. S. (2022). Post-Traumatic Stress Disorder and Difficulties in emotion regulation among university students under the COVID-19 condition. *Journal of Evidence-Based Psychotherapies*, *22*(1), 87–101. doi:10.24193/jebp.2022.1.5

Foslein, L. (2021). *5 things emotionally intelligent managers do differently*. Retrieved from: https://www.fastcompany.com/90705760/5-things-emotionally-intelligent-managers-do-

Fox, A. S., & Shackman, A. J. (2017). (in press). The central extended amygdala in fear and anxiety: Closing the gap between mechanistic and neuroimaging research. *Neuroscience Letters*.

FreudS. (1921). Retrieved from: https://www.bartleby.com/288

García-Bullé, S. (2019). *What is social intelligence and why it should be taught in schools*. Retrieved from: https://observatory.tec.mx/edu-news/social-intelligence

Gardner, H. (1983). *Frames of mind*. Basic Books.

Gilar-Corbi, R., Pozo-Rico, T., Sánchez, B., & Castejón, J. L. (2019). Can emotional intelligence be improved? A randomized experimental study of a business-oriented EI training program for senior managers. *PLoS One*, *14*(10), e0224254.

Giuliani, N. R., & Gross, J. J. (2009). Reappraisal. In D. Sander & K. R. Scherer (Eds.), *Oxford Companion to the Affective Sciences*. Oxford University Press.

Goërtz, Van Herck, Delbressine, Vaes, Meys, Machado, Houben-Wilke, Burtin, Posthuma, Franssen, van Loon, Hajian, Spies, Vijlbrief, van 't Hul, Janssen, & Spruit. (2020). *Persistent symptoms 3 months after a SARS-CoV-2 infection: the post-COVID-19 syndrome?* doi:10.1183/23120541.00542-2020

Goleman, D. (2011a). *The Brain and Emotional Intelligence New Insights*. More Than Sound.

Goleman, D. (2014). *What makes a leader? Why Emotional Intelligence matters*. More Than Sound LLC.

Goleman, D. (2015). How to be emotionally intelligent? *The New York Times*. Retrieved from: https://www.nytimes.com/2015/04/12/education/edlife/how-to-be emotionally-intelligent.html?_r=1

Goleman, D. (2022). *A model to solve workplace problems*. LinkedIn.

Goleman, D., Boyatzis, R., & McKee, A. (2002). *Primal leadership. Realizing the power of emotional intelligence*. Harvard Business School Press.

Görgens-Ekermans, G., & Roux, C. (2021). Revisiting the emotional intelligence and transformational leadership debate: (How) does emotional intelligence matter to effective leadership? *SA Journal of Human Resource Management Menslikehulpbronbestuur, 19*(0), a1279. .v19i0.1279 doi:10.4102/sajhrm

Groves, K., Enrue, M. P., & Shen, W. (2010). Managerial emotional intelligence, affective organizational commitment, and emotional labor. *Journal of Leadership Review.*, *30*, 31–35.

Groves, K. S., & LaRocca, M. A. (2011). An empirical study of leader ethical values, transformational and transactional leadership, and follower attitudes toward corporate social responsibility. *Journal of Business Ethics*, *103*, 511–513. doi:10.100710551-011- 0877-y

Hall, J. (1604). *Two guides to a good life: The genealogy of virtue and the nathomy of sinne. London, England*. Laggard.

Hasan, A., & Masoudi, A. (2011). *Jean Piaget: Intellectual development.* Retrieved from https://www.sk.com.br/sk-piage.html

Hatfield, E., Cacioppo, J., & Rapson, R. (1994). *Emotional contagion (Studies in emotion and social interaction).* Cambridge University Press.

Hines, T. E. (2011). *Leadership: A balancing act.* Wiley Online Library. doi:10.1002/cc.468

Humphrey, R., Kellett, J., & Sleeth, R. (2006). Empathy and the emergence of task and relations leaders. *The Leadership Quarterly, 17*(2), 146–162.

Institute for Health and Human Portal. (2021). *The meaning of emotional intelligence.* https://www.ihhp.com/meaning-ofemotional-intelligence

Izard, C. E. (2006). *Experts' definitions of emotion and their ratings of its components and characteristics* [Unpublished manuscript]. University of Delaware.

Jacobs, I., Wollny, A., Seidler, J., & Wochatz, G. (2021). A trait emotional intelligence perspective on schema modes. *Scandinavian Journal of Psychology, 62*(2), 227–236. https://doi.org/10.1111/sjop.12670

Jiang, L., Qiao, K., & Li, C. (2021). Distance-based functional criticality in the human brain: Intelligence and emotional intelligence. *BMC Bioinformatics, 22*, 1–17. https://doi.org/10.1186/s12859-021-03973-4

Judge, T. A., & Piccolo, R. F. (2004, October). Transformational and transactional leadership: A meta-analytic test of the relative validity. *The Journal of Applied Psychology, 89*(5), 755–757.

Karasneh, A. A.-F., & Al-Momani, E. A. (2020). The influence of emotional intelligence on leadership styles: A profile of executives in ASEZA organization. *Journal of Public Affairs.* doi:10.1002/pa.2069

Karimi, L., Leggat, S. G., Bartram, T., Afshari, L., Sarkeshik, S., & Verulava, T. (2021). Emotional intelligence: Predictor of employees' wellbeing, quality of patient care, and psychological empowerment. *BMC Psychology, 9*, 1–7. https://doi.org/10.1186/s40359-021-00593-8

Kim, H., & Kim, T. (2017). Emotional intelligence and transformational leadership: A review of empirical studies. *Human Resource Development Review, 16*(4), 377–393. https://doi.org/10.1177/1534484317729262

Kumar, S. J., & Iyer, V. R. (2012). Emotional intelligence and quality of work-life among employees in the educational institutions. *SIES Journal of Management, 8*(2), 55.

Lai, F.-Y., Tang, H.-C., Lu, S.-C., Lee, Y.-C., & Lin, C.-C. (2020). Transformational Leadership and Job Performance: The Mediating Role of Work Engagement. *SAGE Open, 10*(1). doi:10.1177/215824401989908

Lange, S. (1922). *The emotions volume*. Retrieved from: archive.org/details/emotionsvolumei007644mbp

Lawrence, J. A., & Parkes, R. (2021). How do you be your best self in a vexatious world? *Nephrology Nursing Journal, 48*(3), 261-265. .2021.48.3.261 doi:10.37526/1526-744X

Lazarus, R. S. (1991). *Emotion and adaptation*. Oxford University Press.

Lee, C., Abdullah, A., Ismail, A., & Alizydeen, N. (2011). How democratic leaders empower teachers job satisfaction? The Malaysian Case. *International Journal of Business and Science*, *2*(10), 251.

Lubbadeh, T. (2020). Emotional intelligence and leadership–the dark and bright sides. Publishing House of Rzeszow University of Technology.

Maden, C. (2012). Transforming public organizations into learning organizations: A conceptual model. *Public Organization Review*, *12*, 73. doi:10.100711115-011-0160-9

Mateiaş, D. E. (2018). Emotional intelligence as a dimension of transformational leadership advancement. *Revista de Psihologie*, *64*(1), 39.

Mayer, J. D., & Salovey, P. (1990). Emotional Intelligence. *Imagination, Cognition and Personality*, *9*(3), 185.

McRae, K., Misra, S., Prasad, A. K., Pereira, S. C., & Gross, J. J. (2011). Bottom-up and top- down emotion generation: Implications for emotion regulation. *Social Cognitive and Affective Neuroscience*, *7*(3), 253–262. doi:10.1093can/nsq103

Moroña, M., & Biolik-Moroń, M. (2021). Trait emotional intelligence and emotional experiences during the COVID19 pandemic outbreak in Poland: A daily diary study. *Personality and Individual Differences*, *168*(110348), 1.

Ng, R., Lai, P., Brown, T. T., Järvinen, A., Halgren, E., Bellugi, U., & Trauner, D. (2018). Neuroanatomical correlates of emotion-processing in children with unilateral brain lesion: A preliminary study of limbic system organization. *Social Neuroscience*, *13*(6), 688–700. https://doi-org.ezproxy.umgc.edu/10.1080/17470919.2017.1386126

Nye, J. (2008). *The powers to lead.* Oxford University Press.

O'Connor, P. J., Hill, A., Kaya, M., & Martin, B. (2019). The measurement of emotional intelligence: A critical review of the literature and recommendations for researchers and practitioners. *Frontiers in Psychology*, *10*, 1–19. doi:10.3389/fpsyg.2019.01116

Osman, I. A. E. (2020). The relationship between emotional intelligence (EI) and middle management leadership styles (transformational and transactional leadership) among managerial personnel at khartoum state: Women and change. *The Ahfad Journal, 37*(1), 3-13. Retrieved from https://www.proquest.com/scholarly-journals/relationship-between-emotional-intelligence-ei/docview/2436886656/se-2?accountid=9708

Owens, R. G. (2004). *Organizational behavior in education: Adaptive leadership and school reform* (8th ed.). Pearson.

Petrides, K. V. (2017). Intelligence, emotional. *Reference Module in Neuroscience and Biobehavioral Psychology*, *1*, 1–16.

Petrides, K. V., Furnham, A., & Mavroveli, S. (2001). *Trait emotional intelligence. Moving forward in the field of EI.* Oxford University Press.

Piaget, J. (1950). *The psychology of intelligence*. Routledge.

Polychroniou, P. (2008). Transformational Leadership and Work Motivation in modern organizations: A critical analysis. Advances in Management, 1(3), 9-12.

Purhaghshenas, S. H., & Esmatnia, M. (2012). Learning organizations. *Interdisciplinary Journal of Contemporary Research in Business*, *4*(7), 243–245.

Raman, G., Peng, N. L., & Chi, C. I. (2021). The mediating effect of leaders' idealized influence on the relationship between leaders' emotional intelligence and job performance among academics in Malaysian public research universities. *International Journal of Management Studies*, *28*(1), 1-24. https://doi.org/10.32890/ijms.28.1.2021.10544

Rinfret, N., Laplante, J., Lagacé, M. C., Deschamps, C., & Privé, C. (2020). Impacts of leadership styles in health and social services: A case from Quebec exploring relationships between emotional intelligence and transformational leadership. *International Journal of Healthcare Management, 13*, 329–339. https://doi-org.ezproxy.umgc.edu/10.1080/20479700.2018.1548153

Robinson, G. (2020). Organizations are people: Reflections, obstacles and learning opportunities. *The Learning Organization, 27*(6), 541–554.

Rolls, E. T. (2005). *Emotion explained.* Oxford University Press.

Ross, E. D. (2021). Differential hemispheric lateralization of emotions and related display behaviors: Emotion-type hypothesis. *Brain Sciences, 11*(8), 1034. https://doi.org/10.3390/brainsci11081034

Saarimäki, H., Glereanb, E., Smirnov, D., Mynttinenb, H., Jääskeläinen, L., Sams, M., & Nummenmaa, L. (2022). Classification of emotion categories based on functional connectivity patterns of the human brain. *NeuroImage. 247.* https://doi.org/10.1016/j.neuroimage.2021.118800

Sadri, G. (2012). Emotional intelligence and leadership development. *Public Personnel Management, 41*(3), 535–538.

Salovey, P., & Mayer, J. D. (1990). Emotional intelligence. *Imagination, Cognition and Personality, 9*, 185–211.

Schutte, N. S., Malouff, J. M., Simunek, M., McKenley, J., & Hollander, S. (2002). Characteristic emotional intelligence and emotional wellbeing. *Cognition and Emotion, 16*(6), 769-85.

Sefela, F. (2017). *The relationship between emotional intelligence and change oriented leadership style of managers and the OCB of non-academic staff members at a tertiary institution in the Western Cape* [Master's thesis]. University of the Western Cape. http://hdl.handle.net/11394/6438

Senge, P. (1992). *The fifth discipline field book: Strategies and tools for building a learning organization.* Doubleday Currency.

Sharpcloud. (2021). *Identify new opportunities in a dynamic business environment.* https://www.sharpcloud.com/blog/identify-business-opportunities-in-a-dynamic-environment

Smith, R., Bajaj, S., Dailey, N. S., Alkozei, A., Smith, C., Sanova, A., Lane, R. D., William, D., & Killgore, S. (2018). Greater cortical thickness within the limbic visceromotor network predicts higher levels of trait emotional awareness. *Consciousness and Cognition*, *57*, 54.

Spearman, C. (1927). *The abilities of man*. Macmillan.

Stockard, M. (2006). *An integrated phenomenological study of teachers' perceptions of a professional learning community utilizing Senge's five disciplines of a learning organization* [Published Doctoral dissertation]. Georgia Southern University.

Supramanian, K., Shahruddin, R., & Sekar, M. (2022). Emotional intelligence using ability model in context of nursing and its impact on end-stage renal disease patients: A narrative review. *International Journal of Current Research and Review*. doi:10.31782/IJCRR.2021.131609

Thorndike, E. L. (1920). Intelligence and its use. *Harper's Magazine, 140*, 227–235.

Thurstone, I. L. (1938). *Primary mental abilities. Psychometric Monographs*. University of Chicago Press.

Valenzano, A., Scarinci, A., Monda, V., Sessa, F., Messina, A., Monda, M., Precenzano, F., Mollica, M. P., Carotenuto, M., Messina, G., & Cibelli, G. (2020). The Social Brain and Emotional Contagion: COVID-19 Effects. *Medicina, 56*(12), 640. doi:10.3390/medicina56120640

Votaw, K. Q. (2022). The Hard Results Of Soft Leaders. *Personal Excellence*, *27*(1), 37.

Waterford, F. (2019). *3 Types of focus every leader needs to cultivate*. Retrieved from: https://www.linkedin.com/pulse/3-types-focus-every-leader-needs-cultivate-felicity-waterford/

Wechsler, D. (1997). *Wechsler adult intelligence scale*. Harcourt Assessment. Retrieved from http://www.sjdm.org/dmidi/Wechsler_Adult_Intelligence_Scale.html

Chapter 9
Impacts on Helping Professionals' Preparation During the COVID-19 Lockdown Through the Perspectives of Two Professional Program Educators

Crystal Walcott
Indiana University-Purdue University Columbus, USA

Deborah S. Judge
https://orcid.org/0000-0002-9256-4182
Western Governors University, USA

ABSTRACT

The authors of this chapter share insights into the impacts of COVID-19 on teacher education and nurse education. The narrative is written from the perspective of two professional educators who live and work in rural to suburban areas of Southern Indiana. The chapter includes a historical perspective on the state of nurse and teacher education from pre-COVID-19 through post-COVID-19 environments. Particular attention is paid to the commonalities of these two helping professions and the effects on building the dispositions of empathy and caring in teacher and nurse candidates during the COVID-19 shutdown.

DOI: 10.4018/978-1-6684-4358-3.ch009

INTRODUCTION

As large sectors of the "helping professions," nursing and teaching have many dispositional aspects in common. In fact, one of the primary goals shared by nurse and teacher educators is the importance of instilling in their students a set of characteristics conducive to forming helping relationships with those they serve. In particular, educators strive to instill elements of empathy and caring in their professional students. These characteristics of helping relationships rely heavily on human interaction and are typically fostered in students within environments that combine safe practice spaces, such as labs and simulations, and within authentic settings, such as healthcare facilities and schools. But what happens when these environments are interrupted, as in the case of the national shutdown in response to the COVID-19 pandemic? This was a question that educators across the nation quickly pondered as they attempted to provide seamless instruction to their students as the nation adjusted to a new normal defined by lockdown restrictions. In this chapter, two educators, one nurse educator, and one teacher educator, discuss the lessons learned from the COVID-19 lockdown as related to educating future nurses and teachers as helping professionals.

COVID-19 ENVIRONMENTS

Every sector of the United States economy was directly impacted by the COVID-19 lockdown. In some industries, employees were furloughed or terminated from employment. The government scrambled to implement financial contingency plans to support those affected by the loss of jobs. However, essential workers were relied upon to keep the country running while keeping themselves and the public safe in light of the highly contagious nature of the virus. In this chapter, consideration of changes to two essential environments, the Education Service Sector and the Healthcare Industry, are shared. Specifically, the authors consider these environments in a southern Indiana, suburban to a rural setting in which the authors live and work as professional program educators (nursing education and teacher education). To roughly define "southern Indiana," consider a map of the state of Indiana cut by a line formed by I-70. The area of the state south of I-70, discounting a radius of roughly 20 miles centered in Indianapolis to account for the vast suburban area extending from the center of the capital city, is considered in this context to be the area defined as "southern Indiana." The living environments in southern Indiana are considered suburban and rural with the largest city being Evansville (population 120,000) located in the extreme southwest corner of the state bordering Kentucky and nearly bordering Illinois.

Healthcare Environments in Suburban and Rural Southern Indiana Pre-COVID

In healthcare, there is always one certainty - evolution. The profession is constantly evolving. The commonality that never changes is the dedication of healthcare professionals to provide care and support for all people. The driving force of new healthcare innovations is technological advancement. The evolution of healthcare technology used to record and communicate healthcare information has been embraced by some but not by others. Mandated in 2009 by the Centers for Medicare and Medicaid, electronic health records provide access to better record keeping and communication (Badalucco, 2015).

Healthcare workers who have been around for a decade or more are familiar with paper charting. In many facilities, night shift nurses were charged with writing admission care plans and daily updating each patient's care plan. Charts were maintained for all records, inclusive of nurse notes, doctor notes, health history, lab values, and any test results. All records were kept inside the nurse's station in an orderly fashion until the patient was discharged and records were transferred to the medical records department for confidential safekeeping.

During the times of paper charting, technology continued to advance. The typewriter was replaced with a word processor, leading eventually to an industry dependent upon computer technology. Growth in reliance on computers coincided with the advancement of telephones. Well-experienced nurses might recall the rudimentary phone system with rotary dials, wall-mounted phones, and even "party lines" when callers waited their turn to make and receive phone calls. These environments restricted healthcare professionals from effectively communicating critical information. Later, private lines were designed with long cords that allowed nurses to move within the room while communicating with stakeholders. By the time the cordless phone was released, healthcare professionals were able to move within a larger space to gather information during a phone conversation. Even within the last few years, cell phone technology continues to evolve from the bag phones released in the 1990s to the handheld devices of today. Still, a one-year-old phone is nearly obsolete.

Computers have rocketed the technology world with the development that parallels the cell phone. Those first computers were big, bulky, slow, and frustrating; but they provided insight into a healthcare future that most people could not fathom. Computers today can be held in your hands and within seconds can retrieve any information you might need or desire. The first version of the Apple iPhone which supported computer capabilities was released in the early 2000s and soon followed by other companies as technology continued to expand (Cantu, 2016). Technology has advanced healthcare and continues to provide new opportunities for creative thinkers to discover ways for increased accessibility to healthcare resources.

Prior to the outbreak of COVID-19, nursing environments in southern Indiana were heavily supported by technology. Most were modern facilities where patients received a variety of services including trauma centers, urgent care, surgery centers, dental care, laboratory and diagnostics, physical and occupational therapies, pharmaceuticals, prenatal care, nutritional support, mental health, substance abuse centers, acute care hospitals, and long-term care. Technology utilized in southern Indiana facilities included telehealth, smart beds, portable monitors, wearable sensors/devices, self-monitoring kiosks, robotics, and healthcare information technology (HIT), such as electronic health records. Student nurses, through their clinical experiences, have access to all these technologies. In addition, preceptors model the appropriate use of the technology to optimize the patient experience.

An example of the importance of technology in the education of nurses is found in a particular school of nursing in southern Indiana. This university provided opportunities for thirty-two faculty members to train in a new online learning management system (LMS). The purpose of the project was to meet the needs for a common user interface in online programs. The participants included 57% full-time and 29% part-time faculty, as well as 14% academic advisors with teaching responsibilities. Before the training, 93% of the participants had not been taught using the specific LMS. The training included several sessions as well as training with a support person or recorded webinars. After teaching in the new LMS for six weeks, the faculty were surveyed. 7% responded that they did not feel competent in using the system while 57% reported that their performance was adequate but believed they would need additional training prior to implementing the system with their students. The remaining participants (36%) agreed they were doing well and utilizing many of the available features, requiring no additional training. Outcomes included decreasing the workload and faculty stress levels associated with implementing new technology. (Judge & Murray, 2017).

Education Environments in Suburban and Rural Southern Indiana Pre-COVID

Prior to the COVID-19 lockdown, Southern Indiana learners had a mixed bag of options for face-to-face and virtual learning. Of course, in public K-12 education traditional face-to-face learning in brick-and-mortar schools was the predominant choice of learners and parents/guardians. A few public charter schools existed, along with online public schools. Homeschooling was on the rise. The percentage of Indiana K-12 students withdrawing from traditional school to enter a homeschooling environment hovered around 1% per year leading up to the pandemic shutdown

(Blackburn, 2021). In terms of student success, the 2019 federal graduation rate for Indiana was 86.46% (Indiana Department of Education, 2021).

In higher education, options in 2019 included public and private brick and mortar 4-year institutions and a strong community college network. The community college structure in Indiana is unique. Unlike other states that have a variety of community college options. There is ONE primary option for a somewhat traditional community college experience in the state. The community college operates out of 40 locations throughout the state and serves commuters rather than resident students. In addition, student choice included a growing number of online offerings from within and outside of the state of Indiana. In terms of student success, the EcO Network (n.d.) reported that in 2017 in the southeastern Indiana counties they serve, 36% of the population had educational credentials beyond a high school diploma. Compared to the state-wide average of 44.9% of residents having formal education beyond high school (Marcus, 2002), the southern region of the state clearly has room to grow in educational attainment.

In teacher education, clinical placements in K-12 schools are critical to the success of pre-service teachers. In these clinical placements, some teacher candidates make their first contact, in the role of instructor, with young students. In addition, these pre-service teachers make their first connections to classroom teachers in the role of a mentee, and sometimes colleague. The state of Indiana (n.d), along with CAEP (Council for the Accreditation of Educator Preparation, 2022), mandate several elements of accredited teacher education programs. For instance, the state requires that all teacher education programs offer a 120-credit hour option. In addition, a major component of the CAEP teacher education standards are opportunities for impactful experiences in K-12 classrooms in the forms of field placements throughout the professional program and a culminating student teaching placement. In southern Indiana, most of these placements occur in rural and suburban public schools because of the lack of urban placement opportunities.

The importance of clinical experiences for preservice teachers cannot be overstated. In fact, Linda Darling-Hammond (2014), a well-known educational researcher, lists clinical experiences as one of the teacher education program features that appear to matter across several different preparation programs that were studied. She suggests that at least 30 weeks of supervised practicum and or student teaching experiences that are closely aligned to coursework is one of the features that help preservice teachers refine the ideas about good teaching that were embedded in them while participating as students in the same system in which one day they will teach. Perhaps even more important, pre-service teachers collaborate with practicing teachers and pick up on professional dispositions from witnessing authentic teacher-student interactions within their clinical experience classroom. Cooperating teachers in the field serve as important role models for those on the cusp of entering the teaching profession.

Healthcare Environments in Suburban and Rural Southern Indiana Mid-COVID

Toward the end of the year 2019, pneumonia of unknown origin was reported in China and rapidly spread around the world, soon to be identified as COVID-19. The virus produced several different symptoms, creating health complications, especially for vulnerable populations such as older people and at-risk patients. Rapid changes were noted in healthcare when the COVID-19 pandemic began. Hospitals began to overflow, there was a shortage of resources including ventilators used to provide oxygenation for life sustainment, hospitals were on divergent status, elective surgeries were canceled, and surgery departments and all available spaces were transformed into patient rooms within the hospitals.

Long-term care facilities closed their doors to outside visitors and limited entrance to employees only. New admissions were on a waiting list. Patients were restricted to their rooms and ate meals in isolation. Families "visited" through a closed exterior window. Many residents suffered increased depression from the isolation and reports of increased death rates were noted.

The public noticed changes at the hospitals, such as no visitor policies, no guests allowed to accompany a patient for outpatient services, and lockdowns that required a screening process for entry. Federal, state, and local agencies adopted policies and guidelines that were modified frequently as the pandemic escalated. Nurses and healthcare workers were constantly in the news and touted as heroes. They were on the frontlines, putting not only themselves in danger, but their families as well. Many healthcare workers have reported living in a basement or a separate section of their home to protect their loved ones from the unforgiving and unfamiliar virus. Their long work hours required wearing protective equipment which included masks, gloves, and gowns. The restrictive equipment placed increased physical demands on the healthcare workers, not to mention the mental and emotional toll on their daily work schedule. Infection control was the highest priority as the virus was spreading rapidly across the United States.

The nursing profession is founded on caring for others and nurse educators role model this basic philosophy within every aspect of teaching their students. When the pandemic hit, nurse educators were faced with multiple concerns that impacted relationships within the academic institution, among clinical partners, and most importantly, the student population. These relationships were affected in different ways, but all were connected to and impacted nurse educators.

Concerns were also evident within universities that prepare nurses. Financial responsibilities always seem to rise to the top of any given crisis and the pandemic was not any different. It was understandable that during an unprecedented time, each institution and organization attempted to provide the best services, care for employees, and meet the needs of those being served to the best of their abilities. Colleges and universities were scrambling to determine how to meet academic as well as accreditation requirements when campus shutdowns prevented face-to-face learning.

Clinical partnerships are at the core of any professional preparation program. At the beginning of the pandemic, many clinical healthcare partners shut their doors to visitors, and this included nursing students. The unknown put fears into healthcare organizations and rather than utilizing students as additional helping hands (and learning opportunities for all healthcare workers), some chose to halt the clinical partnerships during this time. For those who continued clinical rotations and allowed student involvement, the tense environment affected the experiences of the students and created additional stress for preceptors.

Nurse educators attempt to prepare their students for all types of situations. A pandemic consisting of a new virus that has more questions than answers is not something that can be planned. However, the foundations of nursing which consist of caring, and empathy is basic training for nursing students. It was apparent during the Covid crisis that our nurse educators have utilized their expertise in teaching the basic skills as our nurses have proven to be more than competent in providing these basic skills.

Education Environments in Suburban and Rural Southern Indiana Mid-COVID

At the K-12 level, Indiana school closures were imposed by the state on March 20, 2020. Later, on April 2, 2020, Indiana's 1,921 public schools were notified that schools would remain closed for the duration of the 2019-20 academic year ("Map: Coronavirus, "2020). Indiana released a school reopening plan in June of 2020 that included a July 2020 school re-opening option ("School responses," n.d.). The plan contained masking and distancing mandates, as well as an option for students to continue their education with online learning in many school corporations. Since March 20, 2020, modifications to the COVID response plan were made as CDC guidance changed, affecting masking, distancing, and quarantine policies throughout the state. In terms of student success during mid-COVID, the Indiana High School graduation rate increased to 87% in 2020 before dipping to 85.75% in 2021 (Lawson, 2021). Of major concern is the mid-pandemic learning loss across grades K-12. Pass rates on standardized tests dropped from 10% to 50% for some schools. Less known are the lasting effects of pandemic learning loss on students. One researcher posited that "the learning loss of this age cohort will still be evident in the 2060 Census. The only question is how big that loss will be and what compensating factors, like resilience and grit, will take the place of classroom proficiency in these students" (Hicks, 2022).

In higher education, most universities were on spring break in 2020 when COVID-19 delays or closures were announced. Most decided to extend spring break for an additional week to allow for decision-making to occur (USI[1], IU[2], UE[3], Hanover[4], Franklin[5], IvyTech[6]) and moved to virtual learning for the remainder of the spring semester. Timelines for the return to face-to-face instruction varied. In general, face-to-face instruction returned to southern Indiana in the spring of 2021 with a variety of masking and distancing policies in place.

In teacher education, not only were pre-service teachers impacted by the closures of their universities, but they also had to deal with their face-to-face field experiences and student teaching placements shifting due to local public schools transitioning to online instruction (eLearning). Teacher Education Programs (TEPs) in southern Indiana was not alone. Throughout the nation, TEPs were scrambling to find suitable alternatives to face-to-face clinical experiences.

At the author's university, online video repositories such as Annenberg Learner[7], The Teaching Channel[8], Reading Rockets[9], and yes, even YouTube[10], among others, were used to provide virtual teaching observation examples for preservice teachers removed from face-to-face field placements during the spring semester of 2020. Faculty within the teacher education program quickly identified relevant video clips and created a variety of activities around them, including discussion forums, reflection prompts, quizzes, and written assignments. The matter of supporting student teachers with quality experiences was more complex.

By the fall semester of 2020, field placements had been negotiated with an online public k-12 school. Preservice teachers were matched with an online instructor and attended online classes one day per week. In addition, the preservice teachers met online with small groups of students to offer to tutor students who struggled with their assignments. Although the experience was far better than ones had by institutions relying on pre-recoded video of teacher in lieu of in-person field placements, the online public-school environment was not conducive to helping the pre-service teachers build relationships with the cooperating teachers under which they served.

The following spring (2021) schools in the southern portion of the state began returning to face-to-face instruction. Preservice teachers were placed in schools where visitors were allowed in buildings. They were subject to masking and distancing requirements determined by the school of placement. As the student-teacher population was dispersed over several areas of the state, the one-size-fits-all model would not suffice when schools began to close for face-to-face learning in the Spring of 2020. The districts where these student teachers were placed varied greatly in terms of their ability to quickly put courses online. In some districts, learning management systems (LMS) were already in use to provide access to student course information to parents and to provide access to course materials to students. In these districts, students were well versed with the LMS tools, and the transition was smoother than in less technologically savvy districts. In many cases, the student teachers were able to take the lead in launching the online class meetings. In most cases, the launch included the student teacher and the classroom teacher working together to make the transition. In isolated cases when the student-teacher was placed in a remote school in a rural district, many students had no internet access. Classroom teachers worked together to prepare printed learning packets and the two of them would spend time driving to students' homes to deliver the packets. In these cases, personal communication was limited to the written word with an occasional phone call to home.

Not much changed in the following semester (Fall 2020) for student teachers. Although better prepared for a semester of online instruction, schools in which student teachers were placed struggled with technical difficulties, including a lack of high-speed internet in homes of isolated rural students. Student teachers were able to negotiate a more active role in online instruction given the improved preparation for the semester. As with field experience, by the Spring of 2021, some schools were returning to face-to-face instruction, allowing student teachers to complete their teaching in classrooms while adhering to mask and distancing mandates. One-on-one interactions between teachers and students in the masked and distanced classrooms provided a different perspective of authentic teacher-student interactions, one where pre-service teachers rarely witnessed teachers seated near students as they worked.

THE IMPORTANCE OF TECHNOLOGY IN THE PREPARATION OF NURSES AND TEACHERS

Nurses and teachers are required to maintain credentials and willingly pledge to embrace the mission and values of their professional license. It is their responsibility to remain current in their practice and implement evidence-based practices. As technology advances worldwide in both education and healthcare, it is expected for these professionals to encompass digital learning skills within their environments. Both professions must remain competent in their fields and, as professionals, must uphold confidence to utilize technical skills with students and patients. Maintaining digital literacy is essential and it is vital to prepare our future teachers and nurses to evolve as technology is interminable.

Higher Education Technology

Technology has progressed throughout the years and while many educators have sought to learn and advance their professional skills, others chose to remain in their comfort zones within the classroom. Technology has been described by some educators as overwhelming, disruptive, and time-consuming during the initial learning phase. At the same time, for those faculty who chose to become and remain informed on the benefits and functions of technology, numerous time-saving benefits have been realized. LMS systems, such as Blackboard, Canvas, Google Classroom, provide multiple functions including data management, assignment delivery, feedback options, and interaction through discussion forums that just a decade ago could not be housed in one standalone system.

Technology in the Healthcare Professions

Technology is restructuring how healthcare is delivered. As life expectancy is increasing, medicines are being developed at a rapid speed, and medical diagnoses are identified at earlier stages, the healthcare profession must not only stay current with technology but must advance with the newest, most innovative technology challenges.

Health information technology (HIT) includes different types of software such as medical management (administrative and clinical aspects), electronic health records (patient's medical records), E-Prescribing (electronic prescriptions), remote patient monitoring (wearable devices to monitor blood pressure, blood glucose, heart activity, etc.), master patient index (connects patients records from more than one database), patient portal (patients have access to their records, can schedule appointments, and pay bills), urgent care applications (provides health information and allows patient immediate access to care), and medical billing (insurance claims, payments) (Software Suggest, 2020).

The advances in technology promote immediate communication between providers and patients, increase the efficiency of the healthcare organizations, reduce medical errors, and are cost savings. As technology advances healthcare processes will continue to reform and provide improvement in patient care.

Growth of Telehealth

Routine in-person office visits were reduced as a recent study (April 2019 through April 2020) revealed a decrease of healthcare office visits by 1520.8 visits per 10,000 people. During this same time-period telemedicine intensified by 4081%

(Whaley, et al., 2020). Telehealth has been an option in many areas although not widely adopted until the onset of the pandemic.

Telehealth is convenient, affordable, comfortable, safe, and used for several types of healthcare including common medical problems, chronic disease, acute healthcare, mental health, substance abuse, physical and occupational therapy, pediatrics, dermatology, and pre-natal. The convenience includes no wait times, doctors with availability (often 24/7), no travel time which can be an important factor for rural areas as well as cost-saving due to rising gas prices. The visits are usually covered under most medical plans and the cost is similar to an in-person office visit. Meeting in the comfort of the patient's home makes it especially appealing, especially if an illness prevents traveling. And of course, the safety of the patient's home reduces the opportunity for spreading a disease or contracting a contagious virus.

It is important for providers to establish trust and learn how to provide empathy during telehealth visits. Without face-to-face contact, providers must rely on their visual and auditory skills to assure the patient is understood. Telehealth allows providers an opportunity to view a small part of the patient's surroundings, which can give insight into an overall assessment of the patient's needs. It is also imperative for providers to be prepared and on time for appointments as it can be frustrating for a patient waiting, not knowing if they missed the appointment or what is wrong if the provider is not online at the designated time.

Appropriate communication with patients who choose to utilize telehealth services is important for successful outcomes. Communication might include technical support (equipment needed such as a webcam, microphone, internet capabilities, checking equipment prior to appointment time), creating an appropriate environment (noise reduction, distraction-free), and how to take notes or check the online system for follow up directions.

Telehealth offers a different approach to healthcare, allowing providers to see their patients in a home environment. It can be a pleasant experience for both parties and create a relationship of trust and mutual respect. As rural communities continue to work toward delivering broadband internet to all residents, telehealth becomes a more realistic opportunity to promote healthcare across all regions.

The utilization of telehealth requires professionals to continue providing patient experiences as similar as possible to in-person office visits. Learning how to communicate a caring attitude, provide empathy to the patient, and develop a relationship within the confines of a computer screen can be daunting. Nurse educators must begin teaching how to adapt these soft skills to the world of technology as technology continues to advance within healthcare.

Technology in Nursing Education

In the healthcare profession, education that includes hands-on training has always been a critical part of the learning process. Education that included classroom, clinical, lab, simulation, and on-site clinical rotations has been part of the process for many years and proved to provide successful learning opportunities for students. However, as technology advanced and online education became increasingly prominent, educators began to be more creative and implement some form of online education in their programs. Of course, there have always been those who either lack the ambition or simply refuse to make changes that require extensive learning on the part of the educator. As the future evolves, healthcare professionals and educators must move forward with technology and innovative advancements to remain current in their field.

The beginning of the COVID-19 pandemic was met with utter confusion by healthcare workers due to the many unknown factors regarding the highly contagious infection. The guidelines that reduced physical contact and initiated mask mandates impacted education in numerous ways. Social distancing, economic factors, illness, and even death of family and friends resulted in diminished support and any type of pre-pandemic normalcy. The pandemic was an unprecedented challenge that forced academia to delve into innovative solutions and utilize technology to be effective in sustaining educational endeavors. Faculty were no longer in a voluntary role of choosing the desired teaching method but rather were required to transition into an online learning environment. While concerns included student ability to meet required skills and knowledge of the course, synchronous versus asynchronous learning, and how to implement appropriate assessments, the emergent need to move online education during the pandemic provided an opportunity for educators to embrace technology and develop new evidence-based practices to support student learners who migrated toward the online educational opportunities.

Technology in Educational Services

The growth in affordability and availability of technology devices has opened a door to formal and informal virtual learning and communication of learning performance to stakeholders. The Ed-Tech market is one that realized some of the biggest growth during the pandemic. COVID-19 lockdowns served as catalysts to the EdTech market. In fact, "K-12 districts in the U.S. alone spent $35.8 billion in 2020 on hardware, software, curriculum resources, and networks, representing a more than 25 percent increase over 2019" (Paykamian, 2021). Experts predict the investment in EdTech ventures will continue to be on the rise for the remainder of the decade.

Growth of eLearning

Prior to March of 2020, Google Classroom had roughly 50 million users. In one month alone, users increased to over 100 million (De Vynck & Bergen, 2020). The onset of the COVID-19 pandemic lockdown immediately changed the face of eLearning across the globe. Google Classroom is a free service offered to teachers that allows for the communication between teachers and students and the sharing of multiple types of classroom materials and assignments.

Indiana's transition to eLearning began well before the onset of COVID-19 (Digital Learning Collaborative, 2018). In 2011, the Indiana Department of Education (IDOE) unveiled its "Flex Program." This program allowed school districts to design district-level instructional plans that include the option of virtual learning. Early adopters of this plan had up to a full decade of preparation before the COVID-19 pandemic. By 2014, Indiana released an official eLearning policy related to school closure days due to weather. Currently, schools may be eligible to seek approval for their eLearning plans by meeting a set of criteria, including internet access, support for students with special needs, and technical support

Out of the approximately 2,500 public and private schools in Indiana, 229 schools or school districts were approved as eLearning schools in 2017. That number was up to 278 in 2018 (Loughlin, 2018). Comparing these numbers with 100% eLearning beginning in March of 2020, one realizes how few schools were prepared to quickly launch an eLearning environment in response to the shutdown. The data provided is not disaggregated by the area of the state. Anecdotal data indicates that due to the rural nature of the southern portion of Indiana and limited internet access in remote regions, even if schools were approved for eLearning a portion of the student body would not be able to participate.

In higher education, before the pandemic one-third of college students had taken online classes (Gallager & Palmer, 2020). Prior to 2020, over half of U.S college students participating in online programs were 2-year degree seekers (Francis et al.,2019). In Indiana, several colleges and universities known for traditional face-to-face instruction have initiated significant online presence. These include Purdue Global[11], Indiana University Online[12], Notre Dame College Online[13], Butler Online[14], and Ball State Online[15], among others. Of the nearly 425,000 college students in Indiana, around 172,000 (or 40%) are enrolled in exclusively online programs (NIVSTATS, n.d.). Although the data is not disseminated by region of the state, it is assumed that a similar percentage of college students in southern Indiana are enrolled in online programs.

Technology and eLearning in Teacher Education

College students pursuing a degree in K-12 education are positioned in a unique space between technology and eLearning expectations in higher education and K-12 schools. This involves the use of multiple LMS platforms and online learning expectations. Not only do teacher candidates need to know how to integrate technology into their instruction; but they need to be able to anticipate technological hiccups as they teach lessons in their clinical experiences. Many K-12 schools maintain firewalls that block out tools that teacher candidates use at their universities. Sometimes teacher candidate emails do not easily transfer from the university setting to the K-12 setting due to firewalls.

HELPING RELATIONSHIPS

As stated in the introductory paragraph, the career areas of nursing and teaching are among the helping professions and share many dispositional characteristics. The American Psychological Association (n.d.) defines helping relationships as ones "in which at least one of the parties intends to promote the growth, development, maturity, or improved functioning of the other." The two characteristics considered in this discussion of helping relationships are empathy and caring. These two ideas are relational in nature and require the implementation of specific capabilities including cognition, behaviors, emotions, and morals to comprehend and react to the needs of others. Quite simply, there is an art to the ability to care and empathize. And without a propensity toward the dispositions of caring and empathy, the quality of the instructional experience suffers. "Empathy is central to great teaching and when it is lacking, ideas narrow, compassion wanes, and the great health of the teacher-family-child relational dynamic diminishes," as Carly Andrews, principal of Bosque Middle School in Albuquerque, New Mexico posited (Hoerr, p. 38).

Empathy

The concept of empathy stems from "the beginnings of philosophical thought" (Stotland et al, 1978, p. 11). The term itself has existed for over 100 years, dating back to Titchner who first coined the term. "Empathy" is associated with a wide variety of definitions. Cuff et al (2016) explicated the complexity of the term in their article, *Empathy: A review of the concept.* These researchers acknowledge the existence of a minimum of forty-three separate definitions used in English-language research articles investigating the concept of empathy.

An example of the complexity of the term comes from the shift in the definition that Carl R. Rogers, a 20th Century Psychologist, and a founding psychotherapy researcher, outlined in the transformation of his working definition. Rogers recognized a shift in the definition of the meaning of the term, "empathy," from the late 1950s to the mid-1970s. Early on, Rogers considered empathy as a state of being. Rogers posited in 1959 that "the state of empathy, or being empathic, is to perceive the internal frame of reference of another with accuracy and with the emotional components and meanings which pertain thereto as if one were the person, but without ever losing the 'as if' condition. Thus, it means to sense the hurt or the pleasure of another as he senses it and to perceive the causes thereof as he perceives them, but without ever losing the recognition that it is as if I were hurt or pleased and so forth. If this 'as if' quality is lost, then the state is one of identification" (Rogers, 1959). By 1975, Rogers' definition of empathy transitioned from a state of being to a process in which there is the active participation of the empathizer upon the other person. The process includes multiple facets, including laying "aside from the views and values you hold for yourself in order to enter another's world without prejudice" (Rogers, 1975).

For the purpose of this chapter, the act of becoming empathetic in a helping relationship is recognized as a complex process in which the empath is able to identify, from an emotional standpoint, with another to the point that the empath displays compassion for the one in need. The authors also recognize that studies show that the lack of empathy in pre-professional students could be a result of multiple factors including the lack of a proper empathetic role model (Dempsey & Reilly, 2016).

Caring

Caring is an integral part of the education and nursing professions. Both professions exemplify a caring and nurturing attitude for students when role modeling the most important values of their chosen profession. Developing this type of healthy environment creates an opportunity for trust and honest communication to develop between faculty and students. Within this role the use of parenting skills that include providing guidance, support, and positive reinforcement are strengthened with the students through role-play, encouragement, accomplishing tasks, and peer interactions.

Nel Noddings, the foremost researcher on caring in education, addresses the nature of caring in terms of morals and ethics, specifically an ethic of caring. This notion is derived from a feminist perspective. And unlike Kant's definition of moral acts where duty or principle drives moral acts, Noddings' Ethic of Care recognizes moral acts as ones done out of "love and natural inclination" (Noddings, 1988, p. 219). There is a clear connection between an ethic of care and empathy where the natural inclination is closely aligned with an empathetic relationship between the person caring and the person being cared for.

In order for a caring relation to thrive, *the carer* must be attentive and receptive towards the *cared for*'s experiences (another connection to empathy). As the *carer* attends and is receptive to the *cared for* a shift called "motivational displacement" (Noddings, 2012, p. 772) occurs. In this shift the *carer's* motivation is derived from the desire to meet the *cared for*'s needs rather than her own. Inevitably, the *carer* must reply or provide feedback to the *cared-for*; but when doing so the *carer* chooses to react in a way that will maintain the caring relationship. Importantly, a caring relationship does not rely upon the implicit needs of the *cared-for*; but rather on the explicit needs expressed by the *cared-for*.

In this context, it is clear that helping professions of teaching and nursing rely heavily on individuals whose dispositions include qualities of empathy and caring. This is not to claim that to serve as a nurse or a teacher one must be an empathetic and caring individual. However, to maintain a helping relationship between the *carer* (nurse, teacher) and the *cared for* (patient, student) elements of empathy and caring must be in play.

EMPATHY AND CARING IN NURSE AND TEACHER EDUCATION DURING COVID-19 LOCKDOWN

Meeting basic needs and developing a helping relationship have always been the foundation for both educators and healthcare professionals. Developing these dispositions in student nurses and student teachers is difficult even outside the throws of the COVID-19 shutdown. No matter how many dispositional empathic and caring factors college instructors model, such as "offering help, showing compassion, showing interest, caring about the individual, giving time, listening, and getting to know students" (Haskell-McBee & Wescott, 2008, p. 36), instructors are up against competing factors, such as university policies on teaching loads, research, committee work, and service work which requires attention and time.

Research shows that online instruction has its own set of shortcomings, even when a national lockdown is not involved. Some studies have found that online students are more likely to drop courses than their counterparts who attend face-to-face instruction (Xu & Jaggers, 2013). Other studies claim that "online learners received lower course grades, lower pass rates, and higher withdrawal rates than their classmates in face-to-face courses" (Park & Yun, 2017). Yet other studies show the contrary (Driscoll et al., 2012). Studies that consider online students' motivation, achievement, and withdrawal rates are mixed at best.

Given these outcomes, instructors in any higher education setting will struggle to support students in light of the quick transition to online instruction that resulted from the COVID-19 shutdown. In the preparation of caring professionals, before caring and empathy can be modeled in an online learning environment, the pre-professional students need to be proficient enough in the technology and logistics of the online learning environment to be engaged in the learning interaction rather than in the logistics of remotely connecting with their classmates and instructor. Unfortunately, for brick-and-mortar schools, the distance learning curve was so great that much of the effort was on overcoming the lack of experience in teaching in the online environment rather than increasing student engagement with appropriate online tools.

In teacher education, practices closely linked to caring and empathy, such as Universal Design for Learning (CAST, 2008), Social Constructivism (Vygotsky, 1978), Gradual Release (Fisher & Frey, 2003), Hands-on Learning (Ekwueme et al., 2015), and Student-Centered teaching (Santi & Gorghiu, 2017) are at the top of any educators' list of best teaching practices. However, these learning frameworks are difficult at best to implement in an online teaching environment. In asynchronous online teacher education environments, immediate feedback from students is not an option. Instructors rely on student written assignment submissions, and perhaps discussion board postings, to determine the level of understanding of key concepts developed by the students.

Of utmost concern, the modeling of empathy and caring by instructors is limited. When considering the importance of perception in Rogers' definition of empathy, and the importance of attention and reception in Noddings' definition of caring (both shared earlier in the chapter), it is clear that modeling care and empathy is also compromised in the asynchronous online learning environment, simply due to the fact that it is difficult for any instructor to find ways in which to attend to all students in an online learning environment when students are completing assignments posted by instructors in advance without any opportunities to interact with classmates or the instructor. The situation is slightly improved in a synchronous online environment; however, instructors still struggle in connecting with students in a meaningful way due to inequities such as poor internet connections, student-muted microphones and webcams, and limited ability to visualize all students in a large class on one computer screen.

As online learning developed, educators attempted to adapt to a new method of displaying the critical behaviors that form the foundation for both professions. When the pandemic started and campus closures began, effective educators had to quickly acclimate and gain confidence in learning new strategies to model caring behaviors in a synchronous and asynchronous online environment. These efforts resulted in varied outcomes. One research study conducted at a four-year university included twenty experienced K-12 certified teachers who were graduate students. The participants had a minimum of one year of teaching experience with an average of seven years. Data were collected four times throughout the year (two were prior to COVID-19 closures and two were afterward) and consisted of journals, shared learning experiences, and peer discussion forums. The qualitative content analysis framework was based on existing relationships and relationships during remote learning. The results were not surprising that both before and during the transition to remote learning the primary concerns of the educators included relational bonds and authentic care. As remote learning continued, educators' concerns grew regarding maintaining connections. Three discoveries from the research consisted of (a) a need to repair and (re)prioritize teacher-student relationships, (b) educators' efforts to re(connect) with students by using authentic care practices, and (c) challenges to relationship-building due to the digital divide (Miller, 2021, p.121).

PANDEMIC IMPACT

Nurse Faculty

A recent study evaluated the experiences of academic nurse educators during the COVID-19 pandemic. The study included fourteen participants who were interviewed and completed questionnaires. The data analysis revealed five themes which included: *riding a roller coaster, figuring it out, giving and getting help, seeing silver linings, and feeling loss* (Gazza, 2022). *Riding a Roller Coaster* theme revolved around the sudden unplanned changes that continued to evolve daily into additional changes. This included moving from their work environment to working remotely from home. The faculty with little online teaching knowledge relayed the emotions and frustrations associated with the excessive issues experienced daily. What was surprising was the online experienced faculty who admitted similar frustrations. The next theme, *Figuring It Out*, accurately describes the daily hindrances the faculty faced due to teaching in a new environment, lack of face-to-face and immediate assistance, and searching for new resources and avenues to provide learning experiences for students to replace the previous ones. One faculty shared how they struggled to figure out how to hold a virtual classroom and spent hours practicing with family members so the faculty would have confidence and students would not perceive the faculty as not being competent to teach the course. *Giving and Getting Help*—this theme acknowledged the tremendous amount of support required to assist all faculty, most specifically those without online teaching experience. Identified supporters included peers with expertise, technology experts, administrators, publishers, vendors, and instructional designers. The theme titled *Seeing Silver Linings* speaks to the optimistic outlook of dedicated nurse professionals. In the midst of the chaos, frustration, and learning how to be efficient in their jobs within a remote setting, the nurse educators noted positive outcomes such as opportunities to provide more self-care, work-life balance, and an increased amount of student emails that were more personal and compassionate than they might have been with in-person verbal communication. Lastly, *Feeling Loss* was recognized for the sincere feeling of loss of connection with their students, peers, and the academic institution. One faculty reported how much she missed stopping by a colleague's office just to chat and with working remotely she felt as if she was bothering them if she reached out. The most commonly mentioned was the loss of interaction with students within the clinical environment. The faculty stated they missed the live conversations with students and teaching moments that are not the same in a virtual classroom. (Gazza, 2022). Nurse educators are flexible, have inner strength, and their caring and compassionate attitudes reflect the professional role model representative of the profession. The Gazza (2022) study revealed some very important information that can be utilized moving forward for nurse educators to improve remote working environments post-pandemic.

A group of experienced nurse educators published a recent article on diverse learning strategies used to support critical thinking during the pandemic (Dozier, et al., 2021). Critical thinking skills should begin early in nursing education and active learning activities should be implemented throughout the program to promote the learner's level of knowledge in the application of practice. Designing a higher level of thinking activities will lead the student to understand how to analyze and problem solve. The classroom setting offers diversity for numerous types and levels of active learning but when redirected to an online learning environment there are limitations. The limitations include a lack of faculty's perception of student understanding and a lack of student motivation and engagement. The pandemic closure of clinical sites led to adapting clinical and classroom simulations into virtual simulations. The faculty provided scenarios that incited the students to utilize their clinical reasoning skills, prioritize care, and implement organization of patient care. The simulations met course objectives and students were provided a level of competency based on the individual simulation. Student reviews were mixed, some thought the simulation was extremely helpful while others voiced frustration. Another active learning strategy the faculty implemented during the pandemic was employing the collaboration between students using Google Docs. Students were placed in groups and worked in partnership to develop presentations where faculty and students could follow up afterward and post questions for peer feedback. The faculty implemented numerous other online teaching strategies that were restructured to meet online learning outcomes and report that overall, the learning tools were reviewed positively by the students. (Dozier, et al., 2021)

A population of 45 practicing nurses and midwives in Ghana with a minimum of three years of experience as clinical preceptors participated in a recent study that pursued an interest in participants' experience, perceptions, and challenges teaching with technology. The participant's age range was 31-40 years and 33 of them had 11-20 years of experience as practicing nurses (Enyan, et al., 2022). The researchers obtained baseline data, then provided a two-day training including both face-to-face and virtual content on the use of technology in their roles as clinical educators or preceptors. The participants were assigned a group of students and were to implement the learned technology within the clinical groups. Six weeks after the initial training the participants were given the same questionnaire and additional interviews were conducted in which only seven of the 45 preceptors participated in this portion of the study. Results of the study included perceptions of using technology in the clinical setting was useful, improved teaching, quicker to accomplish objectives, improvement in self-efficacy, and increased intention to utilize technology. The interviews resulted in the identification of three key themes related to the use of technology: *strength, constraints, and future approach*. The theme of *strength* acknowledged the acceptance and positive outlook the participants shared regarding implementing technology within clinical teaching, noting it was both helpful to preceptors and students if access was available. Other strengths included convenience, teaching without interruptions, and the ability to review material multiple times if recorded. The *constraints* identified in the clinical environment included poor internet connectivity, demonstration and evaluation of clinical skills, student attendance or lack of attention, preceptors' busy schedule, ineffective collaboration, and time constraints. *Future approach* was identified as a theme by participant comments supporting technology to advance in the clinical education setting and providing personal suggestions for future initiatives. Ideas included training for faculty on multiple online platforms, supporting students with internet packages including appropriate data bundles, and teaching regarding online resources (Enyan, et al., 2022). While in-person clinicals have been the pillar of nursing education for many years, it is vital that educators utilize diverse teaching tools, obtain technical competency, and stay up to date with technological advancements.

Teacher Education Faculty

A coveted aspect of teaching in-person courses is the opportunity for teacher education faculty to debrief with each other. Whether a brief conversation in the hallway or at a faculty meeting, instructors value the opportunity to reflect upon student concerns or successes. As a result of the COVID-19 shutdown, teachers lacked this opportunity to find empathy from colleagues during this challenging and, some would argue, traumatizing time. A natural assumption is that practicing instructors with experience in face-to-face teaching alone would be concerned that students would not perceive them as effective online instructors. This is not always the case. In fact, some instructors believed that they were perceived by students as better online teachers than in-person teachers - a result that improved their self-efficacy as an instructor (Cutri et al., 2020).

The induction by fire to online teaching for college instructors used to teaching in-person courses included quickly designed professional development and support by university IT professionals. At the author's institution located in rural/suburban southern Indiana, some structures for helping faculty move courses from face-to-face to online environments existed prior to the pandemic onset. For instance, the university offered training to faculty using principles from the Quality Matters[16] standards to maximize students' online learning experiences. However, support such as this takes time. The pandemic forced an overnight solution.

Similarly, as described in Borup & Evmenova (2019), information technology support at the author's institution included providing access to and training in technologies, such as Zoom, Kaltura, Google Tools, and others. Because of the university's rural/suburban context, several of the faculty and students did not have access to high-speed internet in their homes. The university set up hotspots in the parking lot, as well as arranged for home hotspots for faculty to use to connect with students. The quick response to technology professional development needs resulted in the university developing enhanced professional development around online teaching in the summer of 2020. What resulted was a comprehensive system to help faculty transition to online teaching and learning in the fall semester.

The teacher education program at the author's institution already had in place several hybrid and online courses that put the teaching unit ahead of the learning curve in transitioning to online learning. The comprehensive training offered prior to the fall semester allowed teacher educators to expand upon the technologies used in the unit to streamline their work. For instance, enhancements to the LMS embedded elements of the Quality Matters standards into shells that guided faculty to include information and features that improved upon the students' learning experiences.

The pandemic forced a change in the way in which teacher education faculty viewed the online learning experience. This was not a result of a natural evolution. It was forced. And, as a result, the institution has a well-trained faculty using cutting-edge technologies. If anything can be held up as a positive example of the pandemic lockdown, it is the strengthening of faculty knowledge around online learning at the author's institution.

Student Population

Students of professional nursing and teaching programs must be devoted to their academic activities to pursue their dreams of becoming licensed professionals. While students who pursue a career in nursing and teaching have many characteristics, the most notable is their inherent sense of caring for all human beings. There is not a particular age for students who desire to be a nurse or teacher, they range from 18- to 70 years old, and it is not uncommon for these students to attend school full-time, work a full-time job, and many of them have families. It would be appropriate to note that juggling life with so many activities take dedication, organization, and a goal to be successful.

Students have different learning needs and for many students, this includes a classroom setting with live lectures. When this type of learning was no longer available, and students were forced into online learning, anxiety, and additional fear of the unknown (technology) escalated. Students who had led busy lives working, attending school, completing clinical hours, and providing care for their families were then thrust into a home isolation atmosphere that required new skills, and the ability to focus and multitask simultaneously and they quickly became overwhelmed.

The decision for moving to an online education forum was a critical turning point for academic institutions to continue meeting educational needs but it affected many students and created additional hardships and stressors. For students living close to a town, the internet was easily accessed within city limits by sitting in parking lots outside of buildings that provided free guest access if they did not have available internet services at home. But for many rural students, internet access was a barrier to education and created additional isolation during a time when depression and mental health became an area of concern.

Moving to the online format for education was not exclusive to higher education. Students who had young school-age children were forced into adapting their lives to assist their children at home with their education via an online format. This created additional stressors which included internet capabilities, access, and usage, multi-asking the children's learning environment with their own educational requirements, not to mention the planning and executing meals and clean-up throughout the day. Additional barriers to the home learning environment were a lack of appropriate learning space (quiet, distraction-free), and meeting the needs of all family members within the designated home space.

One recent study in Belgium reviewed the experiences of the student nurses who were in clinical sites during the pandemic. Areas discussed included awareness, internal and external support, and self-efficacy of competencies (Ulenaers, et al., 2021). The survey was completed by 665 students (3809 were sent out) and consisted of structured questions and open-ended questions that allowed participants to share their additional thoughts. While clinical placement resulted in several site changes for some students, 54.74% of the students were fortunate to remain in clinical settings. Most students were not actively involved with COVID-19 patients, and some stated they were not aware of any infected patients that may have been within their assigned units. The most startling data was that 8.72% of the students had no idea if they provided care for any patients with COVID-19. The survey results provided information that the pandemic has increased student awareness of the risk of infection and the critical importance of hand hygiene and personal protective equipment such as masks and, gloves. Regarding self-efficacy, students felt most confident in identifying patients with oxygen deprivation and least confident regarding care for patients on ventilators. The students recognized the support received from their clinical placement supervisor and admitted that communication was especially vital during the pandemic. There were gaps acknowledged by the students which included creating regular interactions with their clinical placement supervisors, additional psychosocial support, recognition of the challenging work environment, the need for space to decompress, and flexibility with assignments and tests due to the increased demands of clinical, study time and online classes. The students also identified issues with meeting clinical objectives due to multiple clinical site changes and preceptors who were overwhelmed with their workloads and unable to spend time with the students. Other areas students identified needing support included lack of recognition for their contributions at the clinical sites, lack of or delayed communication regarding infected patients or staff and not being included in offerings at the clinical sites for virus testing or supportive services. While most nursing students viewed the clinical experiences as challenging and enriching, others felt insecure, unsafe, and extremely stressed. Students did admit there were fewer learning opportunities and due to the work overload of the preceptors, the students were not monitored closely which resulted in decreased opportunities to practice clinical skills and meet course objectives (Ulenaers, et al., 2021).

Lessons in Caring and Empathy for Post-Pandemic Professional Program Educators

The pandemic allowed professional program educators to take a much-needed deep breath and consider the individual situations of our students, many of which were highly aligned with the notions of caring and empathy. Assignment due dates were modified for students who contracted the virus or had responsibility for the care of immuno-compromised family members. Faculty and students met by phone in cases where students had no access to personal technology devices or high-speed internet access. Faculty struggled to engage with students beyond the limited opportunities provided by a poorly developed LMS system.

What are the lessons learned from the unique context of forced online learning in terms of care and empathy for the helping professions in absence of faculty/student face-to-face interactions? The pandemic provided a shared context for students and faculty to develop empathetic relationships. Even though face-to-face exchanges were not possible, texts, emails, and even phone calls served as vehicles in which faculty and students shared their struggles in living through the pandemic. Emotions, which are critical in developing empathetic relationships, were high for both students and faculty struggling through the realities of the pandemic. If faculty did not hear from their students when expected, they tended to reach out to ask if all was well. This is a practice that should continue post-COVID. COVID shutdown or not, faculty should check in on students beyond the scheduled class time whenever student engagement is lacking. The pandemic fallout necessitated that faculty attend and be receptive to student needs. In many instances, the pandemic created new caring relationships where none existed before due to the magnitude of the shared experience the COVID-19 shutdown provided.

How can faculty continue to nurture empathetic and caring relationships post-pandemic? First, faculty should take the time to introduce themselves to students in a meaningful way. Share something of a personal nature in which students may identify - not deep dark secrets, but perhaps personal interests, hobbies, family structure, etc. Brief bios shared as pre-recorded video introductions posted to the LMS or a short verbal summary offered at the first in-person class meeting will open a door to allow students to see faculty as people willing to share part of their lives beyond the classroom with them. Second, pay attention to student engagement in class. Reaching out to a disengaged student through an email or a message shows students that faculty are attending to their needs. Often times the LMS provides data that reveal the time spent online by each student. Use the data provided to check in on student engagement in the course. Third, be responsive to students' needs by following up on students' questions or comments. Whether referring a student to a tutoring session or providing additional study resources to the student, taking action on a student's concern provides building blocks for the generation of a caring relationship.

Professional program educators must model the professional characteristics and behaviors (especially caring and empathetic dispositions) that are expectations of the professional student, whether the modality of instruction is online, face-to-face, or a combination of both. It should come as no surprise, that true to their core beliefs and value of all human beings, nurse and teacher educators elevated their passion for their professions by adapting to the rapid and innovative deviations from normal teaching practices related to the pandemic, to develop and implement new best practices to promote student success. The educators combined empathy with fortitude and ensured professional students obtained a high level of educational resources, experiences, and learning opportunities to sustain their current academic level and excel them to degree completion. Student success is not only measured by the course completion but a successful pass rate on the NCLEX (The National Council Licensure Examination is a nationwide examination for the licensing of nurses in the United States, Canada, and Australia) that provides a license to practice nursing and the PRAXIS, the teacher licensing assessment required by most states to become a licensed K-12 teacher.

Professional students experience anxiety and fear in a normal educational environment or clinical setting. A global pandemic escalates that fear and anxiety, and educators must act responsibly to provide compassion and empathy to support students' emotional and mental well-being. Educators must be dedicated to futuristic opportunities to promote optimal learning for our students. Technology is key. Open-mindedness, actively learning, and using new tools as they become available will strengthen teaching capabilities as education continues to move toward remote learning environments.

CONCLUSION

What remains clear regarding education: educators are resilient and strive to utilize every resource available to promote student learning and competency; educators have sincere compassion and empathy for students and their personal struggles, and technology is the future. As "helping professionals" forming relationships with our students, providing safe practice spaces for learning, and role modeling characteristics essential for our professions are vital for student success. Providing empathy, acknowledging students' personal struggles, and bestowing resources to meet individual learning needs will nurture the relationship. In turn, the modeling of these caring and empathetic practices will serve to instill in pre-service caring professionals the importance of these dispositions in their future practice.

The need for healthcare and education continues to be a priority and due to the global pandemic, telehealth and eLearning have become increasingly utilized services. Technology is not something to dread or refrain from learning but rather should be embraced as a challenge and dedication to our student's future success and well as the continued advancement of global healthcare. Continued research is needed to investigate the impact of online instruction on the development of dispositions of caring and empathy in pre-service helping professionals.

REFERENCES

American Psychological Association. (n.d.). *APA dictionary of psychology.* Retrieved March 20, 2022 from https://dictionary.apa.org/helping-relationship

Badalucco, T. (2015, November). The battle to move U.S. health care from paper to digital far from over. *U.S. News and World Report.* Retrieved from https://www.usnews.com/news/articles/2015/11/05/the-battle-to-move-us-health-care-from-paper-to-digital-far-from-over

Blackburn, A. (2021, August 5). *State numbers show number of kids withdrawing to homeschooling nearly doubled last school year.* Fox 59. https://fox59.com/news/state-numbers-show-number-of-kids-withdrawing-to-homeschooling-nearly-doubled-last-school-year/

Borup, J., & Evmenova, A.S. (2019). The effectiveness of professional development in overcoming obstacles to effective online instruction in a college of education. *Online Learning, 23*(2).

Cantu, D. V. (2016). The evolution of the telephone. *Children's Technology & Engineering, 21*(1), 14–17.

CAST. (2008). Universal design for learning guidelines version 1.0. Wakefield, MA: Author.

Centers for Disease Control and Prevention. (2020). Impact of the COVID-19 pandemic on emergency department visits—United States, January 1, 2019–May 30, 2020. *MMWR Morb Mortal Wkly Rep., 69*(23), 699–704. .mm6923e1 doi:10.15585/mmwr

Council for the Accreditation Educator Preparation. (2022). *2022 CAEP standards.* https://caepnet.org/standards/2022-itp/introduction

Cuff, B. M., Brown, S. J., Taylor, L., & Howat, D. J. (2016). Empathy: A review of the concept. *Emotion Review, 8*(2), 144–153. doi:10.1177/1754073914558466

Cutri, R. M., Mena, J., & Whiting, E. F. (2020). Faculty readiness for online crisis teaching: Transitioning to online teaching during the COVID-19 pandemic. *European Journal of Teacher Education, 43*(4), 523–541. Advance online publication. doi:10.1080/02619768.2020.1815702

De Vynch, G., & Bergen, M. (2020, April 9). *Google Classroom users doubled as quarantines spread.* Bloomberg. https://www.bloombergquint.com/business/google-widens-lead-in-education-market-as-students-rush-online

Dempsey, C., & Reilly, B. A. (2016). Nurse engagement: What are the contributing factors for success? *Online Journal of Issues in Nursing*, *21*(1).

Digital Learning Collaborative. (2018, September 25). *E-learning days*. DLC. https://www.digitallearningcollab.com/elearning-days

Dozier, A. L., Gilbert, B. G., Hughes, V. W., Mathis, D. P., & Jenkins, L. J. (2021). The Use of active learning strategies during the COVID-19 pandemic to promote critical thinking. *The ABNF Journal*, *32*(1), 12–16.

Driscoll, A., Jicha, K., Hunt, A. N., Tichavsky, L., & Thompson, G. (2012). Can online courses deliver In-class results?: A comparison of student performance and satisfaction in an online versus a face-to-face introductory sociology course. *Teaching Sociology*, *40*(4), 312–331. doi:10.1177/0092055X12446624

Eco Attainment Network. (n.d.). *EcoNetwork*. https://educationcoalition.com/eco-network/

Ekwueme, C. O, Ekon, E. E. & Ezenwa-Nebife, E. E. (2015). The impact of hands-on-approach on student academic performance in basic science and mathematics. *Higher Education Studies, 5*(6).

Enyan, N. I. E., Boso, C. M., Ofori, A. A., Agyare, D. F., Aboh, I. K., Dike, N. O., Agyei, D. D., & Obiri-Yeboah, D. (2022). Clinical preceptorship in Ghana in the era of COVID-19 pandemic: An interventional study. *BMC Nursing*, *21*(1), 1–12. doi:10.118612912-022-00838-w PMID:35287665

Fisher, D., & Frey, N. (2003). Writing instruction for struggling adolescent readers: A gradual release model. *Journal of Adolescent & Adult Literacy*, *46*, 396–407.

Francis, M. K., Wormington, S. V., & Hulleman, C. (2019). The costs of online learning: Examining differences in motivation and academic outcomes in online and face-to-face community college developmental mathematics courses. *Frontiers in Psychology, 10*. doi:10.3389/fpsyg.2019.02054

Gallager, S., & Palmer, J. (2020, September 29). *The pandemic pushed universities online. The change was long overdue*. Harvard Business Review. https://hbr.org/2020/09/the-pandemic-pushed-universities-online-the-change-was-long-overdue

Gazza, E. A. (2022). The Experience of being a full-time academic nurse educator during the COVID-19 pandemic. *Nursing Education Perspectives*, *43*(2), 74–79. doi:10.1097/01.NEP.0000000000000933 PMID:35113079

Hammond, L. D. (2014). Strengthening clinical preparation: The holy grail of teacher education. *Peabody Journal of Education, 89*(4), 547–561. doi:10.1080/0161956X.2014.939009

Haskell-McBee, R. & Westcott, P.M. (2008). Daring to care. *Action in Teacher Education, 29*(5-6), 49-65. doi:10.1080/01626620.2008.10519434

Hicks, M. (2022, February 7). *Opinion: Some surprising lessons from COVID learning loss.* Indiana Economic Digest. https://indianaeconomicdigest.net/MobileContent/Default/Major-Indiana-News/Article/OPINION-Some-surprising-lessons-from-COVID-learning-loss/-3/5308/109729

Hoerr, T. (2017). *The formative five: Fostering grit, empathy, and other success skills every student needs.* ASCD.

Indiana Department of Education. (n.d.). *Indiana educator standards.* https://www.in.gov/doe/educators/educator-licensing/indiana-educator-standards/

Judge, D. S., & Murray, B. (2017). Student and faculty transition to a new online learning management system. *Teaching and Learning in Nursing, 12*(4), 277–280. doi:10.1016/j.teln.2017.06.010

Lawson, H. (2021, January 15). *Indiana graduation rates increase in 2020.* IN.gov. https://www.in.gov/doe/about/news/january-15/#:~:text=Indiana%27s%20federal%20graduation%20rate%20for%202019%20was%2 086.46%20percent

Lawson, H. (2021, December 19). *Nearly 40% of students graduate with honors.* IN.gov. https://www.in.gov/doe/about/news/indiana-department-of-education-announces-2021-graduation-rate/

Loughlin, S. (2018, October 30). More schools in Indiana adopt elearning. *Tribune-Star.*

Map: Coronavirus and school closures. (2020, March 6). *EdWeek.* Retrieved March 30, 2022, from https://www.edweek.org/leadership/map-coronavirus-and-school-closures-in-2019-2020/2020/03

Marcus, M. J. (2002). Indiana makes progress on educational attainment gap. *Indiana Business Review, 77*(2).

Miller, K. E. (2021). A light in students' lives: K-12 teachers' experiences (re) building caring relationships during remote learning. *Online Learning, 25*(1), 115–134. doi:10.24059/olj.v25i1.2486

NIVSTATS. (n.d.). *Indiana colleges student population.* Retrieved March 30, 2022, from https://www.univstats.com/states/indiana/student-population/

Noddings, N. (1988). An ethic of caring and Its implications for instructional arrangements. *American Journal of Education, 96*(2), 215–230. doi:10.1086/443894

Noddings, N. (2012). The caring relation in teaching. *Oxford Review of Education, 38*(6), 771–781. doi:10.1080/03054985.2012.745047

Park, S., & Yun, H. (2017). The influence of motivational regulation strategies on online students' behavioral, emotional, and cognitive engagement. *American Journal of Distance Education, 32*(1), 43–56. doi:10.1080/08923647.2018.1412738

Paykamian, B. (2021, March 29). *What's driving growth in the ed-tech market?* Government Technology. https://www.govtech.com/education/k-12/whats-driving-growth-in-the-ed-tech-market.html

Rogers, C. R. (1975). Empathic: An unappreciated way of being. *The Counseling Psychologist, 5*(2), 2–10. doi:10.1177/001100007500500202

Rogers. C.R. (1959). A theory of therapy, personality and interpersonal relationships as developed in the client-centered framework. In S. Koch (Ed.), *Psychology: A study of a science, vol III. Formulations of the person and the social context.* New York: McGraw Hill.

Rohatgi, K. (2021). Building resilience in nursing students during the Pandemic. *Journal of Teaching and Learning with Technology, 10*, 58–63.

Sahin, I., & Shelley, M. (Eds.). (2020). *Educational practices during the COVID-19 viral outbreak: International Perspectives.* ISTES Organization.

Santi, E., & Gorghiu, G. (2017). The student-centered learning model in John Dewey's progressive conception. *Studia Universitatis Babeş-Bolyai Psychologia-Paedagogia., 62*(2), 77–86. doi:10.24193ubbpsyped.2017.2.04

School responses in Indiana to the coronavirus (COVID-19) pandemic. (n.d.). *Ballotpedia.* https://www.loc.gov/item/lcwaN0016948/

Software Suggest. (2020, January 5). *8 Types of Health Information Technology & Healthcare Software System.* https://www.softwaresuggest.com/blog/types-of-health-information-technology-and-healthcare-software/#

Ulenaers, D., Grosemans, J., Schrooten, W., & Bergs, J. (2021). Clinical placement experience of nursing students during the COVID-19 pandemic: A cross-sectional study. *Nurse Education Today, 99*. Advance online publication. doi:10.1016/j.nedt.2021.104746

Vygotsky, L. S. (1978). *Mind in society: The development of higher psychological processes*. Harvard University Press.

Whaley, C. M., Pera, M. F., Cantor, J., Chang, J., Velasco, J., Hagg, H. K., Sood, N., & Bravata, D. M. (2020). Changes in health services use among commercially insured US populations during the COVID-19 pandemic. *JAMA Network Open, 3*(11), e2024984. doi:10.1001/jamanetworkopen.2020.24984 PMID:33151319

Xu, D., & Jaggars, S. S. (2013). The impact of online learning on students' course outcomes: Evidence from a large community and technical college system. *Economics of Education Review, 37*, 46–57. doi:10.1016/j.econedurev.2013.08.001

ENDNOTES

1 https://www.usi.edu/news/releases/2020/03/march-12-covid-19-update/

2 https://news.iu.edu/stories/2020/03/iu/releases/15-extends-spring-break-remote-teaching-remainder-of-semester.html

3 https://www.evansville.edu/coronavirus/update-2020-03-11.cfm

4 https://www.hanover.edu/about/news/2020/hanovers-12-day-transformation/

5 https://franklincollege.edu/coronavirus/coronavirus-update-april-23/

6 https://www.nwitimes.com/news/education/ivy-tech-to-extend-spring-break-move-classes-to-virtual-learning-in-coronavirus-response/article_865b6ccf-1f98-5cfa-9d0b-55fb73c007e5.html

7 https://www.learner.org/

8 https://www.teachingchannel.com/

9 https://www.readingrockets.org/

10 www.youtube.com

11 https://www.purdueglobal.edu/

12 https://online.iu.edu/

13 https://online.notredamecollege.edu/

14 https://online.butler.edu/

15 https://www.bsu.edu/academics/collegesanddepartments/online

16 https://www.qualitymatters.org/qa-resources/rubric-standards

Chapter 10
Physician Perspectives Regarding the Use of Electronic Health Records in Public Health Disease Reporting:
COVID-19 Reflections

Michelle Stewart
https://orcid.org/0000-0002-1700-3553
Western Governor's University, USA

ABSTRACT

The collection of timely and accurate healthcare data is imperative to successful disease reporting in public health. The devastating effects of fragmented and inaccurate disease reporting became readily apparent during COVID-19. The purpose of this chapter is to review public health disease reporting practices for consideration of incorporation of electronic health records into disease reporting. The perspectives of physicians were gathered through a case study focused on disease reporting and interoperability with electronic health records. Findings from this chapter may prove useful in public health policy review and for decision-making in meaningful use legislation and health information technology integration. Findings from this research may also serve as the catalyst for additional research focused on public health industry perspectives, various disease reporting mandates, and the incorporation of health information technology into public health disease reporting processes.

DOI: 10.4018/978-1-6684-4358-3.ch010

INTRODUCTION

Physicians and facilities are responsible for reporting communicable and chronic diseases to the appropriate public health agencies. Despite the existence of reporting processes since the late 1800s, the reporting methods vary greatly, which often leads to significant gaps in data and information that can be used to facilitate public health activities (CDC, 2017; Wu, 2014). Reporting public health data and information sharing are critical components in improving national health outcomes. This information is used to develop healthcare policies and foster initiatives to enhance population health (Herwehe et al., 2012). Recent reflections on the progression of the COVID-19 pandemic revealed many of the disastrous effects that delays in the transmission of critical public health information can have on overall population health.

Passage of a variety of health care initiatives has increased the opportunity to incorporate electronic health records (EHRs) into patient care and public health activities nationwide. Implementation of Meaningful Use guidelines and financial incentives by the Centers for Medicare and Medicaid Services (CMS) has led to increased collaboration between electronic health systems and public health agencies (Calman, Hauser, Lurio, Wu, & Pichardo, 2012; CMS, 2013; Dixon et al., 2017; Wu, 2014). Despite these regulations and the potential for timely and accurate public health information to inform population health activities, many physicians still report disease information through manual methods (Birkhead, 2017; Mandl, 2014; Richards et al., 2017). Manual reporting of communicable disease data has proved problematic, as it has led to significant gaps in collected data and underreporting of public health information as a whole (CDC, 2017; Markiewicz et al., 2012). These lapses in information collection make it difficult to identify disease trends and public health emergencies, which inhibit activities aimed at proactively mitigating public health risks.

This chapter aims to review public health disease reporting practices for future consideration of incorporating electronic health records into public health syndromic surveillance. The objective will be accomplished by reviewing a case study that focuses on physicians' perspectives as they relate to the use of electronic health records in mandatory public health disease reporting. Implications for leaders across the health care and public health continuums will be explored, and future considerations for research will be discussed.

BACKGROUND

Public Health Surveillance

Public health surveillance is defined as "the systemic and routine collection of data to help identify and monitor health risks over time" (Scutchfield & Keck, 2009, p. 796). Disease reporting is a critical component of public health surveillance functions. According to Smith, Hadler, Stanbury, Rolfs, and Hopkins (2013), public health surveillance aims to provide information that can be used for health action and to guide public health policy and programs. To that end, information from disease reporting is used to detect epidemics, detect health problems, and estimate the scope and magnitude of these results. Information derived from public health disease reporting can also identify and measure trends to characterize diseases and outbreaks in the population.

According to the CDC (2014), there are two surveillance categories: active and passive. Active surveillance involves health agencies initiating contact with the health care providers for disease reports and is often more complete than passive surveillance reports. This type of reporting is commonly used in situations where outbreaks have occurred, and information must be disseminated to the health care community. Passive surveillance is the most common type of disease surveillance, and it occurs when health care providers initiate and transmit reports to local health departments and public health agencies (CDC, 2014). Individual states have enacted various legislation that designates a list of diseases that health care providers and facilities must report to the state (LaMorte, 2017). Examples of passive disease surveillance are physician-initiated reporting of influenza-like illness cases (ILI), positive coronavirus tests, and sexually transmitted illness diagnoses. There have been noted disadvantages to passive surveillance, including incomplete reports and variability in reporting notifiable diseases (CDC, 2014; LaMorte, 2017).

Processes used to report diseases to public health agencies vary among health care facilities. Manual methods involve completing paper forms followed by transmission to public health agencies through a scan, email, fax, and courier (Richards et al., 2017). Electronic forms are also used in manual processes and are often printed out or emailed. Documents are initiated and completed by the health care provider and then sent to the appropriate public health agencies. These methods are categorized as passive reporting.

Electronic methods of reporting to public health are also used for disease reporting. Electronic laboratory reporting and electronic health records are health information technologies that are increasingly used to transmit relevant disease reports to public health agencies (Wu, 2014). The degree of interoperability of electronic methods with public health agencies varies greatly and continues to progress (Richards et al.,

2017). Potential for automated disease reporting to public health agencies through interoperability with electronic health records exist, and a few states are piloting these systems.

Importance of Public Health Disease Reporting

Disease reporting is a critical public health activity that requires timely action to elicit appropriate responses to maintain population health. Over the years, public health agencies have broadened their foci to chronic illnesses; however, contagious diseases continue to be closely monitored through public health surveillance processes to protect population health (Calman et al., 2012). Communicable disease outbreaks and epidemics often reinforce the need for continuous monitoring and reporting, and pandemics such as a measles outbreak reported in ten states in 2019 and the worldwide COVID-19 pandemic.

An additional concern is that reportable disease conditions are historically underreported due to a lack of knowledge associated with the reporting process, such as what to report, to whom, and how (Klompas et al., 2012; Birkhead, Klompas, & Shaw, 2015; Konowitz, Petrossian, & Rose, 1984). Requirements for reporting vary significantly among states, which leads to confusing and complex reporting requirements (DeSalvo et al., 2021). During pandemics, health care resources are focused on saving patients versus navigating complex reporting systems and requirements.

Physician's Role in Disease Reporting for Public Health

Physicians have direct contact with patients during health care encounters and are often seen as the initial collector of data that can be shared with public health agencies. The onus of precise reporting requirements and disease reporting information falls on the public health agencies, and physicians can forward the necessary information to health departments and government agencies (IOM, 2007). Public health agencies disseminate information to physicians in the case of an outbreak, epidemic, or pandemic that may pose risks to population health. A strong collaborative relationship between health care providers and public health agencies is critical to maintaining a healthy community and population.

Physicians are also challenged with aligning with the requirements of their public health departments. Governance and operations of public health functions are distributed across 59 state and territorial health departments (DeSalvo et al., 2021). Public health reporting mandates vary by individual state, and each state has drafted legislation to address the need for reporting communicable diseases (CDC, 2017; Revere et al., 2017). The complex nature of reporting requirements and guidelines can lead to a shared lack of reporting or inaccurate data. The lack of large-scale coordination to share data leads to missed opportunities for public health actions (Madhavan et al., 2021).

USING ELECTRONIC HEALTH RECORDS TO STREAMLINE DISEASE REPORTING

There are emerging technologies that now allow for direct integration and automated disease reporting, such as the Electronic Medical Record Support for Public Health (ESP) system, Electronic Case Reporting System (eCR), and the system created during the Primary Care Information Project; however, they are not widely used among health care and public health organizations (Calman et al., 2012; Klompas et al., 2012; Richards et al., 2017).

The Electronic Medical Record Support for Public Health

The Electronic Medical Record Support (ESP) system is an EHR-based public health platform that facilitates automated reporting of communicable diseases. It was developed through a collaborative project funded by the Centers for Disease Control and Prevention (CDC). The Harvard Center of Excellence in Public Health Informatics and the Massachusetts Department of Public Health developed the platform (Klompas et al., 2012). ESP scans EHRs, extracts information related to public health, and directly transmits this information to the appropriate public health agency (Klompas et al., 2012). It is compatible with various EHR systems and public health agencies.

Health care providers do not manually analyze cases for information of public health importance, as this is done through the use of detailed algorithms inherent in the platform. This extraction method also allows for the collection of relevant demographic data that can assist in identifying trends among populations that may warrant public health intervention. This relatively new technology is currently only operational in three states: Massachusetts, Ohio, and Texas (Klompas et al., 2012).

Massachusetts Department of Public Health Network (MDPHnet)

The state of Massachusetts has developed and implemented a distributed health data network designed to utilize EHR information for disease reporting, known as the Massachusetts Department of Public Health Network (MDPHnet). MDPHnet utilizes complex algorithms to identify diseases and illnesses like influenza-like symptoms and diabetes. Using these algorithms, the public health department can utilize standard International Classification of Diseases codes (ICD) or a prefabricated drop-down list that consists of high-priority conditions. The system permits stratification of the results by demographics, time, and ICD codes.

The software is a combination of two systems used in other areas of the country for electronic reporting: Electronic Medical Record Support for Public Health (ESPnet) and PopMedNet. The system is unique in that it allows for the identification of reportable diseases to public health agencies while protecting patients' personally identifiable information. Health care facilities can view what information is being queried and either approve or deny the release of information. In addition, it reduces the need for self-initiation of most reports by health care providers as the public health department can query the necessary information for public and population health activities (Vogel, Brown, Land, Platt, & Klompas, 2014).

New York City Department of Health and Mental Hygiene (NY DOHMH)

New York City's Department of Health and Mental Hygiene is one of the world's largest public health agencies (Calman et al., 2012). In 2003, NY DOHMH partnered with over 2500 primary health care providers to facilitate public health reporting through a Primary Care Information Project (PCIP). This project allowed for automatic public health data exchange between primary care providers working in the Institute for Family Health and the NY DOHMH, thus making the Institute for Family Health one of the first ambulatory clinics to participate in surveillance reporting through the use of electronic health records.

From the success of the PCIP, the NYC Macroscope was established (Kaye et al., 2014). NYC Macroscope is the first municipal chronic disease EHR surveillance system developed in the United States. Lessons learned from the implementation include the need for standardized language in data collection and reporting, leadership buy-in, and consideration of restrictions on primary care physicians to manage the information in the initial stages (Newton-Dame et al., 2016).

Electronic Case Reporting

Electronic Case Reporting (eCR) involves the automated generation of case reports followed by automated transmission of the reports to the appropriate public health agency (Mac Kenzie et al., 2016). A partnership led by the Robert Woods Johnson Foundation known as Digital Bridge aims to bring high-level leaders from health care, specialists in EHRs, and public health agencies together with a goal of implementing eCR among health care systems (Cooney et al., 2018). In the pilot, the objectives focus on integrating codes associated with reportable diseases into EHR systems, having the EHR systems automatically generate case reports, transfer of the data to a cloud-based system, and eCR to increase its adoption and implementation among health care facilities and public health agencies and to fulfill the objectives outlined in Meaningful Use legislation.

CASE STUDY OF PHYSICIAN PERSPECTIVES

Diseases meeting the reporting criteria under public health surveillance are underreported (Gibbons et al., 2014; Markiewicz et al., 2012). There is strong evidence of chronic underreporting of communicable diseases throughout the literature (Black, Hulkower, Suarez, Patel & Elliott, 2019; Calman et al., 2012; Markiewicz et al., 2012; Revere, Hills, Dixon, Gibson & Grannis, 2017). Physicians are mandated through state legislation to report a selective list of diagnoses to public health agencies for surveillance. Despite the mandate, the reporting compliance rate varies between 9 and 99 percent (Shapiro et al., 2011; Mac Kenzie et al., 2016). Many variables contribute to the underreporting of diseases at the health care level. For example, physicians are often unaware of reporting processes, lack reporting capabilities, or consider disease reporting a time burden (Gibbons et al., 2014; Richards, 2017).

EHRs have the potential to facilitate public health disease reporting, yet physicians still report information to public health agencies through traditional passive methods resulting in laborious work and requiring initiation by physicians (Birkhead, Klompas & Shah, 2015; Calman et al., 2012; CDC, 2017; Mandl, 2014). In addition, physicians are inundated with daily patient care activities and other administrative tasks that may hinder the disease reporting process (Janati, Hosseiny, Gouya, Moradi & Ghaderi, 2015; Richards et al., 2017). Revere et al. (2017) stated that physicians have reported that lack of awareness of reporting protocol, lack of awareness of legal mandates for disease reporting, lack of significant penalties for reporting noncompliance, and the need to protect patient privacy are all perceived barriers to reporting disease occurrences to the appropriate public health agencies.

Methods

An exploratory case study explored physicians' perceptions of and processes in reporting disease information to public health agencies. Twelve physicians who reported diseases to public health agencies and possessed at least one year of experience utilizing electronic health records were recruited to participate in the study. Three research questions focused on perceptions of processes related to mandated disease processes and descriptions of policies and processes used in mandated disease reporting to public health were used to guide this study:

R1. What are physicians' perceptions of current processes related to mandated disease reporting to public health agencies?
R2. Based on physicians' responses and documented state public health reporting processes, what are current organizational processes used to report mandated disease to public health agencies?
R3. Based on physicians' responses and documented state public health reporting policies, what are current organizational policies used to perform mandated disease reporting to public health agencies?

Data were collected through online questionnaires with a representative sample. The overarching goal of the questionnaire was to collect data on the perspectives of physicians as they related to the processes of public health disease reporting and their perspectives for potential automated EHR public health disease reporting. A vetted audience panel obtained from Qualtrics consisting of twelve physicians who reported diseases to public health agencies participated in the study. The nature of the study involved anonymous responses, and the participant's state of medical practice was the only demographic information collected for use in triangulation with published public health disease reporting practices in identified states. Five respondents resided in Texas, three in California, one in Michigan, one in Wisconsin, one in Florida, and one in New York. Table 1 was used to align participants with their state of residence for clarity in data analysis.

Table 1. Summary of Participant's State of Medical Practice

Respondent	State of Medical Practice
01	Texas
02	California
03	Michigan
04	Wisconsin
05	California
06	Florida
07	Texas
08	Texas
09	Texas
10	Texas
11	California
12	New York

Electronic questionnaires consisting of five selection questions and eight free text questions were used during collection. The first question consisted of an Informed Consent Form in which the participants were presented with details of the study and the options to accept or decline the terms of the study. The following three questions focused on the eligibility criteria of the participants. Skip logic was employed to ensure only eligible participants engaged in the questionnaire based on participant responses to the individual eligibility questions and informed consent acceptance. The final selection question involved the participants' selection of their state of medical practice. This question was intended to gather demographic information for further research into individual state public health reporting practices. An extensive review of publicly available policies and documents related to public health disease reporting was conducted to achieve the study's goals.

Theoretical Considerations

The study aligned with two theories related to data collected during the study: the Classical Diffusion of Innovation Theory (DOI) (Rogers, 1983) and the Technology Acceptance Model (TAM) (Davis, 1989). The theories served as references during the analysis of participant responses. Both assisted in determining challenges and opportunities of public health disease reporting using EHRs from physician perspectives.

Classical Diffusion of Innovation Theory (Rogers, 1983) is a social science theory with a solid history across disciplines involving the diffusion of innovations. Rogers (1983) described diffusion as the process through which innovations are communicated through social systems and the involvement of new ideas. Rogers (1983) used DOI to posit that an individual's perception of the innovations will predict the actual decision and rate of adoption. Rates of adoption are further categorized into adopter categories.

Critical components of DOI (Rogers, 1983) are the innovation, the adopter, and the social system associated with the innovation (Dearing, 2009). Innovators are the first to adopt an innovation, while early adopters implement the systems following the experiences of innovators. Rogers (1983) recommended additional research of innovations that focused on the acceptability of innovation prior to diffusion while it is being test marketed. This study aimed to explore processes and perspectives related to accepting and adopting new health information technology processes focused on disease reporting in public health.

The Technology Acceptance Model (TAM) (Davis, 1989) was also considered when identifying themes and interpreting the study results. This model has been extended in many different contexts; however, the core of the model posits that individual intention to use a particular technology is based on perceived usefulness and perceived ease of use (Davis, 1989). Perceived usefulness refers to the degree to which an individual believes the technology will enhance their job performance. Perceived ease of use is based on the degree to which the user feels that a system would be free of physical or mental effort. According to Cooper & Zmund (1990), the six stages that must also be considered when diffusing information technology into the organizational environment are initiation, adoption, adaptation, acceptance, routinization, and infusion. The use of TAM is beneficial in determining the acceptance of health information technologies and helped interpret physician perspectives related to the use of EHRs in processes of disease reporting to public health agencies.

Data Analysis

Data analysis included the use of Braun and Clarke's Six-Phase Thematic Analysis Framework. The six stages of the framework include familiarization with data, generation of initial codes, searching for themes, reviewing themes, defining themes, and producing the report. The researcher collected data from an anonymous questionnaire through the Qualtrics platform utilizing an audience panel during familiarization. Attention was given to each response, and the researcher actively read and re-read the responses searching for semantic and latent meanings.

The generation of codes was performed through manual processes. First pass coding focused on each question. Responses to each of the eight open-ended questions from each participant were coded and categorized using words, phrases, and sentences. Word and phrase frequencies were noted during coding, and data saturation was reached as there were no original themes or patterns to review. During the second pass of coding, the researcher examined questions in groups as they aligned with the study's research questions. Questions 7 and 8 were examined together as they related to the processes used and the perspectives on those processes and aligned with RQ 1. Question 9 was reviewed alone as it focused on the ideal workflow process. Questions 11 and 12 were analyzed together as they focused on the roles and responsibilities of the reporters and their perspectives of those roles and aligned with RQ2. Questions 8 and 13 were used to examine the perspectives and impacts of using integrated EHR systems to report to public health agencies automatically. Question 14 was an opportunity for respondents to provide additional information related to processes and policies aligned with RQ3. During the third pass of coding, a theory-driven approach was utilized. Based on the individual responses, coding was used to determine alignment with the Technology Acceptance Model (TAM) components and the Classical Diffusion of Innovations Theory (DOI).

During phases three thru six, overarching themes were discovered and defined. Several subthemes were noted that aligned with four major themes. These were related to perceptions of reporting processes, perceptions of roles in the reporting processes, and privacy concerns with automation. The researcher observed four themes that focused on public health reporting processes and perceptions, automation of public health disease reporting, technology acceptance in accordance with TAM, and adopter categories that align with DOI.

RESULTS AND FINDINGS

Theme 1: Public Health Disease Reporting Processes Are Problematic and Inefficient for Physician Participants

Six of the twelve participants reported that reporting processes are confusing, burdensome, and time-consuming. Two respondents noted mixed perceptions, with Respondent 06 stating that processes "are currently easy but should be easier," and Respondent 09 reported the processes are "slow but overall efficient ."Of the two respondents who reported positive feedback related to the reporting process, Respondent 01 denied being directly involved in the reporting process. Respondent 03 reported utilizing staff assistance to fill out forms and submit them to public health. Two of the participants answered the question in the context of reporting process descriptions. These responses were moved to the question related to the description of organizational processes for continued analysis.

Reporting Processes

Upon further inquiry into the actual description of the organizational processes used to report diseases to public health, seven respondents reported that the physician is ultimately responsible for identifying mandated reportable diseases, reviewing and completing forms, and submitting them to the appropriate public agency. Respondent 06, Respondent 07, and Respondent 12 stated that some reporting involves actual phone calls to appropriate public health agencies. Respondent 03, Respondent 07, Respondent 09, and Respondent 11 stated that paper forms are used in the reporting process. Respondent 05 and Respondent 06 note that forms are often filled out and submitted through electronic means. Respondent 01 stated that the organizational process consists of a physician review of positive results with the reporting tasks performed by axillary staff. Three respondents stated a lack of standard organizational processes and policies and that the state mandates these processes.

Perceptions of Roles and Responsibilities

All twelve respondents reported some responsibility in mandated public health disease reporting. Upon examining perceptions related to these roles and responsibilities, nine respondents indicated a negative perception, one respondent had mixed perspectives, and two respondents appeared neutral in their responses. Of the nine respondents who reported a negative perception of their reporting role, the common themes were that the processes were inefficient, time-consuming, and disruptive to the workflow. Respondent 08 reported diverse perspectives in that "most of the current process have some automation which is enjoyable; however, the manually reported aspects are time-consuming."

Theme 2: The Automation of Public Health Disease Reporting Processes Is Acceptable by Physician Participants

Ten of twelve respondents believed that integrated EHRs in automated public health disease reporting would streamline work processes and allow for improved data and reporting compliance.

Physician Involvement

Although most respondents agreed with automation through integrated EHRs, six of the twelve physician respondents highlighted a desire to be involved in the disease reporting submission process. Respondent 01, Respondent 02, Respondent 03, Respondent 04, Respondent 07, and Respondent 08 described the ideal process as involving automated identification of the reportable disease, completion and collection of the appropriate information for mandated reporting through the EHR, verification of information, and then submitted by the physician through the use of an automated button within the EHR system.

Privacy Concerns

Despite agreeing that EHRs would be helpful in the public health disease reporting process, three of twelve participants stated concern over the privacy of patient health information by using automation. Respondent 02 reported being "concerned about the health privacy of it being completely automated," which aligned with a previous response indicating a desire to be involved in the verification and submission process of the public health disease report. Respondent 03 revealed being "concerned about a centralized process...would be likely a major data breach waiting to happen", while Respondent 09 stated that the system "would have to be HIPAA compliant."

Theme 3: Based on Components of the Technology Acceptance Model, Physician Participants Have Positive Perceptions Related to Integrated Electronic Health Record Public Health Disease Reporting and Perceive the Process as Easy to Use and Useful in Reporting Practices

Three facets of TAM were used to analyze responses within the questionnaire: attitude towards the technology, perceived ease of use, and perceived usefulness. Nine of the twelve respondents indicated having a positive attitude regarding using technology to enhance public health disease reporting processes. Common responses that indicated a positive attitude towards technology use included "streamlining of work processes," "quick and easy," and "easier with improved efficiency ."Respondent 02, Respondent 03, and Respondent 05 presented diverse perspectives on using technology to report diseases to public health. Respondent 02 conceded that using EHRs to report to public health automatically would result in better resource allocation; however, presented concerns about patient privacy throughout the process. Respondent 03 envisioned the ideal public health disease reporting processes as automatic extraction from EHRs. However, it maintained that this type of technology would not work well with the current system used within the organization. Respondent 05 also reported privacy concerns with the use of technology but reported the ideal reporting process as involving telehealth and EHR use.

Perceived ease of use was drawn through latent analysis of the responses. Based on the researcher's analysis, nine of the twelve respondents perceived integrated EHRs in automated public health disease reporting as easy to use. Respondent 07 stated that it would "decrease physician burden and reduce paperwork ."Eight of twelve respondents reported that the ease of automated EHR public health disease reporting would streamline processes. Two of the twelve respondents appeared to have both positive and negative perspectives on the ease of use involving technology. Respondent 03 reported that EHR use would be "seamless" and would enhance the process but indicated that it might be challenging to maintain the privacy of patient information.

Respondent 06's response was not considered in the analysis as the researcher could not make a connection between the response and the perceived usefulness concept. Regarding perceived usefulness, eleven of twelve respondents reported that EHRs and technology in the public health disease reporting process would enhance performance. Eleven of twelve believed that incorporating EHR systems for direct public health reporting would enhance compliance and improve data collection. The overarching theme of the eleven respondents was that the use of integrated EHRs for automatic public health disease reporting would enhance performance and thus is perceived as a helpful technology.

Theme 4: Based on Components of Classical Diffusion of Innovations, Most Physician Respondents Are Receptive to the Inclusion of Technology Innovations in the Public Health Disease Reporting Process

When asked about the vision of the ideal public health disease reporting process, eleven of twelve respondents stated that automation of public health reporting using the EHR was desired. These responses were further coded based on the adopter categories inherent in Roger's Classical Diffusion of Innovation Theory. The first category combined Innovators and Early Adopters and was defined as those physicians who envisioned the ideal workflow as completely automated without physician involvement. A total of five respondents were considered to be Innovators and Early Adopters. Respondent 05, Respondent 09, Respondent 10, Respondent 11, and Respondent 12 indicated that automatic extraction of the reportable disease from the EHR with subsequent submission to the appropriate public health agency without physician involvement would be ideal. Respondent 12 recommended incorporating ICD-10 codes to assist in identifying the disease reporting burden from the EHR.

Six respondents who supported the use of the EHR for disease reporting were classified as Early and Late Majority adopters. Although the respondents supported the use of automation and EHRs in reporting processes, they all preferred to have some involvement in the information sent to the public health agencies. Respondent 01, Respondent 02, Respondent 04, Respondent 07, and Respondent 08 all recommended using a verification button to allow for initial review and, with approval, subsequent reporting to the public agency.

Respondent 06 did not envision technology as part of the ideal reporting process and was the sole laggard based on the adopter categories. One respondent did not perceive technology as necessary in reporting public health diseases. The respondent reported that the ideal process involved the patient filling out the form with the physician reviewing and submitting it to the appropriate public health agency.

Public Health Disease Reporting Processes and Policies by State

To achieve triangulation, public health disease reporting processes and policies for each state of medical practice of physician participants were reviewed. The study participants reported practicing medicine in six distinct states. Information related to each of the represented states is presented below in order of greatest participants to least.

Texas

The researcher obtained information from the Texas Department of Health and Human Services' publicly available website and discovered Texas mandates the reporting of approximately 88 individual diseases (TDHHS, 2020). The list of reportable conditions can be found in a PDF available to the public and mandated reporters. Within the PDF, there are links to the forms used for reporting, the contact phone number for immediate reporting, and a list of 25 additional footnotes for consideration when reporting. The primary source of reporting is based on a form where physicians fill out patient demographic information, the physician's contact information, diagnostic codes of the reportable disease, and a synopsis of the reportable condition that has been identified.

The responses to the questions related to current workflow processes used in mandated disease reporting, perspectives of the reporting workflow process, the respondents' roles in reporting, and perspectives of the reported roles were analyzed based on the mandated reporting policies found in the Texas Department of Health and Human Services website (TDHHS, 2020). All five of the respondents reported organizational public health disease reporting processes that were consistent with the website's requirements. The overarching theme regarding the reporting processes and roles was that forms were filled out manually and then faxed or phoned to the appropriate public health agency. There was no mention of technology in the mandated disease reporting process for any of the Texas respondents.

Five of the five respondents reported that the organizational reporting processes are time-consuming. Respondent 01 and Respondent 10 indicated that an integrated EHR for automated public health disease reporting is preferred. Respondent 09 stated that physicians should not be responsible for reporting and that "as the years go on, more and more tasks and documentation continue to be added to the physician's workflow ."Respondent 07 indicated that remembering to report and relay information to staff on reporting can be difficult at times, which may lead to the "spread of the disease, and it may not be reported ."The responses of the five respondents aligned with Theme 1 that public health disease reporting processes are problematic and inefficient for physician participants. With Theme 2, the automation of public health disease reporting processes is acceptable to physician participants.

California

According to the California Department of Public Health website, 88 diseases must be reported if discovered (CDPH, 2020). The time range for reporting varies from immediate to within seven days of discovery and includes the use of 88 individual forms listed for the health care provider to choose from and fill out. Electronic submission of public health reporting in California is available through an electronic surveillance reporting system known as the California Reportable Disease Information Exchange (CalREDIE). There is an electronic Case Reporting (eCR) system within this electronic system that allows for the submission of reportable diseases to public health agencies through automated extraction from the provider's EHR system (CDPH, 2020).

Respondent 02, Respondent 05, and Respondent 11 reported practicing in California, and two of the three respondents reported using technology in a small role to assist with public health disease reporting. Respondent 05 indicated a tool embedded in the organization's EHR that assists in populating some of the public health forms. Despite this added feature, the respondent stated, "public health agencies forms are usually too long…making the workflow quite formidable". Respondent 11 stated that the organization's current EHR system would provide a warning if a diagnosis may be a potential reportable disease. Respondent 02 did not report using technology in public health disease reporting. It stated that the physician must "report the same condition to multiple people, who do not necessarily talk to each other ." All three respondents alluded to a negative perception of public health disease reporting practices.

Florida

Florida mandates the reporting of 94 communicable diseases (FDOH, 2017). The manual method of reporting includes the Practitioner Disease Reporting Form that lists each of the reportable diseases with the timeframes for reporting. These completed reports must be sent to the appropriate public health agencies and may be submitted electronically through a link provided directly on the Practitioner Disease Reporting Form. The website also provides instructions to contact local health departments for any questions related to the reporting of disease (FDOH, 2017).

Respondent 06 reported practicing in Florida and described the public health disease reporting process as "currently easy but should be made easier ."The organizational process consisted of referencing a list of diseases that must be reported and then computer or phone transmission of the report to public health. Based on the provided responses, it was difficult to ascertain the extent of technology use in the public health disease reporting process beyond the mention of computer submission of reports.

Michigan

According to the Department of Public Health Website in Michigan, there are approximately 76 diseases that are reportable to local public health departments

(MDHHS, 2020). A list of forms used for reporting could not be located on the website; however, a synopsis of the required information was found. There was a note to indicate that specific local health departments may require additional information. Michigan also utilizes a web-based communicable disease reporting system known as the Michigan Disease Surveillance System (MDSS) (MDHHS, 2020).

Despite the presence of the electronic reporting system, Respondent 03 described the organizational process as the manual completion of public health forms, followed by faxing and calling. The respondent initially described the process as "fairly simple and quick" but later stated that the ideal process would be automated.

New York

The state of New York lists approximately 76 total diseases that must be reported to public health agencies (NYDOH, 2020). Thirty-one out of the 76 diseases require an immediate phone call to report using the 24/7 Provider Access Line (PAL). The maximum time frame for all other diagnoses is 24 hours. A single form from the New York State Division of Epidemiology is used to report diseases through the manual method. Electronic submission of diseases that do not require immediate reporting can be accomplished through a portal entitled NYCMED. Providers must register for an account to be eligible for electronic submission (NYDOH, 2020).

The respondent from New York reported that the organizational process used in public health disease reporting was cumbersome and "requires physically calling the public health agency ."Respondent 12 also stated that this responsibility can be easy to forget during a busy workday and that "an electronic submission should be possible ."The ideal process entailed "automated public health reporting through the entry of billing codes consistent with diseases requiring mandated reporting ."These statements aligned with a positive attitude toward technology, positive perceived usefulness, and the potential for the respondent to be an innovator or early adopter.

Wisconsin

Wisconsin has a list of approximately 90 mandated reportable diseases that are divided into two categories based on the required reporting timeline. (WDHS, 2020) Category I consists of approximately 27 diseases and requires an immediate phone call from the reporter with a faxed, mailed, or electronically submitted a written report. Category II consists of approximately 63 diseases required to be faxed, mailed, or electronically submitted within 72 hours of diagnosis. According to the website, Wisconsin's Electronic Disease Surveillance System (WEDSS) is used in approximately 75% of the reporting cases (WDHS, 2020).

Respondent 04 reported practicing medicine in Wisconsin and stated no specified organizational process on mandated disease reporting. No training or instruction was provided. Each "provider is left to their own regard" for reporting to public health. When asked about perspectives of the current process, the respondent indicated, "it is time-consuming and burdensome ."According to the respondent, the "confusion and burden of disease reporting often detracts from patient care and likely results in less reporting and less meaningful data ."Based on the provided responses, it is difficult to ascertain if any technology is used in the respondent's organization for mandated public health disease reporting. However, the respondent did report that EHRs are not used.

Limitations

One of the study's limitations was the use of an electronic questionnaire to solicit responses from the participants. The nature of delivery for the questionnaire did not allow the researcher to perform member-checking to obtain additional information from participants through non-verbal cues and other verbal cues that may have led to clarifying follow-up questions. Due to the pandemic crisis of COVID-19, securing permissions for research sites proved problematic for the researcher, thus leading to the use of a vetted audience panel obtained through Qualtrics.

Another limitation of the study was a limited geographical reach due to the use of a vetted audience panel. The participants were anonymous to the researcher and were recruited through Qualtrics. The responding participants came from six of fifty states, with only two states having multiple participants. Forty-four states did not have representation, and perspectives may vary from those of the participants in the study.

RELEVANCE OF FINDINGS AND IMPLICATIONS

The study's findings may be used by health care and public health industry leaders to obtain a deeper understanding of physician perspectives related to public health reporting processes and physician perspectives related to the incorporation of automated public health reporting using EHRs. Industry leaders may be able to incorporate the end-user perspectives into decision-making related to compliance with public health reporting legislation and enhanced health information technology in public health reporting practices. The overall goal is that the findings of this study will be used to improve public health processes and the quality of data related to diseases.

Physicians are primary reporters of public health data, and it is imperative to transmit complete and correct information used in public health surveillance (Abdulrahim et al., 2019; Gostin, 1996; Marier, 1977). Public health agencies rely on reported data to inform the public, develop policy, and request funding for health care initiatives (Mac Kenzie et al., 2016). Incorporation of technology into the public health disease reporting process can reduce reporting burdens, increase reporting compliance rates, and streamline workflow practices (Black et al., 2019). As the world navigates through the COVID-19 pandemic, the need for timely, accurate, and complete data is evident. Federal, state, and local public health agencies rely on this data to implement timely critical public health activities and legislation.

Information revealed in the study was instrumental in determining the status of electronic public health reporting by physicians. It may also be valuable in identifying areas for improvements in EHR interoperability with public health databases and reporting processes. Health care leaders may benefit from the results of this study through insight into physicians' perspectives as they relate to public health reporting using EHRs. Based on the Technology Acceptance Model (TAM) (Davis, 1989) and the Classical Diffusion of Innovations Theory (DOI) (Rogers, 1983), these perspectives may be used to predict the intention of use. They may also help decide how to incorporate EHRs into the disease reporting process to achieve public health reporting compliance.

Physicians must perceive the introduced technological application as easy and useful in practice (Davis, 1989; Dornan et al., 2019). These insights may be useful to health care leadership and public health professionals as health information technology are expanded and implemented.

RECOMMENDATIONS

Development of Healthcare Leadership and Electronic Health Record Vendor Workgroup

It is recommended that health care, public health, and electronic health record vendor leadership form a workgroup to develop software that allows for integrating electronic health records into public health disease reporting. Most of the participants expressed some dissatisfaction with public health reporting policies and processes as they currently exist. The reporting processes were viewed as "time-consuming," "burdensome," and "inefficient ."Several participants viewed disease reporting as a disruption to their workday and as interruptions to patient care. Most participants viewed public health disease reporting as essential to overall population health despite negative perceptions. A few participants noted that it can be challenging to remember to report and that some physicians do not report due to the burden. Accurate and timely public health reporting is critical to public health activities and decision making; thus, the lack of reporting has a potential negative population health effect (Dixon et al., 2017).

Based on these findings, the workgroup must strive to improve EHRs that allow for automated public health disease reporting in all states. The addition of this software has the potential to streamline operations and improve compliance with disease reports. During the working sessions, policies and processes must be evaluated for alignment with state mandates to determine the best approach to achieve workflow process efficiency and compliance with state mandates.

Implementation of Education and Training Initiatives

The development of organizational and educational programs focused on public health disease reporting for physicians and end-users is recommended for immediate implementation. The mandate of public health disease reporting has existed since the mid-1700s; however, there is strong evidence that a lack of education and knowledge on how to report diseases to public health has led to a consistent lack of reporting or underreporting (Birkhead et al., 2015; Dixon et al., 2017; Dornan et al., 2019). This sentiment was also present in the study findings, where one participant acknowledged that there are physicians who simply do not report due to a lack of knowledge of processes. Another participant indicated a lack of formal training and that physicians are left on their regard to decipher the public health reporting process.

The importance of public health disease reporting to overall population health can not be understated. A lack of reporting due to a lack of education can be remedied by focusing on solid education and training on reporting processes. Each state has varying policies, procedures, and diseases that must be reported, which may prove complicated for those trying to report and monitor public health disease surveillance (Danila et al., 2015; DeSalvo et al., 2020). It is recommended that leaders revisit organizational processes and policies related to public health and develop robust training programs designed to train and refresh end users on the importance of and processes involved in disease reporting to public health.

Achievement of Meaningful Use

Government leaders must mandate the achievement of Meaningful Use measures associated with public health, including incentives and consequences for noncompliance. Stage Two of Meaningful Use contains a measure associated with public health reporting. Objective Nine reinforces Stage One, Measure Ten, and requires active engagement with public health agencies to submit electronic public health data from certified EHRs, except where legislation prohibits this practice (CDC, 2016). Based on participant responses, many organizations have not made significant progress towards this goal. It is recommended that incentive payments, followed by monetary fines be implemented to catalyze compliance with this measure. If an exception must be made to an organization, it is strongly recommended that the requesting health care organization be required to demonstrate a straightforward, concise, and practical public health disease reporting process and to report compliance rates associated with public health reporting to the government.

This recommendation has the potential to enhance timeliness, accuracy, and compliance with public health disease reporting guidelines. There is strong evidence of chronic underreporting of communicable diseases throughout the literature aligned with the participant responses to complex reporting processes (Black et al., 2019; Calman et al., 2012; Markiewicz et al., 2012; Revere, Hills, Dixon, Gibson & Grannis, 2017). Concerns with public health disease reporting have been revealed during the coronavirus (COVID 19) pandemic (DeSalvo et al., 2020; Madhaven et al., 2021). According to the Congressional Research Service (2020), there have been immediate concerns about the state of the public health disease reporting surveillance systems regarding the timeliness and accuracy of data. The process of collecting vital information to monitor, manage, and mitigate coronavirus is fragmented and varied. These findings, coupled with the findings of this study, further reveal the dire need for improved and efficient public health reporting. Due to the varied nature of public health disease reporting among the states, it is recommended that the federal government work closely with local and state governments to find pathways to achieve Meaningful Use guidelines, significantly enhancing public health data and reporting processes.

FUTURE RESEARCH CONSIDERATIONS

Future research focused on exploring the perspectives and experiences of the public health professionals may prove beneficial, as they are critical components in the disease reporting process. It is recommended that research is conducted on the perspectives of public health professionals regarding disease reporting processes to ascertain the effectiveness of these processes and procedures.

Multiple health information technologies are available and are being used sporadically. It is potentially beneficial to learn from the lessons of current users to determine the next steps for future users. Additional research into organizations using integrated EHRs and other electronic public health reporting methods is also recommended. This research should also include health information technology use from a public health industry perspective to determine technology integration perspectives and outcomes. This research can be further expanded into public health systems' technological and interoperability capabilities. Significant gaps in technological infrastructure and antiquated systems proved problematic during the COVID-19 pandemic (DeSalvo et al., 2020; Madhaven et al., 2021).

As the COVID-19 scenario unfolds, continued research on public health disease reporting systems, processes, and policies is highly recommended. Studies regarding public health reporting processes among different types of health care organizations may add to the body of knowledge in health care. The need for public health disease reporting is absolute; however, there are significant variances in capturing the data. Accurate and timely data is necessary to gauge the health of the population truly and to ensure the mitigation of potential disease outbreaks and pandemics. Research on disease surveillance in public health has been sporadic, and the COVID-19 pandemic crisis highlights the need for an immediate and robust inquiry into public health practices.

CONCLUSION

Strong, solid policies and processes have become apparent during the recent COVID-19 pandemic. It was surprising and quite alarming to learn that as a nation, we have not incorporated solid disease reporting systems and policies despite mandatory guidelines for public health reporting existing since the late 1700s. The absence of these processes and policies ultimately hinders the integrity of data used to make population and public health decisions. Good public health data is essential when the world has witnessed several critical disease outbreaks that threaten the human population and overall global health.

Data collection and reporting have proven to be significant challenges during the COVID-19 pandemic (DeSalvo et al., 2020). It is disheartening that mandated reporters simply are not reporting due to antiquated, cumbersome, complex processes that place time burdens on their availability for patient care. Processes vary based on state regulations, organizational information systems, types of health care facilities, and even political forces.

Responses from participants supported assumptions that public health disease reporting processes are problematic and that EHR technology integration may greatly enhance these processes from a physician's perspective. Participants admitted their responsibilities and roles in the mandated public health disease reporting processes, although most expressed discontentment and negative perceptions of reporting practices. Public health disease reporting is an essential step in monitoring population health, preventing disease outbreaks, and overall policy decision-making (Birkhead et al., 2015). Disease reporting to public health agencies must be improved to allow for accurate, timely data, allowing for swift, deliberate responses to public health emergencies. This study, coupled with the current coronavirus pandemic, are catalysts for further research into the enhancement and improvement of public health disease reporting practices among all health care organizations.

REFERENCES

Abdulrahim, N., Alasasfeh, I., Khader, Y. S., & Iblan, I. (2019). Knowledge, awareness, and compliance with disease surveillance and notification among Jordanian physicians in residency programs. *The Journal of Health Care Organization, Provision, and Financing Inquiry, 56*. Advance online publication. doi:10.1177/0046958019856508 PMID:31220967

Beglaryan, M., Petrosyan, V., & Bunker, E. (2017). Development of a tripolar model of technology acceptance: Hospital-based physicians' perspective on EHR. *International Journal of Medical Informatics, 102*, 50–61. doi:10.1016/j.ijmedinf.2017.02.013 PMID:28495348

Birkhead, G. (2017). Successes and continued challenges of electronic health records for chronic disease surveillance. *American Journal of Public Health, 107*(9), 1365–1366. doi:10.2105/AJPH.2017.303938 PMID:28787206

Birkhead, G., Klompas, M., & Shah, N. (2015). Public health surveillance using electronic health records: Rising potential to advance public health. *Frontiers in Public Health Services & Systems Research, 4*(5), 25–32. doi:10.13023/FPHSSR.0405.05

Black, J., Hulkower, R., Suarez, W., Patel, S., & Elliott, B. (2019). Public health surveillance: Electronic reporting as a point of reference. *The Journal of Law, Medicine & Ethics, 47*(S2), 19–22. doi:10.1177/1073110519857309 PMID:31298129

California Department of Public Health Division of Communicable Disease Control [CDPH]. (2020). *Reportable diseases and conditions.* https://www.cdph.ca.gov/Programs/CID/DCDC/Pages/Reportable-Disease-and-Conditions.aspx

Calman, N., Hauser, D., Lurio, J., Wu, W. Y., & Pichardo, M. (2012). Strengthening Public Health and Primary Care Collaboration Through Electronic Health Records. *American Journal of Public Health, 102*(11), e13–e18. doi:10.2105/AJPH.2012.301000 PMID:22994274

Centers for Disease Control and Prevention (CDC). (2014). *Introduction to Public Health. In: Public Health 101 Services. Atlanta, GA. US Department of Health and Human Services.* https://www.cdc.gov/training/publichealth101/surveillance.html

Centers for Disease Control and Prevention (CDC). (2016, October 14). Summary of Notifiable Infectious Diseases and Conditions — United States, 2014. *MMWR. Morbidity and Mortality Weekly Reports*. https://www.cdc.gov/mmwr/volumes/63/wr/mm6354a1.htm?s_cid=mm6354a1_w

Centers for Disease Control and Prevention (CDC). (2017). *Meaningful Use*. https://www.cdc.gov/ehrmeaningfuluse/introduction.html

Centers for Medicare and Medicaid Services (CMS). (2013). *Electronic health records*. https://www.cms.gov/Medicare/EHealth/EHealthRecords/index.html?redirect=/ehealthrecords/

Congressional Research Service. (2020). *COVID-19: U.S. Public Health Data and Reporting*. https://crsreports.congress.gov/product/pdf/IN/IN11361

Cooney, M. A., Iademarco, M., Huang, M., MacKenzie, W., & Davidson, A. J. (2018). The Public Health Community Platform, Electronic Case Reporting, and the Digital Bridge. *Journal of Public Health Management and Practice*, *24*(2), 185–189. doi:10.1097/PHH.0000000000000775 PMID:29360697

Cooper, R. B., & Zmud, R. W. (1990). Information technology implementation research: A technological diffusion approach. *Management Science*, *36*(2), 123–139. doi:10.1287/mnsc.36.2.123

Danila, R. N., Laine, E. S., Livingston, F., Como-Sabetti, K., Lamers, L., Johnson, K., & Barry, A. M. (2015). Legal Authority for Infectious Disease Reporting in the United States: Case Study of the 2009 H1N1 Influenza Pandemic. *American Journal of Public Health*, *105*(1), 13–18. doi:10.2105/AJPH.2014.302192 PMID:25393187

Davis, F. D. (1989). Perceived usefulness, perceived ease of use, and user acceptance of information technology. *Management Information Systems Quarterly*, *13*(3), 319–340. doi:10.2307/249008

Dearing, J. W. (2009). Applying Diffusion of Innovation Theory to Intervention Development. *Research on Social Work Practice*, *19*(5), 503–518. doi:10.1177/1049731509335569 PMID:20976022

DeSalvo, K., Hughes, B., Bassett, M., Benjamin, G., Fraser, M., Galea, S., Gracia, J. N., & Howard, J. (2021). *Public health COVID-19 impact assessment: Lessons learned and compelling needs*. National Academy of Medicine. doi:10.31478/202104c

Dixon, B., Zhang, Z., Lai, P., Kiriyik, U., Williams, J., Hills, R., Revere, D., ... Grannis, S. (2017). Completeness and timeliness of notifiable disease reporting: A comparison of laboratory and provider reports submitted to a large county health department. *BMC Medical Informatics and Decision Making*, *17*(87), 1–8. doi:10.118612911-017-0491-8 PMID:28645285

Dornan, L., Pinyopornpanish, K., Jiraporncharoen, W., Hashmi, A., Dejkriengkrikul, N., & Angkurawaranon, C. (2019). Utilization of electronic health records for public health in Asia: A review of success factors and potential challenges. *International Journal of Biomedical Research*, *2019*, 1–9. Advance online publication. doi:10.1155/2019/7341841 PMID:31360723

Florida Department of Health [FDOH]. (2017). *Disease reporting and surveillance*. https://www.floridahealth.gov/diseases-and-conditions/disease-reporting-and-management/disease-reporting-and-surveillance/index.html

Gostin, L. O. (1996). The public health information infrastructure. A national review of the law on health information privacy. *Journal of the American Medical Association*, *275*(24), 1921–1927. doi:10.1001/jama.1996.03530480063042 PMID:8648874

Herwehe, J., Wilbright, W., Abrams, A., Bergson, S., Foxhood, J., Kaiser, M., Smith, L., Xiao, K., Zapata, A., & Magnus, M. (2012). Implementation of an innovative, integrated electronic medical record (EMR) and public health information exchange for HIV/AIDS. *Journal of the American Medical Informatics Association: JAMIA*, *19*(3), 448–452. doi:10.1136/amiajnl-2011-000412 PMID:22037891

Institute of Medicine (IOM). (2007). *Training physicians for public health careers*. The National Academies Press., doi:10.17226/11915

Janati, A., Hosseiny, M., Gouya, M. M., Moradi, G., & Ghaderi, E. (2015). Communicable disease reporting systems in the world: A systematic review article. *Iranian Journal of Public Health*, *44*(11), 1453–1465. https://www.ncbi.nlm.nih.gov/pmc/articles/PMC4703224/ PMID:26744702

Kaye, K., Singer, J., Newton-Dame, R., & Shih, S. C. (2014). Health information technology and the primary care information project. *American Journal of Public Health*, *104*(3), e8–e9. doi:10.2105/AJPH.2013.301787 PMID:24432950

Klompas, M., McVetta, J., Lazarus, R., Eggleston, E., Haney, G., Kruskal, B., Yih, K., Daly, P., Oppedisano, P., Beagan, B., Lee, M., Kirby, C., Heisey-Grove, D., DeMaria, A. Jr, & Platt, R. (2012). Integrating clinical practice and public health surveillance using electronic medical record systems. *American Journal of Public Health*, *102*(S3), S325–S332. doi:10.2105/AJPH.2012.300811 PMID:22690967

Konowitz, P. M., Petrossian, G. A., & Rose, D. N. (1984). The underreporting of disease and physicians' knowledge of reporting requirements. *Public Health Reports, 99*(1), 31-5. https://pubmed.ncbi.nlm.nih.gov/6422492/

LaMorte, W. (2017). *Disease surveillance and data sources: Surveillance for infectious disease.* Retrieved from: http://sphweb.bumc.bu.edu/otlt/MPH-Modules/EP/EP713_Surveillance/EP713_Surveillance3.html#headingtaglink_1

Madhavan, S., Bastarache, L., Brown, J. S., Butte, A. J., Dorr, D. A., Embi, P. J., Friedman, C. P., Johnson, K. B., Moore, J. H., Kohane, I. S. O., Payne, P. R., Tenenbaum, J. D., Weiner, M. G., Wilcox, A. B., & Ohno-Machado, L. (2021). Use of electronic health records to support a public health response to the COVID-19 pandemic in the United States: A perspective from 15 academic medical centers. *Journal of the American Medical Informatics Association: JAMIA, 28*(2), 393–401. doi:10.1093/jamia/ocaa287 PMID:33260207

Mandl, K. D. (2014). Ebola in the United States: EHRs as a public health tool at the point of care. *Journal of the American Medical Association, 312*(23), 2499–2500. doi:10.1001/jama.2014.15064 PMID:25329170

Marier, R. (1977). The reporting of communicable diseases. *American Journal of Epidemiology, 105*(6), 587–590. doi:10.1093/oxfordjournals.aje.a112424 PMID:868863

Markiewicz, M., Bevc, C. A., Hegle, J., Horney, J. A., Davies, M., & MacDonald, P. D. M. (2012). Linking public health agencies and hospitals for improved emergency preparedness: North Carolina's public health epidemiologist program. *BMC Public Health, 12*(1), 141. doi:10.1186/1471-2458-12-141 PMID:22361231

Massachusetts Technology Collaborative. (2012). *MeHI introduces new public health surveillance tool.* https://www/masstech.org

Michigan Department of Health and Human Services [MDHHS]. (2020). *Communicable disease reporting in Michigan.* https://www.michigan.gov/mdhhs/0,5885,7-339-71550_5104_31274-12538--,00.html

Morilla, M., Sans, M., Casasa, A., & Gimenez, N. (2017). Implementing technology in healthcare: Insights from physicians. *BMC Medical Informatics and Decision Making, 17*(92), 92. Advance online publication. doi:10.118612911-017-0489-2 PMID:28655299

New York State Department of Health [NYDOH]. (2020). *Diseases & Conditions: Information About Diseases, Viruses, Conditions, and Prevention.* https://health.ny.gov

Newton-Dame, R., McVeigh, K. H., Schreibstein, L., Perlman, S., Lurie-Moroni, E., Jacobson, L., Greene, C., Snell, E., & Thorpe, L. E. (2016). Design of the New York City Macroscope: Innovations in Population Health Surveillance Using Electronic Health Records. *EGEMS (Washington, DC)*, *4*(1), 1265. doi:10.13063/2327-9214.1265 PMID:28154835

Paul, M., Greene, C., Newton-Dame, R., Thorpe, L., Perlman, S., McVeigh, K., & Gourevitch, M. (2015). The state of population health surveillance using electronic health records: A narrative review. *Population Health Management*, *18*(3), 209–216. doi:10.1089/pop.2014.0093 PMID:25608033

Rajkovic, P., Aleksic, D., Jankovic, D., Milenkovic, A., & Petkovic, I. (2018). Checking the potential shift to perceived usefulness: The analysis of users' response to the updated electronic health record core features. *International Journal of Medical Informatics*, *115*, 80–91. doi:10.1016/j.ijmedinf.2018.04.011 PMID:29779723

Revere, D., Hills, R., Dixon, B., Gibson, P., & Grannis, S. (2017). Notifiable condition reporting practices: Implications for public health agency participation in a health information exchange. *BMC Public Health*, *17*(1), 247. doi:10.118612889-017-4156-4 PMID:28284190

Richards, C. L., Iademarco, M. F., Atkinson, D., Pinner, R. W., Yoon, P., Kenzie, W. R., ... Frieden, T. R. (2017). Advances in Public Health Surveillance and Information Dissemination at the Centers for Disease Control and Prevention. *Public Health Reports*, *132*(4), 403–410. doi:10.1177/0033354917709542 PMID:28609194

Rogers, E. (1983). *Diffusion of Innovations* (3rd ed.). The Free Press.

Scutchfield, F., & Keck, C. (2009). *Principles of public health practice* (3rd ed.). Delmar Cengage.

Shapiro, J., Mostashari, F., Hripcsak, G., Soulakis, N., & Kuperman, G. (2011). Using health information exchange to improve public health. *American Journal of Public Health*, *101*(4), 616–623. doi:10.2105/AJPH.2008.158980 PMID:21330598

Smith, P., Hadler, J., Stanbury, M., Rolfs, R., & Hopkins, R. (2013). "Blueprint Version 2.0": Updating public health surveillance for the 21st century. *Journal of Public Management and Practice*, *19*(33), 231–239. doi:10.1097/PHH.0b013e318262906e PMID:22759985

United States Department of Health and Human Services, Centers for Medicare and Medicaid Services (CMS). (2013). *EHR Incentive Programs: Meaningful Use*. http://www.cms.gov/Regulations-and-Guidance/Legislation/EHRIncentivePrograms/Meaningful_Use.html

Vogel, J., Brown, J. S., Land, T., Platt, R., & Klompas, M. (2014). MDPHnet: Secure, distributed sharing of electronic health record data for public health surveillance, evaluation, and planning. *American Journal of Public Health, 104*(12), 2265–2270. doi:10.2105/AJPH.2014.302103 PMID:25322301

Wisconsin Department of Health Services [WDHS]. (2020). *Disease Reporting*. https://www.dhs.wisconsin.gov/disease/diseasereporting.htm#:~:text=Questions%20concerning%20this%20information%20may,%2C%20608%2D266%2D1120.&text=Category%20I%20diseases%20must%20be,health%20officer%2C%20or%20their%20designee

Wu, L. (2014). *Issue Brief: Health IT for Public Health Reporting and Information Systems*. Academic Press.

KEY TERMS AND DEFINITIONS

Active Surveillance Methods: A form of public health surveillance that involves initiation of reporting information from public health agencies to health care facilities.

Interoperability: The ability for systems to communicate with one another.

Meaningful Use: The use of certified electronic health records technology in a manner that provides for the electronic exchange of health information to improve the quality of care (CDC, 2017).

Manual Reporting: The use of paper-based methods to report diseases to public health agencies.

Passive Surveillance Methods: A form of public health surveillance that involves health care providers initiating and transmitting disease reports to local health departments and public health agencies.

Population Health: The health outcomes of a group of individuals, to include the distributions of outcomes within the group.

Qualtrics: An online survey tool that can be used to create, distribute, and collect data for research.

Syndromic Surveillance: The process of monitoring patient encounters to identify symptoms related to infectious diseases and other medical conditions that may affect public health (Calman et al., 2012).

Chapter 11
Shifting Moral Values

Wiley Hughes
https://orcid.org/0000-0003-2295-1731
Destiny Now, USA

ABSTRACT

Church attendance has been a social staple in American culture since the founding years. Therefore, church attendance has historically been a regular practice for many Americans and people who come from religious families often feel expected to attend church. But this may be changing. This chapter explores current shifting moral values.

SHIFTING MORAL VALUES

Church attendance has been a social staple in American culture since the founding years. This is probably because "religion is an important component of people's lives in the U.S. and about 80% categorize themselves as following a religion (Pew Forum, 2008)." "Religious beliefs are remarkably salient in the United States" according to Sherkat and Ellison (1999). Therefore, church attendance has historically been a regular practice for many Americans and people who come from religious families often feel expected to attend church.

The story is told of a young man who, while attending college, became disenfranchised with the religion in which he was raised and began determining his own belief system. His reflection started in a religion class as he listened to a professor explain about having a drug problem when he was a child. His parents "drug" him to church, he said in jest. This resonated with the young man, and he began to resent being forced to attend services when he was younger. It made him feel that to be accepted by God he must attend church. This made church attendance

DOI: 10.4018/978-1-6684-4358-3.ch011

feel more like a requirement than a choice. Many people feel similarly. Some even feel that whether one attends church determines whether God accepts them as being good, "moral" Christians. Unfortunately, this has become a problem as morality and church attendance have become intertwined in the minds of many. Habitually missing church can be viewed as a lapse of morality as "frequent church attendance is often used as a generic measure of religiousness" (Wright and Wisecarver, 1993).

With many churches hosting their weekly services online, since the COVID pandemic began, some Christians have changed how they attend church. The "new normal" for attending church may be ushering in a shift in moral values. This shift in morality has to do with how people feel about attending church in person now that gathering restrictions have limited church attendance and increased online worship experiences. It can be argued that if attending church in person is the "morally correct" thing for religious people to do some may see the lack of church attendance as a lapse in morality. If this is true then the opportunity given church attendees, due to COVID-19, to stay home and watch church services online has enabled church goers to feel differently about how they attend worship. This must be true as a Gallup report states that "the most dramatic result (of the pandemic in religion) has been the exceedingly quick shift of religious services from in-person to online worship" (Dein, Loewenthal, Lewis, and Pargament 2020), which continues even after the number of COVID-19 cases have started to decline.

Most religious people consider attending church in person a requirement but since the pandemic they have seen that they can miss attending church in person for months and remain spiritually healthy. This is unusual because many leaders in the Christian Church teach that God will not bless the believer who does not attend church. Following this reasoning it could be thought that the restrictions placed on congregations to avoid gathering in person would have a negative effect on church attendees. However, one Gallup Poll research group investigated the impact of religious groups on personal health. They tested the theory that COVID-19 will cause a loss of faith but discovered that "the COVID-19 crisis has enhanced spirituality and religion for many Americans and aided to cope with anxiety and depression (Newport, 2020)." The expectation religious people have of themselves to attend church in person has changed thus shifting morality when it comes to some religious obligations. This has opened many up to the idea that maybe attending church in person is not necessary to be accepted by God and experiencing church online is a viable and acceptable option for their religion. This "new morality" is a direct result of how COVID-19 has impacted religion in America and changed how church goers think about attending church in person. Of course, church attendance is not the only thing people have started to think differently about since the COVID-19 pandemic.

This new morality has spilled over into other areas of daily life. Work and various other social activities have been impacted by this new ideology...this new morality.

Many who once considered it improper or lazy to work at home, instead of driving to their job, clocking in, and sitting in an office or a work vehicle all day, have now begun to be open to working virtually and have told themselves, "It's ok, I'm not lazy or slothful because I do not go somewhere to work. I can be perfectly productive right in my own home." Not feeling condemned but experiencing a sense of achievement and success while at home has caused a shift in moral values concerning traditional work patterns. This thought process has birthed a new culture in society (Rioveros, Reyes-Chua, Navigar, Solina, Lanuza, and Garcia, 2021).

As discussed previously, there is not the same level of expectation to stay committed to a traditional way of doing things when it comes to church. There seems to be a similar pattern when it comes to how people work and use technology. This is not surprising considering how their sense of morality normally impacts all areas of their life. Many people have had to use media for the first time. Even after restrictions were lifted on congregations to attend church in person, many church leaders encouraged their older members to stay at home, many of whom never used the internet or social media. To remain part of their congregation these senior adults were left with little option but to learn how to navigate the internet and watch their church services online. They have been forced to have doctor appointments, counseling sessions, buy groceries, attend classes, and worship exclusively online and it stretched them out of their comfort zones to propel them into a new culture of living. One grandmother, who was encouraged by her granddaughter to create a profile on Facebook, said; "the internet is from the devil." She believed that people who go online are either trying to steal personal information or contaminate their minds with filth. She lacked tolerance for it therefore she was not willing to expend any time to learn to use it. To her, the internet was immoral.

In the new culture, since COVID-19 began, such individuals are not restricted by a lack of knowledge or absence of familiarity with technology because they have had to use it. Grandmothers and grandfathers, who were leery of the internet prior to the pandemic, now have Facebook and Instagram accounts. They are sending and receiving emails. Research also shows that "Conservative and fundamentalist religious groups from all religious traditions are ignoring or even defying the coronavirus rules" (Kranz, Niepel, Botes, and Greiff, 2020). Jesus taught to consider others above oneself, which is arguably the moral thing to do. However, these individuals have likely chosen what some would consider the "immoral low ground" and ignored the social gathering restrictions, possibly exposing everyone to a higher risk of getting COVID. Since many Christians are unwilling to keep the golden rule, some would argue that there has been a shift in moral values, at least among conservative and fundamentalist religious groups.

How exactly has there been a shift in moral values in the wrong direction? That is a question that deserves further examination. Up to this point the discussion has been about how there has been a positive shift in moral values but to endanger public safety in the name of religious freedom sounds like regression. Since the reformation, and before, there have been people who have rebelled against the manipulation of religion and have spoken out against the immoral acts done in the name of God. For example, Martin Luther challenged the Roman Church in 1517 over the matter of indulgences. Indulgences were the commutation for money for the temporal penalty due for sin, the practical satisfaction that was a part of the sacrament of repentance. They were granted on papal authority and Luther posted a list of ninety-five propositions for debate concerned with the question of indulgences. He posted them on the door of the Schlosskirche (Castle Church), Wittenberg, on October 31, 1517. This was the beginning of what later would be referred to as the Protestant Reformation.

Morals have often shifted throughout history and not always in a positive direction. Since COVID-19 began, religion and the organized church in America have gone through some radical changes and, as a result, shown some moral shifts. These shifts have not all been good. One pastor reported that many in his congregation, a church that encouraged its members to feel more comfortable being in a multi-ethnic, multi-racial congregation, decided to go back to their previous church during the pandemic (Hughes, W., Franklin, V., Bloom, D.,2021). He suggests that there was a time (prior to COVID) when many church goers recognized that most churches were too segregated on Sundays. That fact inspired them to try a church that was multicultural but when the scare of the pandemic hit many of these individuals regressed to a religious practice that was more familiar and comfortable, leaving a more diverse church to go back to their previous homogeneous congregation. This may be an example of how some religious people prefer that which is more comfortable and familiar and sometimes reject that which is more progressive and seems to be right and moral.

Since the beginning of the COVID-19 pandemic many people seem to live in survival mode, prioritizing their needs ahead of the needs of others. For example, when the populous was encouraged to combat the spread of COVID by using hand sanitizer it flew off the shelf at grocery stores and pharmacies. Many people seemed to ignore common courtesy and share with others. That would seem like the moral thing to do but instead many overbought some products especially necessities. When everyone needs such items, it seems unethical, immoral, and inconsiderate to overbuy and hoard them. This occurrence during the pandemic is referred to "supply shock," and it took place all over the country as the pandemic began to surge through the nation.

Economists would agree that the pandemic had components of both supply and demand shocks. A supply shock reduces the economy's capacity to produce goods and services. A demand shock reduces consumers' ability to purchase goods and services (Brinca, Duarte, and Faria-e-Castro, 2020). This phenomenon was one of many examples since the start of the pandemic of how people's sense of right and wrong changed when they are under pressure. The pressure many felt was the need to survive and provide for themselves and their families. Morality did not always seem to be top priority. It may not be the first primary inclination of many human beings to consider others equal to themselves. Instead, it seems that many prefer to be selfish, and selfishness can be considered immoral because selfish acts seem to be motivated by an absence of empathy and concern for others.

It should not be forgotten that also during the pandemic acts of social injustice occurred, making national news, and furthering immorality. The murder of black men such as Ahmaud Arbery, George Floyd, and others shined a revealing light on injustices made more acute by the stress on the nation during the COVID-19 pandemic. Although such acts of injustice have occurred throughout history, selfishness brought on by the stresses of the pandemic seemed more obvious and a lack of general morality seemed apparent. A shopper, a single mother of three, was asked what her biggest obstacle was during the pandemic while trying to buy groceries for her family. She explained that her struggle was with the overwhelming apathy and general lack of empathy others seemed to have for those who were hit hardest by COVID. These were the people that worked in restaurants, schools, and churches who were laid off when social distancing and other restrictions were first required by the Center for Disease Control and Prevention. Many carried what seemed like an "every man for himself" attitude. It was like some people did not care at all that there were individuals who were struggling to survive, and basic morality seemed to fly out the window. Those who are disadvantaged suffered more during the pandemic.

Research has proven that moral norms shift in crises such as the COVID-19 pandemic. Because people's perspectives change during crises it has an impact on how they behave. A person's sense of personal safety can trigger their survival instinct and cause them to be inclined to behave in uncharacteristic ways. Although there has been some work done in the area of social justice it seems like the pressures of the pandemic, which included supply shortages and other public health concerns, caused many people to regress in their efforts to work toward equality. This caused the progress made in the area of social justice and morality to stall. It also paved the way for the breakdown of morality toward those who had been victims of discrimination or in some way socially disadvantaged. Research points to the fact that "One of the most morally troubling aspects of public health crises is the disproportionate impact on groups that are socially disadvantaged (DeBruin and Leider, 2020)."

Ultimately it is clear that an appropriate response to the immorality perpetuated during the pandemic is needed and should be grounded on values and morality desired as society moves from normalcy to crisis. Crisis response must have clear mechanisms and include accountability measures that respond to needs fairly and efficiently. The precedent in this pandemic for a renewed sense of morality is profound, which produces the need to develop and implement a clear plan to communicate appropriate behavior to the public. One may ask who the ultimate authority and decision maker is to determine an equal morality for everyone but that is a misguided presupposition. The fact that many people allowed their raw sense of survival during the pandemic to temporarily replace their sense of responsibility, integrity, and normalcy, exposes the tendency human beings possess to forget the instinctive motivation we carry to behave morally and treat others with respect.

Shifting moral values have been seen clearly since the start of the COVID-19 pandemic. From religious practices and worship attendance to how people go to work, the sense of what is morally acceptable has changed. Perhaps there are some aspects of what is considered moral by most people that does not have to do with what is right or wrong but that which is considered acceptable by the individual. However, the obvious selfishness, bigotry, and hatred that surfaced in the nation during the pandemic could be considered a moral crisis. Although it may have appeared prior to COVID that Americans were making progress to stomp out injustice, discrimination, and racism, the pressures of the pandemic exposed the deeper feelings of many people who continue to choose self over others. This is the most alarming shift in moral values, and it is a shift backwards. It is a shift backwards that will continue until something is done to create momentum for a morality shift forward.

That which makes a nation better is that which makes a nation acknowledge and alter its immoral behavior. Morality is not determined by what a person's religion or personal belief system determines. Morality is determined by how one's behavior impacts others. That fact has been seen clearly since the COVID-19 pandemic began. People can change unwanted things in their culture that are undesirable and inhuman. People have experienced a shift in their moral values and continue to live just fine. This means they have the power and ability to work on altering their bias, selfishness, stubbornness, and hatred. Although these may seem to be components of human behavior during crises that are uncontrollable, they are not. Such things are only a product of one's impulses and impulses are often launched from the springboard of moral values. Moral values can shift and if morality can shift, then morality can shift in a positive direction, enabling human beings to do better.

REFERENCES

Brinca, P., Duarte, J., & Faria-e-Castro, M. (2020). *Measuring Sectoral Supply and Demand Shocks during COVID-19.* Working Paper 2020-011B, Federal Reserve Bank of St. Louis. https://research.stlouisfed.org/wp/more/2020-011

DeBruin, D., & Leider, J. (2020). COVID-19: The shift from clinical to public health ethics. *Journal of Public Health Management and Practice*, *26*(4), 306–309. doi:10.1097/PHH.0000000000001204 PMID:32433384

Dein, S., Loewenthal, K., Lewis, C., & Pargament, K. (2020). COVID-19, mental health and religion: An agenda for future research. *Mental Health, Religion & Culture*, *23*(1), 1–9. doi:10.1080/13674676.2020.1768725

Hughes, W., Franklin, V., Bloom, D. (2021). *Battling The Assassins of Your Success*. Academic Press.

Kranz, D., Niepel, C., Botes, E., & Greiff, S. (2020). Religiosity Predicts Unreasonable Coping With COVID-19. *Psychology of Religion and Spirituality*. Advance online publication. doi:10.1037/rel0000395

Newport, F. (2020). *The Religion Paradox*. https://news.gallup.com/opinion/polling-matters/310397/religion-paradox.aspx

Pew Forum. (2008). *U.S. religious landscape survey*. Retrieved from https://www.pewforum.org/

Rioveros, G., Reyes-Chua, E., Navigar, N., Solina, A., Lanuza, M., & Garcia, S. (2021). The Impact of Work From Home Set-up During the COVID-19 Pandemic. *Worldwide Journal of Multidisciplinary Research and Development*. http://wwjmrd.com/upload/the-impact-of-work-from-home-set-up-during-the-covid-19-pandemic_1631623804.pdf

Sherkat, D., & Ellison, C. (1999). Recent Developments and Current Controversies In The Sociology of Religion. *Annual Review of Sociology*, *24*, 363–394.

Wright, L., Frost, C., & Wisecarver, S. (1993). Church Attendance, Meaningfulness of Religion and Depressive Symptomology among adolescents. *Journal of Youth and Adolescence*, *22*(5).

Compilation of References

Abdulrahim, N., Alasasfeh, I., Khader, Y. S., & Iblan, I. (2019). Knowledge, awareness, and compliance with disease surveillance and notification among Jordanian physicians in residency programs. *The Journal of Health Care Organization, Provision, and Financing Inquiry, 56.* Advance online publication. doi:10.1177/0046958019856508 PMID:31220967

Acciarini, C., Brunetta, F., & Boccardelli, P. (2020). Cognitive biases and decision-making strategies in times of change: A systematic literature review. *Management Decision*. doi:10.1108/MD-07-2019-1006

Adolphs, R. (2021). Social Cognitive and Affective Neuroscience. doi:10.1093can/nsw153

Aguilar Yuste, M. (2021). Leadership and Emotional Intelligence during a Crisis. *Revista de Relaciones Internacionales. Estrategia y Seguridad, 16*(2), 47–60. doi:10.18359/ries.5619

Aguilera, D. C. (2000). *Crisis intervention: Theory and methodology* (8th ed.). Mosby.

Ahern, S., & Loh, E. (2020). *Leadership during the COVID-19 pandemic: building and sustaining trust in times of uncertainty*. doi:10.1136/leader-2020-000271

Ahmad, R., Imran, H., & Mehmood, T. (2009). Emotional intelligence as a determinant of self-esteem and depression among adolescents. *Pakistan Journal of Psychology, 40*(1), 77–88.

Al-Dabbagh, Z. S. (2020). The role of Decision-maker in crisis management: A qualitative study using grounded theory (COVID-19 pandemic crisis as A model). *Journal of Public Affairs, 20*(4). Advance online publication. doi:10.1002/pa.2186 PMID:32904918

Alhawamdeh, H. M., & Alsmairat, M. A. K. (2019). Strategic decision making and organization performance: A literature review. *International Review of Management and Marketing, 9*(4), 95–99. doi:10.32479/irmm.8161

Al-Helou, G. (2011). School crises in government secondary schools in the northern west bank directorates. *An-Najah University Journal, 24*(1), 223–262.

Allen, R., & Masters, D. (2020). Artificial intelligence: The right to protection from discrimination caused by algorithms, machine learning, and automated decision-making. *ERA-Forum, 20*(4), 585-598. 10.100712027-019-00582-w

American Psychological Association. (n.d.). *APA dictionary of psychology.* Retrieved March 20, 2022 from https://dictionary.apa.org/helping-relationship

American Society of Civil Engineering. (2017). *Infrastructure Report Card: Drinking Water.* Retrieved from https://www.infrastructurereportcard.org/wp-content/uploads/2019/02/Full-2017-Report-Card-FINAL.pdf

American Society of Civil Engineering. (2021). *Infrastructure Report Card: Drinking Water.* Retrieved from https://infrastructurereportcard.org/cat-item/drinking-water/

American Water Works Association. (2011). *Computer modeling of water distribution systems.* American Water Works Association.

American Water Works Association. (2019). *State of the Water Industry.* Retrieved from https://www.awwa.org/Portals/0/AWWA/ETS/Resources/2019_STATE%20OF%20THE%20 WATER%20INDUSTRY_post.pdf

Andiola, L. M., Downey, D. H., & Westermann, K. D. (2020). Examining Climate and Culture in Audit Firms: Insights, Practice Implications, and Future Research Directions. *Auditing, 39*(4), 1–29. doi:10.2308/AJPT-19-107

An, J., & Dorofeev, M. (2019). Short-term foreign exchange forecasting: Decision-making based on expert polls. *Investment Management & Financial Innovations, 16*(4), 215–228. doi:10.21511/imfi.16(4).2019.19

Anoye, A. B., & Kouamé, J. S. (2018). Leadership challenges in virtual team environment. *International Journal of Scientific & Technology Research, 7*(7).

Ansary, A. S. (2019). *Pros and Cons of Charismatic Leadership.* Retrieved on 2/7/22 from http://careercliff/com/pros-and-cons-of-charismatic-leadership

Antle, M., & Silver, R. (2016). Circadian Insights into Motivated Behavior. In Behavioral Neuroscience of Motivation. Springer International Publishing, AG Switzerland.

Antonakis, J. (2020). Leadership to defeat COVID-19. *Group Processes & Intergroup Relations, 24*, 210–215. doi:10.1177/1368430220981418

Arbix, G., & Comin, A. A. (2020). The pandemic, technology, and work are at the crossroads. *Internet Sectoral Overview, 12*(4), 11-17. https://www.cetic.br/media/docs/publicacoes/6/20210115080952/internet_sectoral_overview_year-12_n_4_artificial_intelligence_and_work.pdf

Ardebili, A. A., & Padoano, E. (2020). A literature review of the concepts of resilience and sustainability in group decision-making. *Sustainability (Basel, Switzerland), 12*(7), 2602. doi:10.3390u12072602

Artinger, F., Petersen, M., Gigerenzer, G., & Weibler, J. (2015). Heuristics as adaptive decision strategies in management. *Journal of Organizational Behavior, 36*(S1), S33–S52. doi:10.1002/job.1950

Avolio, B. J., & Bass, B. M. (2007). *Multifactor Leadership Questionnaire, sample set manual, forms, and scoring key*. Mind Garden.

Ayalew, M., & Ayenew, Z. (2022). Do Paradoxical Virtual Leadership and Emotional Intelligence have Relationships? In Particular from Technology Dependence, Geographical Dispersion, and Human Capital Tensions. *International Journal of Organizational Leadership, 11*(1), 1–25. doi:10.33844/ijol.2022.60616

Baba, M. M. (2020). Navigating COVID-19 with emotional intelligence. *International Journal of Social Psychiatry, 66*(8), 810-820.

Badalucco, T. (2015, November). The battle to move U.S. health care from paper to digital far from over. *U.S. News and World Report.* Retrieved from https://www.usnews.com/news/articles/2015/11/05/the-battle-to-move-us-health-care-from-paper-to-digital-far-from-over

Bahrami, B., Olsen, K., Latham, P. E., Roepstorff, A., Rees, G., & Frith, C. D. (2010). Optimally interacting minds. *Science, 329*(5995), 1081–1085. doi:10.1126cience.1185718 PMID:20798320

Baker, M. (2021). *9 Future of Work Trends Post-COVID-19*. Retrieved on 2.5.22 from https://www.gartner.com/smarterwithgartner/9-future-of-work-trends-post-covid-19

Balconi, M., & Pozzoli, U. (2009). Arousal affect on emotional face comprehension, frequency band changes, in different time intervals. *Physiology & Behavior, 97*, 460–462. doi:10.1016/j.physbeh.2009.03.023 PMID:19341748

Ball, C. (2020). Leadership during the COVID crisis and beyond. *Canadian Journal of Surgery, 63*(4), E370–E371. doi:10.1503/cjs.016020 PMID:32813482

Bande, B., Fernandez-Ferrin, P., Varela-Neira, C., & Otero-Neira, C. (2016). Exploring the relationship among servant leadership, intrinsic motivation, and performance in an industrial sales setting. *Journal of Business and Industrial Marketing, 31*(2), 219–231. doi:10.1108/JBIM-03-2014-0046

Banking on technology: 5 tech trends every bank should be prepared for in 2021. (n.d.). Retrieved Jan 13, 2022, from https://www.businesstoday.in/opinion/columns/story/five-technology-trends-to-shape-indian-banking-and-financial-services-in-2021-285594-2021-01-27

Bar-on, R. (1997). *The Baron emotional quotient inventory (EQ-I) technical manual*. Multihealth Systems.

Bar-On, R., & Parker, D. A. (2000). *The handbook of emotional intelligence*. Jossey-Bass.

Barrios, L. M., Prowse, A., & Vargas, V. R. (2020). Sustainable development and women's leadership: A participatory exploration of capabilities in Colombian Caribbean Fisher Communities. *Journal of Cleaner Production, 264*, 121277. https://doi.org/10.1016/j.jclepro.2020.121277

Bassamboo, A., Moreno, A., & Stamatopoulos, I. (2020). Inventory Auditing and Replenishment Using Point-of-Sales Data. *Production and Operations Management, 29*(5), 1219–1231. doi:10.1111/poms.13153

Bass, B. M. (1990). *Bass and Stogdill's handbook of leadership*. Free Press.

Bass, B., & Riggio, R. (2006). *Transformational Leadership* (2nd ed.). Lawrence Erlbaum Associates, Publisher. doi:10.4324/9781410617095

Batson, C. D., Shaw, L. L., & Oleson, K. C. (1992). Distinguishing affect, mood, and emotion: Toward functionally based conceptual distinctions. *Review of Personality and Social Psychology*, *13*, 294.

Bauer, T., Erdogan, B., Cauglin, D., & Truxillo, D. (2021). *Fundamentals of Human Resource Management: People, Data and Analytics*. Sage Publications.

Bayraktar, E., Yalçınkaya, M., & Eyitmiş, A. M. (2021). Examination of emotional intelligence and problem-solving skills of teachers working in secondary education institutions. *International Online Journal of Education & Teaching*, *8*(4), 2228–2247.

Beglaryan, M., Petrosyan, V., & Bunker, E. (2017). Development of a tripolar model of technology acceptance: Hospital-based physicians' perspective on EHR. *International Journal of Medical Informatics*, *102*, 50–61. doi:10.1016/j.ijmedinf.2017.02.013 PMID:28495348

Bhatti, N., Maitlo, G., Shaikh, N., Hashmi, M., & Shaikh, F. (2012). The impact of autocratic and democratic leadership style on job satisfaction. *International Business Research*, *5*(2), 192–193. doi:10.5539/ibr.v5n2p192

Bhuiyan, A. (2014). *Internet Governance and Global South: Demand for a New Framework*. Palgrave Macmillian.

Bialik, K., & Fry, R. (2021, May 28). *How millennials compare with prior generations*. Pew Research Center's Social & Demographic Trends Project. Retrieved March 13, 2022, from https://www.pewresearch.org/social-trends/2019/02/14/millennial-life-how-young-adulthood-today-compares-with-prior-generations-2/

Bingham, C. B., & Eisenhardt, K. M. (2011). Rational heuristics: The 'simple rules' that strategists learn from process experience. *Strategic Management Journal*, *32*(13), 1437–1464. doi:10.1002mj.965

Birkhead, G. (2017). Successes and continued challenges of electronic health records for chronic disease surveillance. *American Journal of Public Health*, *107*(9), 1365–1366. doi:10.2105/AJPH.2017.303938 PMID:28787206

Birkhead, G., Klompas, M., & Shah, N. (2015). Public health surveillance using electronic health records: Rising potential to advance public health. *Frontiers in Public Health Services & Systems Research*, *4*(5), 25–32. doi:10.13023/FPHSSR.0405.05

Bivens, J. (n.d.). *Updated employment multipliers for the U.S. economy*. Economic Policy Institute. Retrieved March 10, 2022, from https://www.epi.org/publication/updated-employment-multipliers-for-the-u-s-economy/?msclkid=c8d85de2a60f11eca5c7f7fbbbeeab79

Blackburn, A. (2021, August 5). *State numbers show number of kids withdrawing to homeschooling nearly doubled last school year.* Fox 59. https://fox59.com/news/state-numbers-show-number-of-kids-withdrawing-to-homeschooling-nearly-doubled-last-school-year/

Black, J., Hulkower, R., Suarez, W., Patel, S., & Elliott, B. (2019). Public health surveillance: Electronic reporting as a point of reference. *The Journal of Law, Medicine & Ethics*, *47*(S2), 19–22. doi:10.1177/1073110519857309 PMID:31298129

Blanken, R. (2019). *Eight Leadership Styles and When to Use Them.* The Center for Association Leadership. Retrieved on 2/14/22 at https://www.asaecenter.org/resources/articles/an_plus/2019/december/eight-leadership-styles-and-when-to-use-them

Blau, P. M. (1964). Justice in social exchange. *Sociological Inquiry*, *34*, 193–206. doi:10.1111/j.1475-682X.1964.tb0058

Bluefin.com. (n.d.). https://www.bluefin.com/bluefin-news/2021-biggest-data-breaches-so-far/

Bögel, P. M. (2019). Company reputation and its influence on consumer trust in response to ongoing CSR communication. *Journal of Marketing Communications*, *25*(2), 115–136. doi:10.1080/13527266.2016.1166146

Børge Brende, P. (n.d.). *India's roadmap for Success and prosperity in the post-COVID world.* World Economic Forum. Retrieved Dec 16, 2021, from https://www.weforum.org/agenda/2021/11/india-s-future-and-role-in-the-post-covid-19-world/

Borup, J., & Evmenova, A.S. (2019). The effectiveness of professional development in overcoming obstacles to effective online instruction in a college of education. *Online Learning, 23*(2).

Boyatzis, R. E., Goleman, D., & Rhee, K. S. (2002). *Clustering competence in emotional intelligence insights from the emotional competency inventory*. Harvard Business Review.

Bradberry, T., & Greaves, J. (2009). *Emotional Intelligence 2.0.* Simon & Schuster.

Brady, D. (2005, Mar. 28). The Immelt Revolution: He's Turning GE's Culture Upside Down, Demanding Far More Risk and Innovation. *Business Week.*

Branson, L., Steele, N. L., & Sung, C.-H. (2010). When two heads are worse than one: Impact of group style and information type on performance evaluation. *Journal of Business and Behavioral Sciences*, *22*, 75–84.

Brinca, P., Duarte, J., & Faria-e-Castro, M. (2020). *Measuring Sectoral Supply and Demand Shocks during COVID-19.* Working Paper 2020-011B, Federal Reserve Bank of St. Louis. https://research.stlouisfed.org/wp/more/2020-011

Brobst, D. (2012). *Choose Your Battles Wisely.* https://www.donbrobst.com/2012/05/choosing-your-battles-wisely/

Brodbeck, F. C., Kerschreiter, R., Mojzisch, A., & Schulz-Hardt, S. (2007). Group decision making under conditions of distributed knowledge: The information asymmetries model. *Academy of Management Review, 32*(2), 459–479. doi:10.5465/amr.2007.24351441

Brown, C., & Nwagbara, U. (2021). Leading Change with the Heart: Exploring the Relationship between Emotional Intelligence and Transformational Leadership in the Era of Covid-19 Pandemic Challenges. *Economic Insights - Trends and Challenges, 3*(3), 1–12. doi:10.51865/EITC.2021.03.01

Bundy, J., Pfarrer, M. D., Short, C. E., & Coombs, W. T. (2017). Crises and crisis management: Integration, interpretation, and research development. *Journal of Management, 43*(6), 1661–1692. doi:10.1177/0149206316680030

Bureau of Labor Statistics. (n.d.). *Number of quits at all-time high November 2021*. Retrieved on 2/5/22 from https://www.bls.gov/opub/ted/2022/number-of-quits-at-all-time-high-in-november-2021.htm

Burity, J. (2021). The Importance of Logistics Efficiency on Customer Satisfaction. *Journal of Marketing Development and Competitiveness, 15*(3), 26–35. doi:10.33423/jmdc.v15i3.4537

Burke, W. W., & Litwin, G. H. (1992). A causal model of organizational performance and change. *Journal of Management, 18*(3), 523–545. doi:10.1177/014920639201800306

Burton, A., Aughterson, H., Fancourt, D., & Philip, K. E. J. (2022). Factors shaping the mental health and well-being of people experiencing persistent COVID-19 symptoms or 'long COVID': Qualitative study. *BJPsych Open, 8*(2), e72. Advance online publication. doi:10.1192/bjo.2022.38 PMID:35307048

Butler, C., Demski, C., Parkhill, K., Pidgeon, N., & Spence, A. (2015). Public values for energy futures: Framing, indeterminacy, and policymaking. *Energy Policy, 87*, 665–672. doi:10.1016/j.enpol.2015.01.035

California Department of Public Health Division of Communicable Disease Control [CDPH]. (2020). *Reportable diseases and conditions.* https://www.cdph.ca.gov/Programs/CID/DCDC/Pages/Reportable-Disease-and-Conditions.aspx

Călinici, M. S., Călinici, T., Balazsi, R., & Miclea, M. (2020). Development and validation of a new ability-based measure of emotional intelligence: Cluj Emotional Intelligence Scale CEIS. *Cognition, Brain, Behavior: An Interdisciplinary Journal, 24*(4), 335–363. doi:10.24193/cbb.2020.24.18

Calman, N., Hauser, D., Lurio, J., Wu, W. Y., & Pichardo, M. (2012). Strengthening Public Health and Primary Care Collaboration Through Electronic Health Records. *American Journal of Public Health, 102*(11), e13–e18. doi:10.2105/AJPH.2012.301000 PMID:22994274

Campbell, T. (2022). The Future of Leadership in a Remote-Work World. *Journal of AHIMA*. https://journal.ahima.org/the-future-of-leadership-in-a-remote-work-world

Cantu, D. V. (2016). The evolution of the telephone. *Children's Technology & Engineering, 21*(1), 14–17.

Carneiro, J., Martinho, D., Marreiros, G., & Novais, P. (2019). How cognitive and affective aspects can influence the outcome of the group decision-making process. *Expert Systems, 36*(3), e12257. doi:10.1111/exsy.12257

Carroll, J. B. (1993). *Human cognitive abilities: A survey of factor-analytic Studies*. Cambridge University Press. doi:10.1017/CBO9780511571312

Carsten Linz, M. E. C. (n.d.). *How to transform your business model for a post-COVID future.* World Economic Forum. Retrieved Jan 14, 2022, from https://www.weforum.org/agenda/2020/11/transform-business-model-post-covid-future/

CAST. (2008). Universal design for learning guidelines version 1.0. Wakefield, MA: Author.

Cavozas, E. A., & Morin, G. (1994). *Cyberspace and the tha Law: Your rights and duties in the Online World.* MIIT press.

Center for Disease Control and Prevention. (2020). *Water and COVID-19 FAQs*. Retrieved from https://www.cdc.gov/coronavirus/2019-ncov/php/water.html

Center for Disease Control and Prevention. (2021). *National Wastewater Surveillance System (NWSS): A new public health tool to understand COVID-19 spread in a community*. Retrieved from https://www.cdc.gov/healthywater/surveillance/wastewater-surveillance/wastewater-surveillance.html

Centers for Disease Control and Prevention (CDC). (2014). *Introduction to Public Health. In: Public Health 101 Services. Atlanta, GA. US Department of Health and Human Services.* https://www.cdc.gov/training/publichealth101/surveillance.html

Centers for Disease Control and Prevention (CDC). (2016, October 14). Summary of Notifiable Infectious Diseases and Conditions — United States, 2014. *MMWR. Morbidity and Mortality Weekly Reports*. https://www.cdc.gov/mmwr/volumes/63/wr/mm6354a1.htm?s_cid=mm6354a1_w

Centers for Disease Control and Prevention (CDC). (2017). *Meaningful Use.* https://www.cdc.gov/ehrmeaningfuluse/introduction.html

Centers for Disease Control and Prevention. (2020). Impact of the COVID-19 pandemic on emergency department visits—United States, January 1, 2019–May 30, 2020. *MMWR Morb Mortal Wkly Rep.*, *69*(23), 699–704. .mm6923e1 doi:10.15585/mmwr

Centers for Medicare and Medicaid Services (CMS). (2013). *Electronic health records*. https://www.cms.gov/Medicare/EHealth/EHealthRecords/index.html?redirect=/ehealthrecords/

Charoensukmongkol, P., & Phungsoonthorn, T. (2020). The effectiveness of supervisor support in lessening perceived uncertainties and emotional exhaustion of university employees during the COVID-19 crisis: The constraining role of organizational intransigence. *The Journal of General Psychology*, 1–20. doi:10.1080/00221309.2020.1800582

Chen, H., Li, T., & Zhang, C. (2021). Going Too Far Is as Bad as Not Going Far Enough: An Inverted U-Shaped Relationship between Internal Controls and Operational Efficiency. *Journal of International Accounting Research, 20*(2), 25–50. doi:10.2308/JIAR-17-571

Cherdantseva, Y., Burnap, P., Blyth, A., Eden, P., Jones, K., Soulsby, H., & Stoddart, K. (2016). A review of cyber security risk assessment methods for SCADA systems. *Computers & Security, 56*, 1–27. doi:10.1016/j.cose.2015.09.009

Chima, A., & Gutman, R. What it Takes to Lead Through an Era of Exponential Change. *Harvard Business Review*.

Choo, L., Grimm, V., Horváth, G., & Nitta, K. (2019). Whistleblowing and diffusion of responsibility: An experiment. *European Economic Review, 119*, 287–301. doi:10.1016/j.euroecorev.2019.07.010

Clark, J. D., Jones, M. C., & Armstrong, C. P. (2007). The Dynamic Structure of Management Support Systems: Theory Development, Research Focus, and Direction. *Management Information Systems Quarterly, 31*(3), 579–615. doi:10.2307/25148808

Clark, T. (2016). *Leading with Character & Competence*. Berrett-Koehler Publishers, Inc.

Clay, N. (1982) *Paraprofessional utilization and reading achievement gains of third and fourth grade compensatory education students* [Published Dissertation]. University of Michigan, Ann Arbor, MI.

Clore, G. L. (1992). *Cognitive phenomenology: Feelings and the construction of judgment*. Erlbaum.

Collin, K., Herranen, S., Paloniemi, S., Auvinen, T., Riivari, E., Sintonen, T., & Lemmertty, S. (2018). Leadership as an enabler of professional agency and creativity: Case studies from the Finnish information technology sector. *International Journal of Training and Development, 22*(3), 222–232. doi:10.1111/ijtd.12130

Collins, J., & Hansen, M. (2011). *Great by Choice: Uncertainty, Chaos and Luck—Why Some Thrive Despite Them All*. HarperCollins Publishers.

Congressional Research Service. (2020). *COVID-19: U.S. Public Health Data and Reporting*. https://crsreports.congress.gov/product/pdf/IN/IN11361

Conrad, F., Couper, M., Tourangeau, R., & Peytchev, A. (2010, September 1). The Impact or Progress Indicators on Task Completion. *Interacting with Computers, 22*(5), 417–427. doi:10.1016/j.intcom.2010.03.001 PMID:20676386

Contreras, F., Baykal, E., & Abid, G. (2020). E-Leadership and Teleworking in Times of COVID-19 and Beyond: What we Know and Where Do We Go. *Frontiers in Psychology, 1*, 1–10. doi:10.3389/fpsyg.2020.590271 PMID:33362656

Cooney, M. A., Iademarco, M., Huang, M., MacKenzie, W., & Davidson, A. J. (2018). The Public Health Community Platform, Electronic Case Reporting, and the Digital Bridge. *Journal of Public Health Management and Practice*, *24*(2), 185–189. doi:10.1097/PHH.0000000000000775 PMID:29360697

Cooper, R. B., & Zmud, R. W. (1990). Information technology implementation research: A technological diffusion approach. *Management Science*, *36*(2), 123–139. doi:10.1287/mnsc.36.2.123

Corporate Finance Institute. (n.d.). *Ethical dilemma*. https://corporatefinanceinstitute.com/resources/knowledge/other/ethical-dilemma/

Cotterill, S., Bunney, S., Lawson, E., Chisholm, A., Farmani, R., & Melville-Shreeve, P. (2020). COVID-19 and the water sector: Understanding impact, preparedness and resilience in the UK through a sector-wide survey. *Water and Environment Journal: the Journal / the Chartered Institution of Water and Environmental Management*, *34*(4), 715–728. doi:10.1111/wej.12649 PMID:33230399

Council for the Accreditation Educator Preparation. (2022). *2022 CAEP standards*. https://caepnet.org/standards/2022-itp/introduction

Cousar, R.D. (2017). *Counting on change: A quantitative evaluation of the relationship of continuous improvement to organizational performance outcomes at United States public sector clean water utilities*. Academic Press.

Coustasse, A. (2021). Upcoding Medicare: Is Healthcare Fraud and Abuse Increasing? *Perspectives in Health Information Management*, *18*(4), 1–14. PMID:34975355

Covey, S. R. (2004). The 7 habits of highly effective people: Powerful lessons in personal change (25th ann. ed.). New York: Simon & Schuster.

Covid-19 and the automotive industry: Changes, challenges and the future. (n.d.). *Dartmouth Undergraduate Journal of Science*. Retrieved Jan 5, 2022, from https://sites.dartmouth.edu/dujs/2020/11/09/covid-19-and-the-automotive-industry-changes-challenges-and-the-future/

Covid-19 impact on the Indian insurance industry - pwc.in. (n.d.). Retrieved Jan 3, 2022, from https://www.pwc.in/assets/pdfs/services/crisis-management/covid-19/covid-19-impact-on-the-indian-insurance-industry.pdf

Covid-19: How the crisis impacts high tech. (n.d.). Retrieved Dec 16, 2021, from https://www.accenture.com/_acnmedia/PDF-121/Accenture-High-Tech-COVID-19-Impacts.pdf

Cribb, L. C. (2018). *Exploring effective virtual leadership and the use of computer-mediated communication technology on a virtual team* [Master's thesis]. University of Alberta.

Cuff, B. M., Brown, S. J., Taylor, L., & Howat, D. J. (2016). Empathy: A review of the concept. *Emotion Review*, *8*(2), 144–153. doi:10.1177/1754073914558466

Cutri, R. M., Mena, J., & Whiting, E. F. (2020). Faculty readiness for online crisis teaching: Transitioning to online teaching during the COVID-19 pandemic. *European Journal of Teacher Education, 43*(4), 523–541. Advance online publication. doi:10.1080/02619768.2020.1815702

D'Auria, G., & De Smet, A. (2020). *Leadership in a Crises: Responding to the coronavirus outbreak and future challenges*. McKinsey & Company.

D'Aveni, R. A. (1994). *Hypercompetition: Managing the Dynamics of Strategic Maneuvering*. The Free Press.

D'Aveni, R. A., Dagnino, G. B., & Smith, K. G. (2010). The age of temporary advantage. *Strategic Management Journal, 31*(13), 1371–1385. doi:10.1002mj.897

Dagnino, G. B., Picone, P. M., & Ferrigno, G. (2021). Temporary competitive advantage: A State-of-the-Art literature review and research directions. *International Journal of Management Reviews: IJMR, 23*(1), 85–115. doi:10.1111/ijmr.12242

Danila, R. N., Laine, E. S., Livingston, F., Como-Sabetti, K., Lamers, L., Johnson, K., & Barry, A. M. (2015). Legal Authority for Infectious Disease Reporting in the United States: Case Study of the 2009 H1N1 Influenza Pandemic. *American Journal of Public Health, 105*(1), 13–18. doi:10.2105/AJPH.2014.302192 PMID:25393187

Darley, J. M., & Latane, B. (1968). Bystander intervention in emergencies: Diffusion of responsibility. *Journal of Personality and Social Psychology, 8*(4), 377-383. doi:10.1037/h0025589

Das, T. K., & Teng, B. (1999). Cognitive biases and strategic decision processes: An integrative perspective. *Journal of Management Studies, 36*(6), 757–778. doi:10.1111/1467-6486.00157

Dastin, J. (2018, October 11). *Amazon scraps secret AI recruiting tool that showed bias against women*. https://www.reuters.com/article/us-amazon-com-jobs-automation-insight-idUSKCN1MK08G

Dattner, B., Chamorro-Premuzic, T., Buchband, R., & Schettler, L. (2019, April 25). *The legal and ethical implications of using AI in hiring*. Harvard Business Review. https://hbr.org/2019/04/the-legal-and-ethical-implications-of-using-ai-in-hiring

Davidson, E. J., & Chismar, W. G. (2007). The Interaction of Institutionally Triggered and Technology-Triggered Social Structure Change: An Investigation of Computerized Physician Order Entry. *Management Information Systems Quarterly, 31*(4), 739–758. doi:10.2307/25148818

Davis, F. D. (1989). Perceived usefulness, perceived ease of use, and user acceptance of information technology. *Management Information Systems Quarterly, 13*(3), 319–340. doi:10.2307/249008

De Vynch, G., & Bergen, M. (2020, April 9). *Google Classroom users doubled as quarantines spread*. Bloomberg. https://www.bloombergquint.com/business/google-widens-lead-in-education-market-as-students-rush-online

Dearing, J. W. (2009). Applying Diffusion of Innovation Theory to Intervention Development. *Research on Social Work Practice, 19*(5), 503–518. doi:10.1177/1049731509335569 PMID:20976022

DeBruin, D., & Leider, J. (2020). COVID-19: The shift from clinical to public health ethics. *Journal of Public Health Management and Practice, 26*(4), 306–309. doi:10.1097/PHH.0000000000001204 PMID:32433384

Dein, S., Loewenthal, K., Lewis, C., & Pargament, K. (2020). COVID-19, mental health and religion: An agenda for future research. *Mental Health, Religion & Culture, 23*(1), 1–9. doi:10.1080/13674676.2020.1768725

Deloitte. (2022). *Workforce Strategies for a Post-COVID-19 recovery: Workbook.* Retrieved on 1.12.22 from https://www2.deloitte.com/content/dam/Deloitte/uk/Documents/human-capital/deloitte-uk-workforce-strategies-for-post-covid-recovery-workbook.pdf

Dempsey, C., & Reilly, B. A. (2016). Nurse engagement: What are the contributing factors for success? *Online Journal of Issues in Nursing, 21*(1).

DeSalvo, K., Hughes, B., Bassett, M., Benjamin, G., Fraser, M., Galea, S., Gracia, J. N., & Howard, J. (2021). *Public health COVID-19 impact assessment: Lessons learned and compelling needs.* National Academy of Medicine. doi:10.31478/202104c

Dess, G., & Shaw, J. (2001). Voluntary turnover, social capital, and organizational performance. *Academy of Management Review, 26*(3), 446-456.

Dhingra, N., Samo, A., Schaninger, B., & Schrimper, M. (n.d.). *Help Your Employees Find Purpose—or Watch Them Leave.* McKinsey & Company, Retrieved on 1/22/22 from https://www.mckinsey.com/business-functions/people-and-organizational-performance/our-insights/help-your-employees-find-purpose-or-watch-them-leave

Di Fabio, A., & Saklofske, D. H. (2018). The contributions of personality and emotional intelligence to resiliency. *Personality and Individual Differences, 123*, 140–144. doi:10.1016/j.paid.2017.11.012

Digital Learning Collaborative. (2018, September 25). *E-learning days.* DLC. https://www.digitallearningcollab.com/elearning-days

Dimitrias, N., & Psychogios, A. (2016). *Neuroscience for Leaders: A brain-adaptive leadership approach.* Kogan Page Limited.

Dirani, K. M., Abadi, M., Alizadeh, A., Barhate, B., Garza, R. C., Gunasekara, N., & .... (2020). Leadership competencies and the essential role of human resource development in times of crisis: A response to Covid-19 pandemic. *Human Resource Development International, 23*, 380–394. doi:10.1080/13678868.2020.1780078

Dixon, T. (2012). *Emotion: The history of a key word in crisis.* doi:10.1177/1754073912445814

Dixon, B., Zhang, Z., Lai, P., Kiriyik, U., Williams, J., Hills, R., Revere, D., ... Grannis, S. (2017). Completeness and timeliness of notifiable disease reporting: A comparison of laboratory and provider reports submitted to a large county health department. *BMC Medical Informatics and Decision Making*, *17*(87), 1–8. doi:10.118612911-017-0491-8 PMID:28645285

Dolcos, F., Alexandru, D. I., & Dolcos, S. (2011). Neural correlates of emotion-cognition interactions: a review of evidence from brain imaging investigations. *Journal of Cognitive Psychology*. Retrieved from: https://www.tandfonline.com/doi/full/10.1080/20445911.2011.594433

Domingues, J., Vieira, V. A., & Agnihotri, R. (2017). The interactive effects of goal orientation and leadership style on sales performance. *Marketing Letters*, *28*(4), 637–649. doi:10.100711002-017-9436-3

Dorminey, J. W., Fleming, A. S., Kranacher, M.-J., & Riley, R. A. (2010). Beyond the Fraud Triangle. *The CPA Journal*, *80*(7), 16–23.

Dornan, L., Pinyopornpanish, K., Jiraporncharoen, W., Hashmi, A., Dejkriengkrikul, N., & Angkurawaranon, C. (2019). Utilization of electronic health records for public health in Asia: A review of success factors and potential challenges. *International Journal of Biomedical Research*, *2019*, 1–9. Advance online publication. doi:10.1155/2019/7341841 PMID:31360723

Dozier, A. L., Gilbert, B. G., Hughes, V. W., Mathis, D. P., & Jenkins, L. J. (2021). The Use of active learning strategies during the COVID-19 pandemic to promote critical thinking. *The ABNF Journal*, *32*(1), 12–16.

Dressler, G. (2020). *Human Resource Management* (16th ed.). Pearson.

Drigas, A. S., & Papoutsi, C. (2018). A new layered model on emotional intelligence. *Behavioral Sciences (Basel, Switzerland)*, *8*(5), 45. doi:10.3390/bs8050045 PMID:29724021

Driscoll, A., Jicha, K., Hunt, A. N., Tichavsky, L., & Thompson, G. (2012). Can online courses deliver In-class results?: A comparison of student performance and satisfaction in an online versus a face-to-face introductory sociology course. *Teaching Sociology*, *40*(4), 312–331. doi:10.1177/0092055X12446624

Druskat, V., & Wolff, S. (2001). Building the Emotional Intelligence of Groups. In D. Goleman, A. McKee, & S. Achor (Eds.), *Everyday Emotional Intelligence* (pp. 73–93). Harvard Business School Publishing Corporation.

Du Plessis, M., Waglay, A., & Becker, J. (2020). A leader member exchange perspective. *SA Journal of Industrial Psychology*, *46*. Advance online publication. doi:10.4102ajip.v46i0.1762

DuBrin, A. J. (2007). *Fundamentals of organizational behavior (4th ed.)*. Thomson South-Western.

Eco Attainment Network. (n.d.). *EcoNetwork*. https://educationcoalition.com/eco-network/

Edelman, P., & Van Knippenberg, D. (2018). Emotional intelligence, management of subordinate's emotions, and leadership effectiveness. *Leadership & Organization Development Journal. Bradford., 39*(5), 592–607. doi:10.1108/LODJ-04-2018-0154

Education in a post-COVID world-nine ideas for public action. (n.d.). Retrieved Jan 16, 2022, from https://en.unesco.org/sites/default/files/education_in_a_post-covid_world-nine_ideas_for_public_action.pdf

Eichensehr, K. E. (2014). Cyber law of Nations. *The Georgetown Law Journal, 103*, 317.

Ekwueme, C. O, Ekon, E. E. & Ezenwa-Nebife, E. E. (2015). The impact of hands-on-approach on student academic performance in basic science and mathematics. *Higher Education Studies, 5*(6).

El Samea, D. A., & Rashed, A. (2021). The Impact of Customer's Perception of the Practices of Corporate Social Responsibility on Purchase Intention. *Journal of Accounting, Business & Management, 28*(2), 83–103.

Elbanna, S., Thanos, I. C., & Jansen, R. J. G. (2020). A literature review of the strategic decision-making context: A synthesis of previous mixed findings and an agenda for the way forward. *Management, 23*(2), 42–60. doi:10.37725/mgmt.v23i2.4621

Elbarazi, A. S. (2022). Post-Traumatic Stress Disorder and Difficulties in emotion regulation among university students under the COVID-19 condition. *Journal of Evidence-Based Psychotherapies, 22*(1), 87–101. doi:10.24193/jebp.2022.1.5

Engineers, . (n.d.). *Reshoring poised to surge 38% to record high.* Reshoring Initiative® IH2021 Data Report. Retrieved March 20, 2022, from https://reshorenow.org/blog/reshoring-initiative-ih2021-data-report/

Enyan, N. I. E., Boso, C. M., Ofori, A. A., Agyare, D. F., Aboh, I. K., Dike, N. O., Agyei, D. D., & Obiri-Yeboah, D. (2022). Clinical preceptorship in Ghana in the era of COVID-19 pandemic: An interventional study. *BMC Nursing, 21*(1), 1–12. doi:10.118612912-022-00838-w PMID:35287665

Eriksson, T., Bigi, A., & Bonera, M. (2020). Think with me, or think for me? on the future role of artificial intelligence in marketing strategy formulation. *The TQM Journal, 32*(4), 795–814. doi:10.1108/TQM-12-2019-0303

EUM. (2017). *Effective Utility Management: A Primer for Water and Wastewater Utilities.* Retrieved from https://www.epa.gov/sites/default/files/2017-01/documents/eum_primer_final_508-january2017.pdf

Everard, J. (2000). *VirtualmStates: The Internet and the boundaries of the Nation State.* Psycology Press.

Ferguson, S. (2022, March 10). *Understanding America's labor shortage.* U.S. Chamber of Commerce. Retrieved March 15, 2022, from https://www.uschamber.com/workforce/understanding-americas-labor-shortage

Fischer, P., Krueger, J. I., Greitemeyer, T., Vogrincic, C., Kastenmuller, A., Frey, D., Heene, M., Wicher, M., & Kainbacher, M. (2011). The bystander-effect: A meta-analytic review on bystander intervention in dangerous and non-dangerous emergencies. *Psychological Bulletin, 137*(4), 517–537. doi:10.1037/a0023304 PMID:21534650

Fisher, D., & Frey, N. (2003). Writing instruction for struggling adolescent readers: A gradual release model. *Journal of Adolescent & Adult Literacy, 46*, 396–407.

Florida Department of Health [FDOH]. (2017). *Disease reporting and surveillance.* https://www.floridahealth.gov/diseases-and-conditions/disease-reporting-and-management/disease-reporting-and-surveillance/index.html

Flynn, J. (2020). *Smart Strategies for Managing a Dispersed Workforce.* Retrieved on 2.12.22 from https://www.insight.com/en_US/content-and-resources/tech-journal/fall-2020/smart-strategies-for-managing-a-dispersed-workforce.html

Forbes Expert Panel and Forbes Coaches Council AI. (2021, August 27). *Key ways artificial intelligence can improve recruiting in the hiring process.* Forbes. https://www.forbes.com/sites/forbescoachescouncil/2021/08/27/key-ways-artificial-intelligence-can-improve-recruiting-in-the-hiring-process/?sh=78908a412c52

Foslein, L. (2021). *5 things emotionally intelligent managers do differently.* Retrieved from: https://www.fastcompany.com/90705760/5-things-emotionally-intelligent-managers-do-

Four types of company culture: Which one is yours? (2021, November 26). Retrieved Jan 25, 2022, from https://www.personio.com/hr-lexicon/company-culture-types/

Fox, A. S., & Shackman, A. J. (2017). (in press). The central extended amygdala in fear and anxiety: Closing the gap between mechanistic and neuroimaging research. *Neuroscience Letters.*

Francis, M. K., Wormington, S. V., & Hulleman, C. (2019). The costs of online learning: Examining differences in motivation and academic outcomes in online and face-to-face community college developmental mathematics courses. *Frontiers in Psychology, 10.* doi:10.3389/fpsyg.2019.02054

Fred, C. (2019). *The 24-Hour Rule: Leading in a Frenetic World.* Magnusson-Skor Publishing, LLC.

FreudS. (1921). Retrieved from: https://www.bartleby.com/288

Futcher, C. (2019). *Transformational Leadership Advantages and Disadvantages Explored.* Retrieved on 2/7/22 from https://info.cavendishwood.com/blog/transformational-leadership-advantages-and-disadvantages-explored

Gallager, S., & Palmer, J. (2020, September 29). *The pandemic pushed universities online. The change was long overdue.* Harvard Business Review. https://hbr.org/2020/09/the-pandemic-pushed-universities-online-the-change-was-long-overdue

Gallup. (2015). *State of the American Manager.* Retrieved on 2.24.22 from https://www.gallup.com/services/182138/state-american-manager.aspx

Gallup. (2021). *Americans Offer Gloomy State of the Nation Report.* Retrieved on 1/22/22 from https://news.gallup.com/poll/389309/americans-offer-gloomy-state-nation-report

Gamble, J. E., Peteraf, M. A., & Thompson, A. A. (2021). *Essentials of Strategic Management: The quest for competitive advantage* (7th ed.). McGraw-Hill Education.

García-Bullé, S. (2019). *What is social intelligence and why it should be taught in schools.* Retrieved from: https://observatory.tec.mx/edu-news/social-intelligence

Gardner, H. (1983). *Frames of mind.* Basic Books.

Gartner. (2019). Available at: https://www.gartner.com/en/doc/374107-predicts-2019-

Gazza, E. A. (2022). The Experience of being a full-time academic nurse educator during the COVID-19 pandemic. *Nursing Education Perspectives, 43*(2), 74–79. doi:10.1097/01.NEP.0000000000000933 PMID:35113079

Ghadi, M. Y., Fernando, M., & Caputi, P. (2013). Transformational leadership and work engagement; the mediating effect of meaning in work. *Leadership and Organization Development Journal, 34*(6), 532–550. doi:10.1108/LODJ-10-2011-0110

Gigerenzer, G., & Brighton, H. (2009). Homo heuristics: Why biased minds make better inferences. *Topics in Cognitive Science, 1*(1), 107–143. https://doi.org/10.1111/j.1756-8765.2008.01006.x

Gilar-Corbi, R., Pozo-Rico, T., Sánchez, B., & Castejón, J. L. (2019). Can emotional intelligence be improved? A randomized experimental study of a business-oriented EI training program for senior managers. *PLoS One, 14*(10), e0224254.

Giuliani, N. R., & Gross, J. J. (2009). Reappraisal. In D. Sander & K. R. Scherer (Eds.), *Oxford Companion to the Affective Sciences.* Oxford University Press.

Giurge, L., & Bohns, V. (2020). 3 Tips to Avoid WFH Burnout. *Harvard Business Review.*

Glaser, J. (2014). *Conversational Intelligence: How Great Leaders Build Trust and Get Extraordinary Results.* Bibliomotion, Inc.

Global Workplace Analytics. (n.d.). *Work-at-Home After COVID-19 – Our Forecast.* Retrieved on 2/5/22 from https://globalworkplaceanalytics.com/work-at-home-after-covid-19-our-forecast

Goërtz, Van Herck, Delbressine, Vaes, Meys, Machado, Houben-Wilke, Burtin, Posthuma, Franssen, van Loon, Hajian, Spies, Vijlbrief, van 't Hul, Janssen, & Spruit. (2020). *Persistent symptoms 3 months after a SARS-CoV-2 infection: the post-COVID-19 syndrome?* doi:10.1183/23120541.00542-2020

Gohm, C., Corser, G., & Dalsky, D. (2005). Emotional Intelligence under Stress: Useful, unnecessary, or irrelevant? *Personality and Individual Differences, 39*(6), 1017–1028. doi:10.1016/j.paid.2005.03.018

Goleman, D. (2014). *What makes a leader? Why Emotional Intelligence matters.* More Than Sound LLC.

Goleman, D. (2015). How to be emotionally intelligent? *The New York Times*. Retrieved from: https://www.nytimes.com/2015/04/12/education/edlife/how-to-be emotionally-intelligent.html?_r=1

Goleman, D. (2022). *A model to solve workplace problems*. LinkedIn.

Goleman, D. (2011a). *The Brain and Emotional Intelligence New Insights*. More Than Sound.

Goleman, D. (2013). What is Empathy? In D. Goleman, A. McKee, & S. Achor (Eds.), *Everyday Emotional Intelligence* (pp. 293–297). Harvard Business School Publishing Corporation.

Goleman, D., Boyatzis, R., & McKee, A. (2001). Primal Leadership: The Hidden Driver of Great Performance. In D. Goleman, A. McKee, & S. Achor (Eds.), *Everyday Emotional Intelligence* (pp. 25–44). Harvard Business School Publishing Corporation.

Goleman, D., Boyatzis, R., & McKee, A. (2002). *Primal leadership. Realizing the power of emotional intelligence*. Harvard Business School Press.

Gordan, M. (2021). A Review of B. F. Skinner's 'Reinforcement Theory of Motivation.'. *International Journal of Research in Education Methodology.*, *5*(3), 680–688. doi:10.24297/ijrem.v5i3.3892

Görgens-Ekermans, G., & Roux, C. (2021). Revisiting the emotional intelligence and transformational leadership debate: (How) does emotional intelligence matter to effective leadership? *SA Journal of Human Resource Management Menslikehulpbronbestuur, 19*(0), a1279. .v19i0.1279 doi:10.4102/sajhrm

Gostin, L. O. (1996). The public health information infrastructure. A national review of the law on health information privacy. *Journal of the American Medical Association, 275*(24), 1921–1927. doi:10.1001/jama.1996.03530480063042 PMID:8648874

Green, S. (2022, February 2). *What is candidate experience & why you should be paying more attention to it*. https://www.innoflow.io/2022/02/02/what-is-candidate-experience-why-you-should-be-paying-more-attention-to-it/

Green, M. T., & Puetzer, M. (2002). The value of mentoring: A strategic approach to retention and recruitment. *Journal of Nursing Care Quality, 17*(1), 67–74. doi:10.1097/00001786-200210000-00008 PMID:12369749

Groves, K. S., & LaRocca, M. A. (2011). An empirical study of leader ethical values, transformational and transactional leadership, and follower attitudes toward corporate social responsibility. *Journal of Business Ethics*, *103*, 511–513. doi:10.100710551-011- 0877-y

Groves, K., Enrue, M. P., & Shen, W. (2010). Managerial emotional intelligence, affective organizational commitment, and emotional labor. *Journal of Leadership Review.*, *30*, 31–35.

Grubb, M., Hourcade, J.-C., & Neuhoff, K. (2014). *Planetary Economics: Energy, climate change, and the three domains of sustainable development*. Routledge.

Gruszka, A., Matthews, G., & Szymura, B. (2010). *Handbook of Individual Differences in Cognition: Attention, Memory, and Executive Control*. Springer.

Guha, T. (2021, February 4) *AI is transforming how and where we work*. Accenture. https://www.accenture.com/us-en/blogs/business-functions-blog/how-ai-is-transforming-work

Gupta, N., & Beehr, T. (1979). Job Stress and Employee Behaviors. *Organizational Behavior and Human Performance, 23*(3), 373–387. doi:10.1016/0030-5073(79)90004-7 PMID:10242304

Gürçay, B., Mellers, B. A., & Baron, J. (2014). The Power of Social Influence on Estimation Accuracy. *Journal of Behavioral Decision Making, 28*, 250–261. doi:10.1002/bdm.1843

Hadley, C. N., Pittinsky, T. L., Sommer, S. A., & Zhu, W. (2011). Measuring the efficacy of leaders to assess information and make decisions in a crisis: The C-LEAD scale. *The Leadership Quarterly, 22*(4), 633–648. https://doi.org/10.1016/j.leaqua.2011.05.005

Hall, J. (1604). *Two guides to a good life: The genealogy of virtue and the nathomy of sinne. London, England*. Laggard.

Hammond, L. D. (2014). Strengthening clinical preparation: The holy grail of teacher education. *Peabody Journal of Education, 89*(4), 547–561. doi:10.1080/0161956X.2014.939009

Hasan, A., & Masoudi, A. (2011). *Jean Piaget: Intellectual development*. Retrieved from https://www.sk.com.br/sk-piage.html

Haskell-McBee, R. & Westcott, P.M. (2008). Daring to care. *Action in Teacher Education, 29*(5-6), 49-65. doi:10.1080/01626620.2008.10519434

Hastie, R., & Dawes, R. M. (2010). *Rational Choice in an Uncertain World: The Psychology of Judgment and Decision Making* (2nd ed.). Sage.

Hatfield, E., Cacioppo, J., & Rapson, R. (1994). *Emotional contagion (Studies in emotion and social interaction)*. Cambridge University Press.

Hauer, M. P., Kevekordes, J., & Haeri, M. A. (2021). Legal perspective on possible fairness measures: A legal discussion using the example of hiring decisions. *Computer Law & Security Review, 42*, 105583. doi:10.1016/j.clsr.2021.105583

Hauser, A., Eggers, F., & Güldenberg, S. (2020). Strategic decision-making in SMEs: Effectuation, causation, and the absence of strategy. *Small Business Economics, 54*(3), 775–790. https://doi.org/10.1007/s11187-019-00152-x

Heflin, C., Jensen, J., & Miller, K. (2014). Understanding the economic impacts of disruptions in water service. *Evaluation and Program Planning, 46*, 80–86. doi:10.1016/j.evalprogplan.2014.05.003 PMID:24950018

Heinegg, V., & Heintschel, W. (2012, June). *Legal Implications of Territorial Soverignty in Cyberspace*. Academic Press.

Heinzman, J. R. (2016). *Divided You Fall, Creating an Appreciative Environment*. Kendall Publishing.

Helfat, C. E., & Peteraf, M. A. (2015). Managerial cognitive capabilities and the microfoundations of dynamic capabilities. *Strategic Management Journal, 36*(6), 831–850. https://doi.org/10.1002/smj.2247

Hellriegel, D., & Slocum, J. W. (2007). *Organizational behavior* (11th ed.). Thomson South-Western.

Hernandez, R. (2022). Will the COVID-19 pandemic reduce potential output? *Monthly Labor Review*. Retrieved on 2/5/22 from https://www.bls.gov/opub/mlr/2022/beyond-bls/pdf/will-the-covid-19-pandemic-reduce-potential-output

Héroux, S., Fortin, A., & Goupil, C. (2020). Adherence to expense report approval control: An application of the theory of planned behavior. *Journal of Applied Accounting Research, 21*(3), 397–413. doi:10.1108/JAAR-12-2018-0198

Herwehe, J., Wilbright, W., Abrams, A., Bergson, S., Foxhood, J., Kaiser, M., Smith, L., Xiao, K., Zapata, A., & Magnus, M. (2012). Implementation of an innovative, integrated electronic medical record (EMR) and public health information exchange for HIV/AIDS. *Journal of the American Medical Informatics Association: JAMIA, 19*(3), 448–452. doi:10.1136/amiajnl-2011-000412 PMID:22037891

Hicks, M. (2022, February 7). *Opinion: Some surprising lessons from COVID learning loss*. Indiana Economic Digest. https://indianaeconomicdigest.net/MobileContent/Default/Major-Indiana-News/Article/OPINION-Some-surprising-lessons-from-COVID-learning-loss/-3/5308/109729

Hines, T. E. (2011). *Leadership: A balancing act*. Wiley Online Library. doi:10.1002/cc.468

Hoerr, T. (2017). *The formative five: Fostering grit, empathy, and other success skills every student needs*. ASCD.

Hughes, W., Franklin, V., Bloom, D. (2021). *Battling The Assassins of Your Success*. Academic Press.

Humphrey, R., Kellett, J., & Sleeth, R. (2006). Empathy and the emergence of task and relations leaders. *The Leadership Quarterly, 17*(2), 146–162.

Huntingford, C., Jeffers, E. S., Bonsall, M. B., Christensen, H. M., Lees, T., & Yang, H. (2019). Machine learning and artificial intelligence to aid climate change research and preparedness. *Environmental Research Letters, 14*(12), 124007. https://doi.org/10.1088/1748-9326/ab4e55

Ilgen, D. R., Hollenbeck, J. R., Johnson, M., & Jundt, D. (2005). Teams in organizations: From input-process-output models to IMOI models. *Annual Review of Psychology, 56*, 517–543.

Impact of covid-19 on Technology. (2021, July 13). Retrieved Jan 16, 2022, from https://bernardmarr.com/impact-of-covid-19-on-technology/

Indiana Department of Education. (n.d.). *Indiana educator standards.* https://www.in.gov/doe/educators/educator-licensing/indiana-educator-standards/

Institute for Business Value. (2020). *COVID-19 and the future of business.* IBM Corporation. https://www.ibm.com/thought-leadership/institute-business-value/report/covid-19-future-business

Institute for Health and Human Portal. (2021). *The meaning of emotional intelligence.* https://www.ihhp.com/meaning-ofemotional-intelligence

Institute of Medicine (IOM). (2007). *Training physicians for public health careers.* The National Academies Press., doi:10.17226/11915

ISAS.NUS.EDU.SG. (n.d.). Retrieved Dec 6, 2021, from https://www.isas.nus.edu.sg/papers/education-during-the-covid-19-pandemic-in-india/

Ivleva, N. P. (2021). Chemical Analysis of Microplastics and Nanoplastics: Challenges, Advanced Methods, and Perspectives. *Chemical Reviews, 121*(19), 11886–11936. doi:10.1021/acs.chemrev.1c00178 PMID:34436873

Izard, C. E. (2006). *Experts' definitions of emotion and their ratings of its components and characteristics* [Unpublished manuscript]. University of Delaware.

Jacobs, I., Wollny, A., Seidler, J., & Wochatz, G. (2021). A trait emotional intelligence perspective on schema modes. *Scandinavian Journal of Psychology, 62*(2), 227–236. https://doi.org/10.1111/sjop.12670

Jadoul, Q., Nascimento, A., Salo, O., & Willi, R. (2020). *Agility in the Time of COVID-19: Changing Your Operating Model in an Age of Turbulence.* McKinsey & Company. https://www.mckinsey.com/business-functions/people-and-organizational-performance/our-insights/agility-in-the-time-of-covid-19-changing-your-operating-model-in-an-age-of-turbulence

Jain, J., Walia, N., & Gupta, S. (2020). Evaluation of behavioral biases affecting investment decision-making of individual equity investors by fuzzy analytic hierarchy process. *Review of Behavioral Finance, 12*(3), 297–314. https://doi.org/10.1108/RBF-03-2019-0044

Janati, A., Hosseiny, M., Gouya, M. M., Moradi, G., & Ghaderi, E. (2015). Communicable disease reporting systems in the world: A systematic review article. *Iranian Journal of Public Health, 44*(11), 1453–1465. https://www.ncbi.nlm.nih.gov/pmc/articles/PMC4703224/ PMID:26744702

January 2021: Educational Resurgence Journal: DYPCOED, Pimpri. (n.d.). Retrieved Dec 14, 2021, from https://coed.dypvp.edu.in/educational-resurgence-journal/jan-2021.aspx

Jarrahi, M. H. (2018). Artificial intelligence and the future of work: human-AI symbiosis in organizational decision making. *Business Horizons, 61*(4), 577–586.

Jasinowski, J. (2013, November 27). *The Magic job multiplier of manufacturing.* HuffPost. Retrieved March 13, 2022, from https://www.huffpost.com/entry/jobs-multiplier_b_4002113#:~:text=In%20some%20advanced%20manufacturing%20sectors%2C%20such%20as%20electronic,not%20sweetness%20and%20light%20on%20the%20manufacturing%20front

Jenks, L. (2019). The Importance of Inventory Control. *Skin Inc*, *31*(4), 58–63.

Jiang, L., Qiao, K., & Li, C. (2021). Distance-based functional criticality in the human brain: Intelligence and emotional intelligence. *BMC Bioinformatics*, *22*, 1–17. https://doi.org/10.1186/s12859-021-03973-4

Ji, B., Zhao, Y., Wei, T., & Kang, P. (2021). Water science under the global epidemic of COVID-19: Bibliometric tracking on COVID-19 publication and further research needs. *Journal of Environmental Chemical Engineering*, *9*(4), 105357. doi:10.1016/j.jece.2021.105357 PMID:33747765

Jones, J. (2022). *Americans Offer Gloomy State of the Nation Report*. Retrieved on 2.2.22 from https://news.gallup.com/poll/389309/americans-offer-gloomy-state-nation-report.aspx?version=print

Jong, W., Dückers, M. L. A., & van der Velden, P. G. (2016). Leadership of mayors and governors during crises: A systematic review on tasks and effectiveness. *Journal of Contingencies and Crisis Management, 24*(1), 46-58. doi:10.1111/1468-5973.12091

Judge, D. S., & Murray, B. (2017). Student and faculty transition to a new online learning management system. *Teaching and Learning in Nursing*, *12*(4), 277–280. doi:10.1016/j.teln.2017.06.010

Judge, T. A., & Piccolo, R. F. (2004, October). Transformational and transactional leadership: A meta-analytic test of the relative validity. *The Journal of Applied Psychology*, *89*(5), 755–757.

Kahai, S. S., Sosik, J. J., & Avolio, B. J. (1997). Effects of leadership style and problem structure on workgroup process and outcomes in an electronic meeting system environment. *Pers. Psychol.*, *50*, 121–146. doi: .1997.tb00903.x doi:10.1111/j.1744-6570

Kahneman, D., & Tversky, A. (1979). Prospect Theory: An Analysis of Decisions Under Risk. *Econometrica*, *47*(2), 263–291. doi:10.2307/1914185

Kao, A. B., & Couzin, I. D. (2014). Decision accuracy in complex environments is often maximized by small group sizes. *Proceedings. Biological Sciences*, *281*, 1–8. doi:10.1098/rspb.2013.3305

Karasneh, A. A.-F., & Al-Momani, E. A. (2020). The influence of emotional intelligence on leadership styles: A profile of executives in ASEZA organization. *Journal of Public Affairs*. doi:10.1002/pa.2069

Karimi, L., Leggat, S. G., Bartram, T., Afshari, L., Sarkeshik, S., & Verulava, T. (2021). Emotional intelligence: Predictor of employees' wellbeing, quality of patient care, and psychological empowerment. *BMC Psychology*, *9*, 1–7. https://doi.org/10.1186/s40359-021-00593-8

Katzenbach, J. R., & Smith, D. K. (1999). *The Wisdom of Teams: Creating High- Performance Organizations*. HarperBusiness.

Kaye, K., Singer, J., Newton-Dame, R., & Shih, S. C. (2014). Health information technology and the primary care information project. *American Journal of Public Health, 104*(3), e8–e9. doi:10.2105/AJPH.2013.301787 PMID:24432950

Kennedy, M. (2016, April 20). *Lead-Laced Water In Flint: A Step-By-Step Look At The Makings Of A Crisis*. NPR. Retrieved from https://www.npr.org/sections/thetwo-way/2016/04/20/465545378/lead-laced-water-in-flint-a-step-by-step-look-at-the-makings-of-a-crisis

Kerr, N. L., Maccoun, R. J., & Kramer, G. P. (1996). Bias in judgment: Comparing individuals and groups. *Psychological Review, 103*, 687–719.

Kerr, N. L., & Tindale, R. S. (2004). Group Performance and Decision Making. *Annual Review of Psychology, 55*, 623–655. doi:10.1146/annurev.psych.55.090902.142009

Kim, H., & Kim, T. (2017). Emotional intelligence and transformational leadership: A review of empirical studies. *Human Resource Development Review, 16*(4), 377–393. https://doi.org/10.1177/1534484317729262

King, A. J., Cheng, L., Starke, S. D., & Myatt, J. P. (2012). Is the true 'wisdom of the crowd' to copy successful individuals? *Biology Letters, 8*, 197–200. doi:10.1098/rsbl.2011.0795

Kipling, R. (1943). *If*. Retrieved on 2.12.22 from https://www.poetryfoundation.org/poems/46473/if--

Kleinbaum, A. M., & Stuart, T. E. (2014). Inside the black box of the corporate staff: Social networks and the implementation of corporate strategy. *Strategic Management Journal, 35*(1), 24–47. https://doi.org/10.1002/smj.2090

Klompas, M., McVetta, J., Lazarus, R., Eggleston, E., Haney, G., Kruskal, B., Yih, K., Daly, P., Oppedisano, P., Beagan, B., Lee, M., Kirby, C., Heisey-Grove, D., DeMaria, A. Jr, & Platt, R. (2012). Integrating clinical practice and public health surveillance using electronic medical record systems. *American Journal of Public Health, 102*(S3), S325–S332. doi:10.2105/AJPH.2012.300811 PMID:22690967

Kokemuller, N. (2022). The Importance of Timing in Leadership. *Small Business Chronicle*. Retrieved on 1/12/22 from https://smallbusiness.chron.com/importance-timing-leadership-51987.html

Konowitz, P. M., Petrossian, G. A., & Rose, D. N. (1984). The underreporting of disease and physicians' knowledge of reporting requirements. *Public Health Reports, 99*(1), 31-5. https://pubmed.ncbi.nlm.nih.gov/6422492/

Kowalczyk, M. (2017). *The Support of Decision Processes with Business Intelligence and Analytics: Insights on the Roles of Ambidexterity, Information Processing, and Advice*. Springer.

Kraemer, H. (2015). *Becoming the Best: Build a World-Class Organization through Values-Based Leadership*. John Wiley & Sons, Inc.

Kranz, D., Niepel, C., Botes, E., & Greiff, S. (2020). Religiosity Predicts Unreasonable Coping With COVID-19. *Psychology of Religion and Spirituality*. Advance online publication. doi:10.1037/rel0000395

Krogerus, M., & Tschäppeler, R. (2017). *The Decision Book: 50 Models for Strategic Thinking*. WW Norton & Company.

Kuckertz, A., Brändle, L., Gaudig, A., Hinderer, S., Morales Reyes, C. A., Prochotta, A., Steinbrink, K. M., & Berger, E. S. C. (2020). Startups in times of crisis – A rapid response to the COVID-19 pandemic. *Journal of Business Venturing Insights, 13*, e00169. https://doi.org/10.1016/j.jbvi.2020.e00169

Kumar, S. J., & Iyer, V. R. (2012). Emotional intelligence and quality of work-life among employees in the educational institutions. *SIES Journal of Management, 8*(2), 55.

Kuntz, J., Davies, B., & Naswall, K. (2019). From transactional to transformational: Exploring the role of leadership style on CEO succession outcomes. *Leadership and Organization Development Journal, 40*(7), 815–827. doi:10.1108/LODJ-01-2019-0012

Lagowska, U., Sobral, F., & Furtado, L. M. G. P. (2020). Leadership under crises: A research agenda for the post-COVID-19 era. *Brazilian Administration Review, 17*(2), 1-5. doi:10.1590/1807-7692bar2020200062

Lai, F.-Y., Tang, H.-C., Lu, S.-C., Lee, Y.-C., & Lin, C.-C. (2020). Transformational Leadership and Job Performance: The Mediating Role of Work Engagement. *SAGE Open, 10*(1). doi:10.1177/215824401989908

LaMorte, W. (2017). *Disease surveillance and data sources: Surveillance for infectious disease.* Retrieved from: http://sphweb.bumc.bu.edu/otlt/MPH-Modules/EP/EP713_Surveillance/EP713_Surveillance3.html#headingtaglink_1

Lange, S. (1922). *The emotions volume.* Retrieved from: archive.org/details/emotionsvolumei007644mbp

Lapointe, E., & Vandenberghe, C. (2018). Examination of the Relationships Between Servant Leadership, Organizational Commitment, and Voice and Antisocial Behaviors. *Journal of Business Ethics, 148*(1), 99–115. doi:10.100710551-015-3002-9

Larsen, D. A., & Wigginton, K. R. (2020). Tracking COVID-19 with wastewater. *Nature Biotechnology, 38*(10), 1151–1153. doi:10.103841587-020-0690-1 PMID:32958959

Latham, G. (2004). The Motivational Benefits of Goal setting. *The Academy of Management Executive, 18*(4), 126-129.

Latham, G., & Locke, E. (2007). New Development in and Directions for Goal-Setting Research. *European Psychologist, 12*(4), 290–300. doi:10.1027/1016-9040.12.4.290

Lawrence, J. A., & Parkes, R. (2021). How do you be your best self in a vexatious world? *Nephrology Nursing Journal, 48*(3), 261-265. .2021.48.3.261 doi:10.37526/1526-744X

Lawson, H. (2021, December 19). *Nearly 40% of students graduate with honors.* IN.gov. https://www.in.gov/doe/about/news/indiana-department-of-education-announces-2021-graduation-rate/

Lawson, H. (2021, January 15). *Indiana graduation rates increase in 2020.* IN.gov. https://www.in.gov/doe/about/news/january-15/#:~:text=Indiana%27s%20federal%20graduation%20rate%20for%202019%20was%2 086.46%20percent

Lawson, E., Bunney, S., Cotterill, S., Farmani, R., Melville-Shreeve, P., & Butler, D. (2021). COVID-19 and the UK water sector: Exploring organisational responses through a resilience framework. *Water and Environment Journal: the Journal / the Chartered Institution of Water and Environmental Management.*

Lawton-Misra, N., & Pretorius, T. (2021). Leading with heart: Academic leadership during the Covid-19 crisis. *South African Journal of Psychology. Suid-Afrikaanse Tydskrif vir Sielkunde, 51*(2), 205–214. doi:10.1177/0081246321992979

Lazarus, R. S. (1991). *Emotion and adaptation.* Oxford University Press.

Lee, C., Abdullah, A., Ismail, A., & Alizydeen, N. (2011). How democratic leaders empower teachers job satisfaction? The Malaysian Case. *International Journal of Business and Science, 2*(10), 251.

Liborius, P. (2014). Who is worthy of being followed? The impact of leaders' character and the moderating role of followers' personality. *The Journal of Psychology, 148*, 347–385. doi:10.1080/00223980.2013.801335

Lilian, S. (2014). Virtual Teams: Opportunities and challenges for e-leaders. *Procedia: Social and Behavioral Sciences, 110*, 1251–1261. doi:10.1016/j.sbspro.2013.12.972

Limited, I. (n.d.). *Navigating the next for financial services.* Infosys. Retrieved Jan 11, 2022, from https://www.infosys.com/industries/financial-services.html

Loayza, A. (2020). Cyber Privateers: Protecting American Intellectual Property from Cyber Theft. *Landslide, 12*(5), 27–32.

Lorenz, J., Rauhut, H., Schweitzer, F., & Helbing, D. (2011). How social influence can undermine the wisdom of crowd effect. *Proceedings of the National Academy of Sciences of the United States of America, 108*, 9020–9025. doi:10.1073/pnas.1008636108

Loten, A. (2022, January 7). *Employers, investors take notice of ai tools to speed job recruitment.* The Wall Street Journal. https://www.wsj.com/articles/employers-investors-take-notice-of-ai-tools-to-speed-job-recruitment-11641599629

Loughlin, S. (2018, October 30). More schools in Indiana adopt elearning. *Tribune-Star.*

Lubbadeh, T. (2020). Emotional intelligence and leadership–the dark and bright sides. Publishing House of Rzeszow University of Technology.

Lu, J., Jiang, X., Yu, H., & Li, D. (2015). Building collaborative structures for teachers' autonomy and self-efficacy: The mediating role of participative management and learning culture. *School Effectiveness and School Improvement, 26*, 240–257. doi:10.1080/09243453.2014.888086

Lumen Learning. (n.d.). *Listening: Three A's of Active Listening*. Retrieved on 2/15/22 from https://courses.lumenlearning.com/ivytech-comm101-master/chapter/chapter-4-three-as-of-active-listening/

MacDonald, M. (2022). *Six Ways Leaders Can Adapt to the Workplace of 2022*. Retrieved on 2.4.22 from https://sloanreview.mit.edu/article/six-ways-leaders-can-adapt-to-the-workplace-of-2022/

Maden, C. (2012). Transforming public organizations into learning organizations: A conceptual model. *Public Organization Review, 12*, 73. doi:10.100711115-011- 0160-9

Madhavan, S., Bastarache, L., Brown, J. S., Butte, A. J., Dorr, D. A., Embi, P. J., Friedman, C. P., Johnson, K. B., Moore, J. H., Kohane, I. S. O., Payne, P. R., Tenenbaum, J. D., Weiner, M. G., Wilcox, A. B., & Ohno-Machado, L. (2021). Use of electronic health records to support a public health response to the COVID-19 pandemic in the United States: A perspective from 15 academic medical centers. *Journal of the American Medical Informatics Association: JAMIA, 28*(2), 393–401. doi:10.1093/jamia/ocaa287 PMID:33260207

Mandl, K. D. (2014). Ebola in the United States: EHRs as a public health tool at the point of care. *Journal of the American Medical Association, 312*(23), 2499–2500. doi:10.1001/jama.2014.15064 PMID:25329170

Manjakkal, L., Mitra, S., Petillot, Y., Shutler, J., Scott, E., Willander, M., & Dahiya, R. (2021). Connected Sensors, Innovative Sensor Deployment,and Intelligent Data Analysis for Online Water Quality Monitoring. *IEEE Internet of Things Journal, 8*(18), 13802–13824. doi:10.1109/JIOT.2021.3081772

Map: Coronavirus and school closures. (2020, March 6). *EdWeek*. Retrieved March 30, 2022, from https://www.edweek.org/leadership/map-coronavirus-and-school-closures-in-2019-2020/2020/03

Marcus, M. J. (2002). Indiana makes progress on educational attainment gap. *Indiana Business Review, 77*(2).

Marier, R. (1977). The reporting of communicable diseases. *American Journal of Epidemiology, 105*(6), 587–590. doi:10.1093/oxfordjournals.aje.a112424 PMID:868863

Markiewicz, M., Bevc, C. A., Hegle, J., Horney, J. A., Davies, M., & MacDonald, P. D. M. (2012). Linking public health agencies and hospitals for improved emergency preparedness: North Carolina's public health epidemiologist program. *BMC Public Health, 12*(1), 141. doi:10.1186/1471-2458-12-141 PMID:22361231

Massachusetts Technology Collaborative. (2012). *MeHI introduces new public health surveillance tool.* https://www/masstech.org

Mateiaş, D. E. (2018). Emotional intelligence as a dimension of transformational leadership advancement. *Revista de Psihologie, 64*(1), 39.

Mather, P. (2020). Leadership and governance in a crisis: Some reflections on COVID-19. *Journal of Accounting & Organizational Change, 16*(4), 579–585. https://doi.org/10.1108/JAOC-08-2020-0123

Maurer, R. (2021, March 26). *The pros and cons of virtual and in-person interviews*. https://www.shrm.org/resourcesandtools/hr-topics/talent-acquisition/pages/pros-and-cons-virtual-in-person-interviews.aspx

Mayer, C.-H., & Louw, L. (2012). Managing cross-cultural conflict in organizations. *International Journal of Cross Cultural Management, 12*(1), 3–8. doi:10.1177/1470595811413104

Mayer, J. D., & Salovey, P. (1990). Emotional Intelligence. *Imagination, Cognition and Personality, 9*(3), 185.

Mayo Clinic. (n.d.a). *COVID-19 (coronavirus): Long-term effects*. Retrieved on November 5, 2021, from https://www.mayoclinic.org/diseases-conditions/coronavirus/in-depth/coronavirus-long-term-effects/art-20490351

Mayo Clinic. (n.d.b). *Resilience Training*. Retrieved 2/12/22 from Mayoclinic.org/tests-procedures/resilience-training/in-depth/resilience/art-20046311

McCleskey, J., & Gruda, D. (2021). Risk-taking, resilience, and state anxiety during the COVID-19 pandemic: A coming of (old) age story. *Personality and Individual Differences, 170*. doi:10.1016/j.paid.2020.110485

McGrath, R. G. (2013). Six signs that your innovation program is broken. *Ivey Business Journal*.

McKeever, A. (2021, September 21). COVID-19 surpasses 1918 flu as deadliest pandemic in U.S. history. *National Geographic*. Retrieved from https://www.nationalgeographic.com/history/article/covid-19-is-now-the-deadliest-pandemic-in-us-history

McKibbin, W. J., & Fernando, R. (2020). *The global macroeconomic impacts of COVID-19: seven scenarios*. doi:10.2139srn.3547729

McKinsey & Company. (2022, April 13). *Covid-19: Implications for business in 2021*. McKinsey & Company. Retrieved April16, 2022, from https://www.mckinsey.com/business-functions/risk-and-resilience/our-insights/covid-19-implications-for-business-2021

McKinsey. (2020). *Psychological safety, emotional intelligence, and leadership in a time of flux*. July 2, 2020. Retrieved on 1/21/22 from https://www.mckinsey.com/featured-insights/leadership/psychological-safety-emotional-intelligence-and-leadership-in-a-time-of-flux

McRae, K., Misra, S., Prasad, A. K., Pereira, S. C., & Gross, J. J. (2011). Bottom-up and top-down emotion generation: Implications for emotion regulation. *Social Cognitive and Affective Neuroscience, 7*(3), 253–262. doi:10.1093can/nsq103

McRaney, D. (2011). *You are not so smart*. Gotham Books.

Meyer, R. K., Dalal, R. S., & Hermida, R. (2010). A review and synthesis of situational strength in the organizational sciences. *Journal of Management, 36*(1), 121–140. doi:10.1177/0149206309349309

Michigan Department of Health and Human Services [MDHHS]. (2020). *Communicable disease reporting in Michigan.* https://www.michigan.gov/mdhhs/0,5885,7-339-71550_5104_31274-12538--,00.html

Miller, K. E. (2021). A light in students' lives: K-12 teachers' experiences (re)building caring relationships during remote learning. *Online Learning, 25*(1), 115–134. doi:10.24059/olj.v25i1.2486

Mitre Media. (2015, September 11). Moody's Research Reports. *By Issuer* [data set]. Retrieved from http://www.municipalbonds.com/bonds/moodys_reports/

Mofijur, M., Fattah, I. R., Alam, M. A., Islam, A. S., Ong, H. C., Rahman, S. A., ... Mahlia, T. M. I. (2021). Impact of COVID-19 on the social, economic, environmental and energy domains: Lessons learnt from a global pandemic. *Sustainable Production and Consumption, 26*, 343-359.

Mohapatra, H., & Rath, A. (2020). IoT-based Smart Water. In IoT Technologies in Smart-Cities: From sensors to big data, security and trust (pp. 63-93). The Institution of Engineering and Technology. doi:10.1049/PBCE128E_ch3

Monks, I., Stewart, R. A., Sahin, O., & Keller, R. (2019). Revealing unreported benefits of digital water metering: Literature review and expert opinions. *Water (Basel), 11*(4), 838. doi:10.3390/w11040838

Montibeller, G., & von Winterfeldt, D. (2015). Cognitive and motivational biases in decision and risk analysis. *Risk Analysis, 35*(7), 1230–1251. https://doi.org/10.1111/risa.12360

Morilla, M., Sans, M., Casasa, A., & Gimenez, N. (2017). Implementing technology in healthcare: Insights from physicians. *BMC Medical Informatics and Decision Making, 17*(92), 92. Advance online publication. doi:10.118612911-017-0489-2 PMID:28655299

Morońa, M., & Biolik-Moroń, M. (2021). Trait emotional intelligence and emotional experiences during the COVID19 pandemic outbreak in Poland: A daily diary study. *Personality and Individual Differences, 168*(110348), 1.

Motlaghzadeh, K., Kerachian, R., & Tavvafi, A. (2020). An evidential reasoning-based leader-follower game for hierarchical multi-agent decision making under uncertainty. *Journal of Hydrology (Amsterdam), 591*, 125294. https://doi.org/10.1016/j.jhydrol.2020.125294

Mukarram, M. (2020). Impact of COVID-19 on the UN sustainable development goals (SDGs). *Strategic Analysis, 44*(3), 253–258. doi:10.1080/09700161.2020.1788363

Mukherjee, M. (2022, January 30). *The lost generation: How unending school closures have crippled the education of India's children.* Retrieved Feb 6, 2022, from https://scroll.in/article/1016178/the-lost-generation-how-unending-school-closures-have-crippled-the-education-of-india-s-children

Nappinai, N. S. (2009). Cybercrime Law in India. Has law kept pace with emerging trends? An Emperical Study. *Journal of International Commercial Law and Technology*, *5*(1), 22–28.

National Infrastructure Advisory Council. (2016). *Water Sector Resilience Final Report and Recommendations*. Retrieved from https://www.cisa.gov/sites/default/files/publications/niac-water-resilience-final-report-508.pdf

Netz, J., Svensson, M., & Brundin, E. (2020). Business disruptions and affective reactions: A strategy-as-practice perspective on fast strategic decision making. *Long Range Planning*, *53*(5), 101910. https://doi.org/10.1016/j.lrp.2019.101910

New York State Department of Health [NYDOH]. (2020). *Diseases & Conditions: Information About Diseases, Viruses, Conditions, and Prevention*. https://health.ny.gov

Newport, F. (2020). *The Religion Paradox*. https://news.gallup.com/opinion/polling-matters/310397/religion-paradox.aspx

Newton-Dame, R., McVeigh, K. H., Schreibstein, L., Perlman, S., Lurie-Moroni, E., Jacobson, L., Greene, C., Snell, E., & Thorpe, L. E. (2016). Design of the New York City Macroscope: Innovations in Population Health Surveillance Using Electronic Health Records. *EGEMS (Washington, DC)*, *4*(1), 1265. doi:10.13063/2327-9214.1265 PMID:28154835

Ng, R., Lai, P., Brown, T. T., Järvinen, A., Halgren, E., Bellugi, U., & Trauner, D. (2018). Neuroanatomical correlates of emotion-processing in children with unilateral brain lesion: A preliminary study of limbic system organization. *Social Neuroscience*, *13*(6), 688–700. https://doi-org.ezproxy.umgc.edu/10.1080/17470919.2017.1386126

Nguyen, A. (2020). On the Clock and at Home: Post-COVID-19 Employee Monitoring in the Workplace. *People & Strategy*, *43*(3), 30–33.

NIVSTATS. (n.d.). *Indiana colleges student population*. Retrieved March 30, 2022, from https://www.univstats.com/states/indiana/student-population/

No right to education in the pandemic? (2020, December 7). Retrieved Jan 1, 2022, from https://countercurrents.org/2020/12/no-right-to-education-in-the-pandemic/

Noddings, N. (1988). An ethic of caring and Its implications for instructional arrangements. *American Journal of Education*, *96*(2), 215–230. doi:10.1086/443894

Noddings, N. (2012). The caring relation in teaching. *Oxford Review of Education*, *38*(6), 771–781. doi:10.1080/03054985.2012.745047

Norheim, O. F., Abi-Rached, J. M., Bright, L. K., Baeroe, K., Ferraz, O. L. M., Gloppen, S., & Voorhoeve, A. (2021). Difficult trade-offs in response to COVID-19: The case for open and inclusive decision making. *Nature Medicine*, *27*, 10–13. https://doi.org/10.1038/s41591-020-01204-6

Nye, J. (2008). *The powers to lead*. Oxford University Press.

O'Connor, P. J., Hill, A., Kaya, M., & Martin, B. (2019). The measurement of emotional intelligence: A critical review of the literature and recommendations for researchers and practitioners. *Frontiers in Psychology, 10*, 1–19. doi:10.3389/fpsyg.2019.01116

Orcutt, M. (2020). An Elegy for Cash: The Technology We Might Never Replace. *MIT Technology Review, 123*(2), 32–37.

Osman, I. A. E. (2020). The relationship between emotional intelligence (EI) and middle management leadership styles (transformational and transactional leadership) among managerial personnel at khartoum state: Women and change. *The Ahfad Journal, 37*(1), 3-13. Retrieved from https://www.proquest.com/scholarly-journals/relationship-between-emotional-intelligence-ei/docview/2436886656/se-2?accountid=9708

Ott, T. E., & Eisenhardt, K. M. (2020). Decision weaving: Forming novel, complex strategy in entrepreneurial settings. *Strategic Management Journal, 41*(12), 2275–2314. https://doi.org/10.1002/smj.3189

Owens, R. G. (2004). *Organizational behavior in education: Adaptive leadership and school reform* (8th ed.). Pearson.

Pacheco-de-Almeida, G. (2010). Erosion, time compression, and self-displacement of leaders in hypercompetitive environments. *Strategic Management Journal, 31*, 1498–1526.

Paddy Padmanabhan, C. (2020, March 27). *How the COVID-19 pandemic is reshaping healthcare with technology*. CIO. Retrieved Dec4, 2021, from https://www.cio.com/article/193198/how-the-covid-19-pandemic-is-reshaping-healthcare-with-technology.html

Park, S., & Yun, H. (2017). The influence of motivational regulation strategies on online students' behavioral, emotional, and cognitive engagement. *American Journal of Distance Education, 32*(1), 43–56. doi:10.1080/08923647.2018.1412738

Parks, R. (2021). *COVID-19 Mental Health Crises: Holistic Understanding and Solutions*. Parks Press.

Pasmore, W., & Mallis, E. (2022). *Re-set Your Organization for the Post Covid Future with Creative Leadership*. Center for Creative Leadership. Retrieved on 1.12.22 from https://www.ccl.org/articles/white-papers/reset-your-organization-for-post-covid-future/

Patel, A., & Reinsch, L. (2003). Companies Can Apologize: Corporate Apologies and Legal Liability. *Business Communication Quarterly, 66*(1), 9–25. doi:10.1177/108056990306600103

Patterson, K., Grenny, J., McMillan, R., & Switzler, A. (2004). *Crucial Confrontations: Tools for Resolving Broken Promises, Violated Expectations, and Bad Behavior*. McGraw-Hill.

Paul, M., Greene, C., Newton-Dame, R., Thorpe, L., Perlman, S., McVeigh, K., & Gourevitch, M. (2015). The state of population health surveillance using electronic health records: A narrative review. *Population Health Management, 18*(3), 209–216. doi:10.1089/pop.2014.0093 PMID:25608033

Pawson, R., & Tilley, N. (1997). *Realistic Evaluation*. Sage.

Paykamian, B. (2021, March 29). *What's driving growth in the ed-tech market?* Government Technology. https://www.govtech.com/education/k-12/whats-driving-growth-in-the-ed-tech-market.html

Pecson, B., Gerrity, D., Bibby, K., Drewes, J. E., Gerba, C., Gersberg, R., Gonzalez, R., Haas, C. N., Hamilton, K. A., Nelson, K. L., Olivieri, A., Rock, C., Rose, J., & Sobsey, M. (2020). Editorial Perspectives: Will SARS-CoV-2 reset public health requirements in the water industry? Integrating lessons of the past and emerging research. *Environmental Science. Water Research & Technology*, *6*(7), 1761–1764. doi:10.1039/D0EW90031A

Peek, N., Sujan, M., & Scott, P. (2020, June 1). Digital Health and care in pandemic times: Impact of covid-19. *BMJ Health & Care Informatics*. Retrieved Nov 16, 2021, from https://informatics.bmj.com/content/27/1/e100166

Peng, J. C., & Chen, S. W. (2021). Servant Leadership and Service Performance: A Multilevel Mediation Model. *Psychological Reports*, *124*(4), 1738–1760. doi:10.1177/0033294120950302 PMID:32838663

Pérez, I. J., Cabrerizo, F. J., Alonso, S., Dong, Y. C., Chiclana, F., & Herrera-Viedma, E. (2018). On dynamic consensus processes in group decision-making problems. *Information Sciences*, *459*, 20–35. https://doi.org/10.1016/j.ins.2018.05.017

Petrides, K. V. (2017). Intelligence, emotional. *Reference Module in Neuroscience and Biobehavioral Psychology*, *1*, 1–16.

Petrides, K. V., Furnham, A., & Mavroveli, S. (2001). *Trait emotional intelligence. Moving forward in the field of EI*. Oxford University Press.

Pew Forum. (2008). *U.S. religious landscape survey*. Retrieved from https://www.pewforum.org/

Phillips-Wren, G., & Adya, M. (2020). Decision making under stress: The role of information overload, time pressure, complexity, and uncertainty. *Journal of Decision Systems*, 1-13. doi:10.1080/12460125.2020.1768680

Piaget, J. (1950). *The psychology of intelligence*. Routledge.

Pieterse, A. H., & de Vries, M. (2013). On the suitability of fast and frugal heuristics for designing values clarification methods in patient decision aids: A critical analysis. *Health Expectations*, *16*(3), e73–e79. https://doi.org/10.1111/j.1369-7625.2011.00720.x

Poch, M., Garrido-Baserba, M., Corominas, L., Perelló-Moragues, A., Monclús, H., Cermerón-Romero, M., Melitas, N., Jiang, S. C., & Rosso, D. (2020). When the fourth water and digital revolution encountered COVID-19. *The Science of the Total Environment*, *744*, 140980. doi:10.1016/j.scitotenv.2020.140980 PMID:32687996

Podsakoff, P. M., MacKenzie, S. B., Moorman, R. H., & Fetter, R. (1990). Transformational leader behaviors, and their effects on followers' trust in leader, satisfaction, and organizational citizenship behaviors. *The Leadership Quarterly, 1*(2), 107–142. doi:10.1016/1048-9843(90)90009-7

Poister, T. H. (2010). The future of strategic planning in the public sector: Linking strategic management and performance. *Public Administration Review, 70*, 246–254. doi:10.1111/j.1540-6210.2010.02284.x

Poll, K., Widen, J., & Weller, S. (2014). Six Instructional Best Practices for Online Engagement and Retention. *Journal of Online Doctoral Education, 1*(1), 56–72.

Polychroniou, P. (2008). Transformational Leadership and Work Motivation in modern organizations: A critical analysis. Advances in Management, 1(3), 9-12.

Porter, M. E. (1980). *Competitive strategy: Techniques for analyzing industries and competitors*. Simon & Schuster.

Potocan, V., & Nedelko, Z. (2021). The Behavior of Organization in Economic Crisis: Integration, Interpretation, and Research Development. *Journal of Business Ethics*, 1–19. doi:10.1007/s10551-021-04928-8

Pratt, M. K., & Florentine, S. (2021, August 20). *Employee retention: 10 strategies for retaining top talent*. https://www.cio.com/article/251060/how-to-improve-employee-retention.html

Pulse of Remote Work: Before & After covid-19. (2022, March 8). Retrieved May 6, 2022, from https://www.pipefy.com/blog/pulse-of-remote-work-before-after-covid-19

Purhaghshenas, S. H., & Esmatnia, M. (2012). Learning organizations. *Interdisciplinary Journal of Contemporary Research in Business, 4*(7), 243–245.

Radraf, A., Ferreira, M., Sosa, J., & Filip, I. (2021). Emergent Crisis of COVID-19 Pandemic: Mental Health Challenges and Opportunities. *Frontiers in Psychology, 12*, 631008. doi:10.3389/fpsyt.2021.631008 PMID:34349675

Raftelis. (2020). *Financial Impact of the COVID-19 Crisis on U.S. Drinking Water Utilities*. Retrieved from https://www.awwa.org/Portals/0/AWWA/Communications/AWWA-AMWA-COVID-Report_2020-04.pdf

Rajkovic, P., Aleksic, D., Jankovic, D., Milenkovic, A., & Petkovic, I. (2018). Checking the potential shift to perceived usefulness: The analysis of users' response to the updated electronic health record core features. *International Journal of Medical Informatics, 115*, 80–91. doi:10.1016/j.ijmedinf.2018.04.011 PMID:29779723

Raju, V. (n.d.). *30 AI-driven tools to optimize your talent acquisition process*. https://www.aihr.com/blog/artificial-intelligence-talent-acquisition/

Raman, G., Peng, N. L., & Chi, C. I. (2021). The mediating effect of leaders' idealized influence on the relationship between leaders' emotional intelligence and job performance among academics in Malaysian public research universities. *International Journal of Management Studies, 28*(1), 1-24. https://doi.org/10.32890/ijms.28.1.2021.10544

Rauf, A. A. (2021). New Moralities for New Media? Assessing the Role of Social Media in Acts of Terror and Providing Points of Deliberation for Business Ethics. *Journal of Business Ethics, 170*(2), 229–251. doi:10.100710551-020-04635-w

Readwrite.com. (n.d.). *Why You Should Allow Employees to do Online Shopping at Work.* https://www.reputationsciences.com/companies-with-the-best-and-worst-reputations-at-the-start-of-2021/

Renukappa, S., Kamunda, A., & Suresh, S. (2021). Impact of COVID-19 on water sector projects and practices. *Utilities Policy, 70*, 101194. doi:10.1016/j.jup.2021.101194 PMID:34873382

Revere, D., Hills, R., Dixon, B., Gibson, P., & Grannis, S. (2017). Notifiable condition reporting practices: Implications for public health agency participation in a health information exchange. *BMC Public Health, 17*(1), 247. doi:10.118612889-017-4156-4 PMID:28284190

Richards, C. L., Iademarco, M. F., Atkinson, D., Pinner, R. W., Yoon, P., Kenzie, W. R., ... Frieden, T. R. (2017). Advances in Public Health Surveillance and Information Dissemination at the Centers for Disease Control and Prevention. *Public Health Reports, 132*(4), 403–410. doi:10.1177/0033354917709542 PMID:28609194

Riles, W. (1975). ECE in California Passes Its First Tests. *Phi Delta Kappan, 57*(1), 3–7.

Rinfret, N., Laplante, J., Lagacé, M. C., Deschamps, C., & Privé, C. (2020). Impacts of leadership styles in health and social services: A case from Quebec exploring relationships between emotional intelligence and transformational leadership. *International Journal of Healthcare Management, 13*, 329–339. https://doi-org.ezproxy.umgc.edu/10.1080/20479700.2018.1548153

Ring, C. (2019). Construction Corner: Don't Let Invoice Fraud Take Your Company Down. *Business Credit, 121*(5), 31.

Rioveros, G., Reyes-Chua, E., Navigar, N., Solina, A., Lanuza, M., & Garcia, S. (2021). The Impact of Work From Home Set-up During the COVID-19 Pandemic. *Worldwide Journal of Multidisciplinary Research and Development.* http://wwjmrd.com/upload/the-impact-of-work-from-home-set-up-during-the-covid-19-pandemic_1631623804.pdf

Robbins, S. P., & Judge, T. A. (2010). *Essentials of organizational behavior* (10th ed.). Pearson, Inc.

Robertson, D. (2020, December 23). Flint Has Clean Water Now. Why Won't People Drink It? *POLITICO.* Retrieved from https://www.politico.com/news/magazine/2020/12/23/flint-water-crisis-2020-post-coronavirus-america-445459

Robinson, G. (2020). Organizations are people: Reflections, obstacles and learning opportunities. *The Learning Organization, 27*(6), 541–554.

Rodrigues, M., Franco, M., Sousa, N., & Silva, R. (2021). COVID-19 and the business management crisis: An empirical study in SMEs. *Sustainability, 13*(11), 5912. doi:10.3390/su13115912

Roelich, K., & Giesekam, J. (2019). Decision making under uncertainty in climate change mitigation: Introducing multiple actor motivations, agency and influence. *Climate Policy, 19*(2), 175–188. https://doi.org/10.1080/14693062.2018.1479238

Rogers. C.R. (1959). A theory of therapy, personality and interpersonal relationships as developed in the client-centered framework. In S. Koch (Ed.), *Psychology: A study of a science, vol III. Formulations of the person and the social context*. New York: McGraw Hill.

Rogers, C. R. (1975). Empathic: An unappreciated way of being. *The Counseling Psychologist, 5*(2), 2–10. doi:10.1177/001100007500500202

Rogers, E. (1983). *Diffusion of Innovations* (3rd ed.). The Free Press.

Rohatgi, K. (2021). Building resilience in nursing students during the Pandemic. *Journal of Teaching and Learning with Technology, 10*, 58–63.

Rolls, E. T. (2005). *Emotion explained*. Oxford University Press.

Roman, A., Van Wart, M., Wang, X., Liu, C., Kim, S., & McCarthy, A. (2019). Defining e-leadership as competence in ICT-mediated communications: An exploratory assessment. *Public Administration Review, 79*(6), 853–866. doi:10.1111/puar.12980

Ross, E. D. (2021). Differential hemispheric lateralization of emotions and related display behaviors: Emotion-type hypothesis. *Brain Sciences, 11*(8), 1034. https://doi.org/10.3390/brainsci11081034

Rottenstreich, Y., & Hsee, C. K. (2001). Money, kisses, and electric shocks: On the affective psychology of risk. *Psychological Science, 12*(3), 185–190.

Rowell, S., & Torres, B. (2022). *Strategic Insights for 2022: Linking DEI and Harassment Prevention Initiatives*. Retrieved on 2.1.12 from https://www.shrm.org/learningandcareer/learning/webcasts/pages/0122rowell.aspx

Rudroff, T., Fietsam, A., Deters, J., Bruant, A., & Kamholz, J. (2020, December 19). Post-COVID-19 Fatigue: Potential Contributing Factors. *Brain Sciences, 10*(12), 1012. doi:10.3390/brainsci10121012 PMID:33352638

Rumelt, R. (2011). *Good Strategy Bad Strategy: The difference and why it matters*. Profile Books.

Russell, P. (2019, September 2). *The Internet of Water*. ENR. Retrieved from https://www.enr.com/articles/47424-data-drives-smarter-water-systems-decisions

Saarimäki, H., Glereanb, E., Smirnov, D., Mynttinenb, H., Jääskeläinen, L., Sams, M., & Nummenmaa, L. (2022). Classification of emotion categories based on functional connectivity patterns of the human brain. *NeuroImage. 247*. https://doi.org/10.1016/j.neuroimage.2021.118800

Sadri, G. (2012). Emotional intelligence and leadership development. *Public Personnel Management, 41*(3), 535–538.

Sagnak, M. (2016). Participative leadership and change-oriented organizational citizenship: The mediating effect of intrinsic motivation. *Eurasian J. Educ. Res.*, *62*, 181–194.

Sahin, I., & Shelley, M. (Eds.). (2020). *Educational practices during the COVID-19 viral outbreak: International Perspectives*. ISTES Organization.

Saliu, U. A., Wankasi, A. J., Eromosele, G. O., & Olukade, A. O. (2018). Leadership Styles and Motivation on Job Performance of Library Personnel in Public University Libraries in North Central Nigeria. *Library and Philosophy Practice*. https://link.gale.com/apps/doc/A580773835/AONE?=anon~585617ba&sid=googleScholar&xid=7ceece62

Salovey, P., & Mayer, J. D. (1990). Emotional intelligence. *Imagination, Cognition and Personality*, *9*, 185–211.

Santi, E., & Gorghiu, G. (2017). The student-centered learning model in John Dewey's progressive conception. *Studia Universitatis Babeş-Bolyai Psychologia-Paedagogia.*, *62*(2), 77–86. doi:10.24193ubbpsyped.2017.2.04

Saragih, R., Pradana, M., Wijaksana, T., Arwiyah, Y., & Fadhillah, N. (2021). Antecedents of Work Performance: The Effect of Leadership Style, Work Motivation and Discipline. *Journal of Management Information and Decision Sciences*, *24*(6), 1–8.

Saura, J. R., Reyes-Menendez, A., & Palos-Sanchez, P. R. (1970, January 1). *The Digital Tourism Business: A systematic review of Essential Digital Marketing Strategies and trends*. IGI Global. Retrieved Dec 17, 2021, from https://www.igi-global.com/chapter/the-digital-tourism-business

Schecter, T., & Gould, M. (2020). Lead from Your Heart: The Art of Relationship-Based Leadership. HTI Institute.

School responses in Indiana to the coronavirus (COVID-19) pandemic. (n.d.). *Ballotpedia*. https://www.loc.gov/item/lcwaN0016948/

Schools in the pandemic: Seeking Education Equity for India's Marginalized Kids. (2021, April 8). Retrieved Jan 2, 2022, from https://news.microsoft.com/en-in/features/schools-in-the-pandemic-seeking-education-equity-for-indias-marginalized-kids/

Schröder, D., & Gilboa Freedman, G. (2020). Decision making under uncertainty: The relation between economic preferences and psychological personality traits. *Theory and Decision*, *89*(1), 61–83. https://doi.org/10.1007/s11238-019-09742-3

Schutte, N. S., Malouff, J. M., Simunek, M., McKenley, J., & Hollander, S. (2002). Characteristic emotional intelligence and emotional wellbeing. *Cognition and Emotion*, *16*(6), 769-85.

Schwenk, C. H. (1986). Information, cognitive biases, and commitment to a course of action. *Academy of Management Review*, *11*(2), 298–310.

Scott, H. (2003, July 1). *Water AMR Market Continues Strong Growth*. WaterWorld. Retrieved from https://www.waterworld.com/home/article/16190548/water-amr-market-continues-strong-growth

Scott, W. R., & Davis, G. F. (2007). *Organizations and organizing rational, natural, and open systems perspectives*. Prentice Hall.

Scutchfield, F., & Keck, C. (2009). *Principles of public health practice* (3rd ed.). Delmar Cengage.

Searcy, C., Morali, O., Karapetrovic, S., Wichuk, K., McCartney, D., McLeod, S., & Fraser, D. (2012). Challenges in implementing a functional ISO 14001 Environmental Management System. *International Journal of Quality & Reliability Management, 29*(7), 779–796. doi:10.1108/02656711211258526

Sefela, F. (2017). *The relationship between emotional intelligence and change oriented leadership style of managers and the OCB of non-academic staff members at a tertiary institution in the Western Cape* [Master's thesis]. University of the Western Cape. http://hdl.handle.net/11394/6438

Senge, P. (1992). *The fifth discipline field book: Strategies and tools for building a learning organization*. Doubleday Currency.

Shapiro, J., Mostashari, F., Hripcsak, G., Soulakis, N., & Kuperman, G. (2011). Using health information exchange to improve public health. *American Journal of Public Health, 101*(4), 616–623. doi:10.2105/AJPH.2008.158980 PMID:21330598

Sharpcloud. (2021). *Identify new opportunities in a dynamic business environment*. https://www.sharpcloud.com/blog/identify-business-opportunities-in-a-dynamic-environment

Shea, C., & Howell, J. (1999). Charismatic Leadership and Task Feedback: A Laboratory Study of their Effects on Self-Efficacy and Task Performance. *The Leadership Quarterly, 10*(3), 375–396. doi:10.1016/S1048-9843(99)00020-X

Sherkat, D., & Ellison, C. (1999). Recent Developments and Current Controversies In The Sociology of Religion. *Annual Review of Sociology, 24*, 363–394.

Siegrist, M., & Sütterlin, B. (2014). Human and nature-caused hazards: The affect heuristic causes biased decisions. *Risk Analysis, 34*(8), 1482–1494.

Sinclair and stabler 1997 have defined the tourism enterprise as a composition: Course hero. (n.d.). Retrieved Nov 6, 2021, from https://www.coursehero.com/file/p244ji8g/Sinclair-and-Stabler-1997

Sirmon, D. G., Hitt, M. A., Ireland, R. D., & Gilbert, B. A. (2011). Resource orchestration to create competitive advantage: Breadth, depth, and life cycle effects. *Journal of Management, 37*(5), 1390–1412. https://doi.org/10.1177/0149206310385695

Smith, P., Hadler, J., Stanbury, M., Rolfs, R., & Hopkins, R. (2013). "Blueprint Version 2.0": Updating public health surveillance for the 21st century. *Journal of Public Management and Practice, 19*(33), 231–239. doi:10.1097/PHH.0b013e318262906e PMID:22759985

Smith, R., Bajaj, S., Dailey, N. S., Alkozei, A., Smith, C., Sanova, A., Lane, R. D., William, D., & Killgore, S. (2018). Greater cortical thickness within the limbic visceromotor network predicts higher levels of trait emotional awareness. *Consciousness and Cognition, 57*, 54.

Software Suggest. (2020, January 5). *8 Types of Health Information Technology & Healthcare Software System.* https://www.softwaresuggest.com/blog/types-of-health-information-technology-and-healthcare-software/#

Soluk, J., Kammerlander, N., & De Massis, A. (2021). Exogenous shocks and the adaptive capacity of family firms: Exploring behavioral changes and digital technologies in the COVID-19 pandemic. *R & D Management, 51*(4), 364–380. https://doi.org/10.1111/radm.12471

Sowby, R. B. (2020). Emergency preparedness after COVID-19: A review of policy statements in the US water sector. *Utilities Policy, 64*, 101058. doi:10.1016/j.jup.2020.101058 PMID:32406402

Spearman, C. (1927). *The abilities of man*. Macmillan.

Stark, A. (2021). Top 10 Work Trends for 2021. *SIOP Member News*. Retrieved on 1.22.22 from https://www.siop.org/Research-Publications/Items-of-Interest/ArtMID/19366/ArticleID/4914/Top-10-Work-Trends-for-2021

Stasser, G., & Stewart, D. (1992). Discovery of hidden profiles by decision-making groups: Solving a problem versus making a judgment. *Journal of Personality and Social Psychology, 63*, 426–434. doi:10.1037/0022-3514.63.3.426

Stasser, G., & Titus, W. (1985). Pooling of Unshared Information in Group Decision Making: Biased Information Sampling During Discussion. *Journal of Personality and Social Psychology, 48*, 1467–1478. doi:10.1037/0022-3514.48.6.1467

Stefan, R., & Carutasu, G. (2020). Explainable machine learning for ethical artificial intelligence based decisions. *Journal of Information Systems & Operations Management,* 151-161.

Stockard, M. (2006). *An integrated phenomenological study of teachers' perceptions of a professional learning community utilizing Senge's five disciplines of a learning organization* [Published Doctoral dissertation]. Georgia Southern University.

Sull, D., Turconi, S., Sull, C., & Yoder, J. (2018). How to develop strategy for execution. *MIT Sloan Management Review, 59*(2), 130–135.

Supramanian, K., Shahruddin, R., & Sekar, M. (2022). Emotional intelligence using ability model in context of nursing and its impact on end-stage renal disease patients: A narrative review. *International Journal of Current Research and Review*. doi:10.31782/IJCRR.2021.131609

Talari, S., Shafie-khah, M., Siano, P., Loia, V., Tommasetti, A., & Catalão, J. (2017). A Review of Smart Cities Based on the Internet of Things Concept. *Energies, 10*(421), 421. Advance online publication. doi:10.3390/en10040421

Tangney, J., Baumeister, R., & Boone, A. (2004). High self-control predicts good adjustment, less pathology, better grades, and interpersonal success. *Journal of Personality, 72*(2), 271–324. doi:10.1111/j.0022-3506.2004.00263.x PMID:15016066

Teaching and learning after covid-19. (n.d.). Retrieved Jan 4, 2022, from https://www.insidehighered.com/digital-learning/blogs/learning-innovation/teaching-and-learning-after-covid-19

Temido, J., Sousa, J., & Malheiro, R. (2014). SCADA and smart metering systems in water companies. A perspective based on the value creation analysis. 12th International Conference on Computing and Control for the Water Industry, CCWI2013. *Procedia Engineering,* (70), 1629–1638. 10.1016/j.proeng.2014.02.180

The role of entrepreneurship in tourism industry: An overview. (n.d.). Retrieved Feb 4, 2022, from https://www.researchgate.net/publication/316240958_The_Role_of_Entrepreneurship_in_Tourism_Industry_An_Overview

Thomas, K. W., Thomas, G. F., & Schaubhut, N. (2008). Conflict styles of men and women at six organization levels. *International Journal of Conflict Management, 14*(2), 1–38. doi:10.1108/10444060810856085

Thorndike, E. L. (1920). Intelligence and its use. *Harper's Magazine, 140*, 227–235.

Thurstone, I. L. (1938). *Primary mental abilities. Psychometric Monographs.* University of Chicago Press.

Tversky, A., & Kahneman, D. (1992). Advances in prospect theory: Cumulative representation of uncertainty. *Journal of Risk and Uncertainty, 5*(4), 297–323. doi:10.1007/bf00122574

U.S. Bureau of Labor Statistics. (n.d.). *Employment projections home page*. U.S. Bureau of Labor Statistics. Retrieved March 10, 2022, from https://www.bls.gov/emp/

U.S. EPA. (2004, September). *Primer for Municipal Wastewater Treatment Systems.* Retrieved from https://www3.epa.gov/npdes/pubs/primer.pdf

U.S. EPA. (2008). *Ensuring a Sustainable Future: An Energy Management Guidebook for Wastewater and Water Utilities.* Retrieved from https://nepis.epa.gov/Exe/ZyPDF.cgi/P1003Y1G.PDF?Dockey=P1003Y1G.PDF

U.S. EPA. (2013). *Drinking water infrastructure needs survey and assessment, fifth report to Congress* (EPA 816-R-13-006). Retrieved from https://water.epa.gov/grants_funding/dwsrf/upload/epa816r13006.pdf

U.S. EPA. (2015, June). *Framework for Comparing Alternatives For Water Quality Surveillance and Response Systems.* Retrieved from https://www.epa.gov/sites/default/files/2015-07/documents/framework_for_comparing_alternatives_for_water_quality_surveillance_and_response_systems.pdf

U.S. EPA. (2015, November). *Dashboard Design Guidance For Water Quality Surveillance and Response Systems.* Retrieved from https://www.epa.gov/sites/default/files/2015-12/documents/srs_dashboard_guidance_112015.pdf

U.S. EPA. (2016, July). *Guidance for Designing Communications Systems For Water Quality Surveillance and Response Systems.* Retrieved from https://www.epa.gov/sites/default/files/2016-09/documents/srs_communications_guidance_081016.pd

U.S. EPA. (2019, March). *Addressing Lead in Drinking Water with the Drinking Water State Revolving Fund.* Retrieved from https://nepis.epa.gov/Exe/ZyPDF.cgi/P100WD7W.PDF?Dockey=P100WD7W.PDF

U.S. EPA. (2021). *Proposed Revisions to the Lead and Copper Rule.* Retrieved from https://www.epa.gov/ground-water-and-drinking-water/proposed-revisions-lead-and-copper-rule

U.S. Equal Employment Opportunity Commission. (2019, November 18). *Dollar General to pay $6 million to settle EEOC class race discrimination suit.* https://www.eeoc.gov/newsroom/dollar-general-pay-6-million-settle-eeoc-class-race-discrimination-suit

U.S. Equal Employment Opportunity Commission. (n.d.). *Prohibited employment policies/practices.* https://www.eeoc.gov/prohibited-employment-policiespractices

Ulenaers, D., Grosemans, J., Schrooten, W., & Bergs, J. (2021). Clinical placement experience of nursing students during the COVID-19 pandemic: A cross-sectional study. *Nurse Education Today, 99.* Advance online publication. doi:10.1016/j.nedt.2021.104746

United States Department of Health and Human Services, Centers for Medicare and Medicaid Services (CMS). (2013). *EHR Incentive Programs: Meaningful Use.* http://www.cms.gov/Regulations-and-Guidance/Legislation/EHRIncentivePrograms/Meaningful_Use.html

Usman, M., Farooq, M., Jellali, S., Charabi, Y., Al-Sadi, A. M., & Al-Badi, A. (2021). COVID-19: The Challenges and Opportunities for Water, Air, Agriculture and Energy Sectors. *International Journal of Agriculture and Biology, 25*(5), 1085–1095. doi:10.17957/IJAB/15.1767

Valenzano, A., Scarinci, A., Monda, V., Sessa, F., Messina, A., Monda, M., Precenzano, F., Mollica, M. P., Carotenuto, M., Messina, G., & Cibelli, G. (2020). The Social Brain and Emotional Contagion: COVID-19 Effects. *Medicina, 56*(12), 640. doi:10.3390/medicina56120640

Valeriani, D., Cinel, C., & Poli, R. (2017). Group Augmentation in Realistic Visual-Search Decisions via a Hybrid Brain-Computer Interface. *Scientific Reports, 7*(1), 7772. https://doi.org/10.1038/s41598-017-08265-7

Van Fleet. (2020). *Quality Time: Productivity Through Time Management.* Information Age Publishing.

van Kleef, G. A. (2014). Understanding the positive and negative effects of emotional expressions in organizations: EASI does it. *Human Relations, 67*(9), 1145-1164. doi:10.1177/0018726713510329

Van Thiel, E. (2021). *Big 5 Personality Test.* Retrieved on 1/25/22 at https://www.123test.com/personality-test/

VanBavel, J., Baicker, K., Boggio, P., Capraro, V., Cichocka, A., Cikara, M., Crockett, M., Crum, A., Douglas, K., Druckman, J., & Drury, J. (2020). Using Social and Behavioural Science to Support Covid-19 Pandemic Response. *Nature Human Behaviour, 4*(5), 460–471. doi:10.103841562-020-0884-z PMID:32355299

Very Well Mind. (n.d.). *What is resilience?* Retrieved 2/12/22 from Verywellmind.com/what-is-resilience-2795059

Vinokur, A. (1971). Review and theoretical analysis of the effects of group processes upon individual and group decisions involving risk. *Psychological Bulletin, 76*(4), 231–250. https://doi.org/10.1037/h0031568

Vogel, J., Brown, J. S., Land, T., Platt, R., & Klompas, M. (2014). MDPHnet: Secure, distributed sharing of electronic health record data for public health surveillance, evaluation, and planning. *American Journal of Public Health, 104*(12), 2265–2270. doi:10.2105/AJPH.2014.302103 PMID:25322301

Voice, C. (2021, March 28). *Covid-19 pandemic has changed the Banking Industry.* InsuranceNewsNet. Retrieved Dec 24, 2021, from https://insurancenewsnet.com/oarticle/covid-19-pandemic-has-changed

Votaw, K. Q. (2022). The Hard Results Of Soft Leaders. *Personal Excellence, 27*(1), 37.

Vygotsky, L. S. (1978). *Mind in society: The development of higher psychological processes.* Harvard University Press.

Walls, E. (2019). The Value of Situational Leadership. *Community Practitioner*, 31–33.

Water Environment Federation. (2021). *Key Safety Information for Covid-19 and Other Biohazards.* Retrieved from https://www.wef.org/globalassets/assets-wef/news-hub/coronavirus/key-safety-information-for-covid-19-and-other-biohazards.pdf

Water Research Foundation. (2020). *Virtual International Water Research Summit on COVID-19.* Retrieved from https://www.waterrf.org/event/virtual-international-water-research-summit-covid-19

Waterford, F. (2019). *3 Types of focus every leader needs to cultivate.* Retrieved from: https://www.linkedin.com/pulse/3-types-focus-every-leader-needs-cultivate-felicity-waterford/

Wechsler, D. (1997). *Wechsler adult intelligence scale.* Harcourt Assessment. Retrieved from http://www.sjdm.org/dmidi/Wechsler_Adult_Intelligence_Scale.html

Weston, M. (2020, June 12). *How covid-19 has impacted organisational culture.* theHRDIRECTOR. Retrieved Dec 26, 2021, from https://www.thehrdirector.com/how-covid-19-has-impacted-organisational-culture/

Whaley, C. M., Pera, M. F., Cantor, J., Chang, J., Velasco, J., Hagg, H. K., Sood, N., & Bravata, D. M. (2020). Changes in health services use among commercially insured US populations during the COVID-19 pandemic. *JAMA Network Open, 3*(11), e2024984. doi:10.1001/jamanetworkopen.2020.24984 PMID:33151319

White, S. K. (2021, September 17). *AI in hiring might do more harm than good.* https://www.cio.com/article/189212/ai-in-hiring-might-do-more-harm-than-good.html

Whitwell, G. (2021). *Resetting the Leadership Agenda Post-COVID-19*. AACSB. Retrieved on 1/12/22 from https://www.aacsb.edu/insights/articles/2021/01/resetting-the-leadership-agenda-post-covid-19

Williams, M. (2020, August 21). Council post: How to scale your business operations fast without burning your money. *Forbes*. Retrieved Jan 16, 2022, from https://www.forbes.com/sites/forbesbusinesscouncil/2020/08/24/how-to-scale-your-business-operations-fast-without-burning-your-money/

Williams, R. Jr, Raffo, D. M., & Clark, L. A. (2018). Charisma as an attribute of transformational leaders: What about credibility? *Journal of Management Development*, *37*(6), 512–524. doi:10.1108/JMD-03-2018-0088

Wisconsin Department of Health Services [WDHS]. (2020). *Disease Reporting*. https://www.dhs.wisconsin.gov/disease/diseasereporting.htm#:~:text=Questions%20concerning%20this%20information%20may,%2C%20608%2D266%2D1120.&text=Category%20I%20diseases%20must%20be,health%20officer%2C%20or%20their%20designee

Wong, N. C., & Shayegan, B. (2019). Patient centered care for prostate cancer-how can artificial intelligence and machine learning help make the right decision for the right patient? *Annals of Translational Medicine*, *7*(Suppl 1), S1–S1. https://doi.org/10.21037/atm.2019.01.13

Wright, L., Frost, C., & Wisecarver, S. (1993). Church Attendance, Meaningfulness of Religion and Depressive Symptomology among adolescents. *Journal of Youth and Adolescence*, *22*(5).

Wu, L. (2014). *Issue Brief: Health IT for Public Health Reporting and Information Systems*. Academic Press.

Xu, D., & Jaggars, S. S. (2013). The impact of online learning on students' course outcomes: Evidence from a large community and technical college system. *Economics of Education Review*, *37*, 46–57. doi:10.1016/j.econedurev.2013.08.001

Yahoo. (n.d.). *Recovery and rebirth for restaurants in a post-pandemic world*. Yahoo! Finance. Retrieved Nov 5, 2021, from https://ca.finance.yahoo.com/news/recovery-rebirth-restaurants-post-pandemic-171100354.html

Yam, K., Lian, H., Ferris, D., & Brown, D. (2017). Leadership Takes Self-Control. Here's What We Know About It. *Harvard Business Review*. Retrieved on 2.11.22 from https://hbr.org/2017/06/leadership-takes-self-control-heres-what-we-know-about-it

You-De, D., You-Yu, D., Kuan-Yang, C., & Hui-Chun, W. (2013). Transformational vs. transactional leadership: Which is better? A study on employees of international tourist hotels in Taipei City. *International Journal of Contemporary Hospitality Management*, *25*(5), 760–778. doi:10.1108/IJCHM-Dec-2011-0223

Youssef, N. (2017). Requirements of decision-makers in confronting crises. *Journal of Legal Studies and Research*, *4*, 109–110.

Zhu, W., Avolio, B. J., & Walumbwa, F. O. (2009). Moderating role of follower characteristics with transformational leadership and follower work engagement. *Group & Organization Management*, *34*(5), 590–619. doi:10.1177/1059601108331242

Zhu, W., Newman, A., Miao, Q., & Hooke, A. (2013). Revisiting the mediating role of trust in transformational leadership effects: Do different types of trust make a difference? *The Leadership Quarterly*, *24*(1), 94–105. doi:10.1016/j.leaqua.2012.08.004

Zhu, X. S., Wolfson, M. A., Dalal, D. K., & Mathieu, J. E. (2021). Team decision making: The dynamic effects of team decision style composition and performance via decision strategy. *Journal of Management*, *47*(5), 1281–1304. https://doi.org/10.1177/0149206320916232

ZuckermanJ. (2021). Zuckermanlaw.com/jury-awards

About the Contributors

Joseph Robert Heinzman is a U.S. Army Veteran, he holds degrees in Engineering and Business (BS), an MBA in Technology Management, and a Doctor of Management in Organizational Development and Change. He has a diverse expertise in areas of engineering, production and operations as well as an extensive process improvement background and instruction. He has been involved in design and development of devices for amusement parks, automotive, and aerospace. Dr. Heinzman has experience at a corporate level in training and mentoring operations processes and has over ten years' experience teaching Organizational Behavior, Leadership, and Management courses. Dr. Heinzman's core research is focused on leadership's appreciative environment and the effect on follower's job satisfaction. Dr. Heinzman's research proved the appreciative environment created by leadership can impact employees' job satisfaction, performance, and retention. Dr. Heinzman is a Certified Manufacturing Technologist, SME Member, SAE Reviewer, Life Member of Alpha Kappa Psi Fraternity, and Chief Editor and Founder of IABS Journal.

Deborah Judge has been a nurse for over 25 years and dedicated the last 15 years to promoting nursing education. Her interests focus on collaboration, addictions, spirituality, disaster simulation, scholarship of teaching and learning, and nursing research. Her scholarly pursuits have led to collaborative efforts with schools of nursing resulting in joint simulations and dissemination of work through presentations and publications. Passion for teaching combined with a passion for serving and relationship building is the basis for her dedication to the advancement of nursing education. Dr. Judge is a member of Sigma Theta Tau International- Eta Chi and Psi Upsilon Chapters, National League for Nursing, and Indiana League for Nursing. Community service includes serving in many different forms within her church as well as supporting the local transitional homes for those with addictions (past board member/president).

Vincent S. Franklin served 25 years in the United States Marine Corps, earning the rank of Master Gunnery Sergeant/E-9. During his service, Dr. Franklin served multiple tours of duty during Gulf War operations, specialized in leadership, leadership development, intelligence analysis, intelligence operations, and the development, management, and training of Marines. Dr. Franklin is a Bronze Star recipient. Since retiring from the Marine Corps, Dr. Franklin has held positions in corporate human resource acquisition, development, management, and leadership; hospital performance improvement; worked as a Deputy Directorate Manager (Deputy G2) of a U. S. Army Command; and served as a Professor at the University level. Dr. Franklin has served as Course Instructor at Western Governors University since June 2019. Dr. Franklin holds a Doctor of Philosophy in Organization and Management (Specializing in Leadership). Dr. Franklin's principal areas of study include (but are not limited to): Leadership; Business management; In-group out-group theory; leader/member exchange theory; Prejudice and stereotyping; organizational cultural development; social group and social identity theory; and ethics within workplaces and work environments. Dr. Franklin is a member of Delta Mu Delta International Honor Society in Business Administration and is a life member of Alpha Phi Alpha Fraternity, Incorporated.

Jim A. McCleskey is a Course Faculty Manager at Western Governors University and an Instructor for the Houston Community College System. He holds a Ph.D. in Organization and Management with an emphasis in Leadership Studies and an MBA in Management and Marketing. His research areas of interest include leadership emergence, attachment, dark leadership, leadership efficacy, emotional labor, and affect in organizations. He is a native Texan who lives in the Houston area. He is a husband, a father of three sons, and a new grandfather. When he is not working, he enjoys reading and martial arts.

* * *

Lezlie Banks has over 14 years of higher education experience. She teaches English, Literature, Business, Human Resources, and Data Analytics, classes. Currently, she facilitates business strategy courses in the College of Business at Western Governors University. Lezlie earned a Bachelor of Science, in English, American Literature, and Language from Eastern Michigan University, a Master of Science in Human Resource Management from Troy University, and a Doctorate of Business Administration from Argosy University. In addition to higher education, Lezlie has 15 years of leadership experience. She earned a Green Belt in Six Sigma; Effective Leadership certification from the Jack Welch Institute; and Diversity, Equity, and Inclusion Certification from the University of South Florida. Her research interests include emotional intelligence and organizational behavior. When not teaching,

Lezlie enjoys traveling and reading. Though originally from Michigan, she now resides in Georgia.

Alma Beciragic graduated from Queens University of Charlotte with a BS; she then completed her PhD in Environmental Sciences and Engineering at the UNC Gillings School of Global Public Health in Fall 2020. As an AWWA, AMTA/USBR, and EPA Science to Achieve Results graduate fellow, Beciragic's research focused on using and developing methods to better micropollutants in water including Contaminants of Emerging Concern (CECs), regulated and unregulated DBPs, and leachates from nanofiltration membranes. She has presented research findings at ACE, MTC, and WQTC. Dr. Beciragic currently serves as a Management Consultant for the water sector of Arcadis North America.

Kimberly Brooks earned a Ph.D. in Education and Leadership specializing in Human Resource Development from Barry University. She also achieved a Master of Human Resource Management degree from Rollins College and a Bachelor of Arts degree in Psychology from the University of Central Florida. Kimberly has over 20 years of human resource and leadership experience in multi-unit organizations. As the President of a multi-unit enterprise, Kimberly led the human resource, operations, marketing, IT, construction, leasing, and accounting departments. Kimberly also has experience as a college professor for both bachelor and master level courses. She also holds the esteemed certification of Senior Professional in Human Resources (SPHR) from the Human Resource Certification Institute (HRCI).

Regina Cousar attended Howard University and Georgia Tech to obtain degrees in Chemical Engineering, and then completed her doctorate at Northeastern University. She continued her educational journey to obtain her MBA in IT Management at WGU in 2021. Before joining the WGU quantitative analysis faculty in 2017, she had several past private and public sector roles over 25 years, including Chief Operating Officer and Continuous Improvement Officer for water utilities. Dr. Cousar now has a greater challenge - focusing on sharing knowledge with students pursuing MBAs at WGU, as well as doctoral students at Vanderbilt University since 2021.

Nabanita Ghosh with 15 years of experience in teaching and learning is currently working as an Asst Prof in Christ University in Bangalore. Having keen interest in research and research activities.

Wiley Hughes serves as Lead Pastor of Destiny Now Church in Fayetteville, NC and owns Southeast Counseling Services LLC, a mental health counseling practice in Fayetteville, Raleigh, and Wilmington NC. He earned a Psy.D. in Pastoral Psychology from Graduate Theological Foundation, a Doctor of Ministry Degree from Oral Roberts University, a master's in Clinical Mental Health Counseling from the University of North Carolina Pembroke, and a Masters in Christian Education from Southeastern Baptist Theological Seminary in Wake Forest NC. His bachelor's is also in education. Dr. Hughes is a licensed clinical mental health counselor and a certified pastoral counselor in the state of North Carolina. He has authored several books and articles in the fields of theology and psychology.

Kevin J. Jones is an associate professor of management at Indiana University – Purdue University Columbus specializing in organizational change and business ethics. His interests include digital ethics including data acquisition and management. He is currently focusing on the ethical implications of Artificial Intelligence (AI) decision-making. Over the course of a 22-year business career, he held management and leadership roles in telecommunications, healthcare, education, and business consulting.

Sunita Kumar is an Assistant Professor at the School of Business and Management, Christ (Deemed to be University), Bangalore. Her doctoral thesis was on brand strategies in retail marketing. She has published articles in various international and national refereed journals. Her areas of research interests include social media marketing, branding, Advertising and neuromarketing. She has chaired various international conferences and was a keynote speaker.

Rebecca Melton serves as Faculty in the College of Business at Western Governors University, teaching courses in Human Resources. She earned a Ph.D. in Business Psychology from the Chicago School of Professional Psychology, and a Masters in Business Administration (MBA) and Bachelors in Psychology from Bowling Green State University. Her research interests include motivation (temporal discounting), leadership, and university disability services. With more than 25 years of experience in Leadership, Operations and Human Resources and 16 years experience in teaching, she holds the SHRM-SCP and SPHR certifications. Further, she is a certified Vocational Expert in the Franklin County Domestic Courts. When not teaching, she enjoys reading, movies and serving on multiple boards (both non-profit and for profit). Becky currently resides in Ohio, with husband Paul and their beloved pets.

Nazish Saba is an entrepreneur. Born and raised in Karachi, Pakistan. Nazish earned a Bachelor and Masters in Pharmaceutical Chemistry. She earned an MBA from Indiana University in 2015. She brings a unique perspective to every discussion, having experienced two very different cultures. When not at work, she spends her time volunteering and exploring the outdoors.

Anubha Srivastava is working as Assistant Professor at School of Law, Christ University, Bangalore. She specialises in the areas of Law, Commerce, and Management.

Michelle Stewart, Instructor Faculty in the College of Health Professions at Western Governors University, teaching graduate courses in Health Leadership. She has earned a Doctorate in Health Administration, a Master of Business Administration, a Masters of Management in Health Care Management, and a Master's in Public Administration. She is a member of the American College of Healthcare Executives (ACHE) where she serves on the Board of Directors for the South Texas Chapter. She has worked in higher education for over twelve years and has over 25 years of health care experience in direct patient care, clinical management, and health care education. When she is not actively teaching or engaged in health industry-related activities, she enjoys spending time with her husband, two teenage sons, and her two loving dogs.

Crystal Walcott, Professor of Mathematics Education, teaches in the Elementary Teacher Education program at Indiana University Purdue University – Columbus. She also serves as the Head of the Division of Education on that campus. Her research focus is secondary analysis of mathematics results from the National Assessment of Educational Progress (NAEP). She served as co-principal investigator on an NAEP secondary analysis grant funded by the National Science Foundation and housed at IU Bloomington. She has received funding on several STEM-focused grants aimed to improve STEM education in Indiana. She currently is serving as PI on a STEM teacher recruitment grant funded by the state of Indiana.

Index

D

E

F

G

H

I

L

M

N

O

P

Q

R

S

T

U

V

W

Premier Reference Source
Stabilizing Currency and Preserving Economic Sovereignty Using the Grondona System

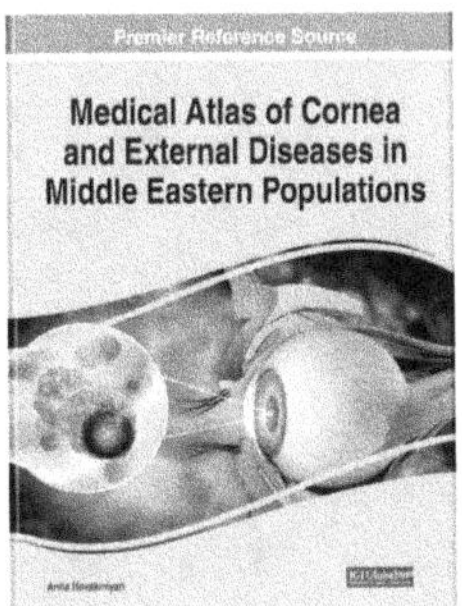

Premier Reference Source
Medical Atlas of Cornea and External Diseases in Middle Eastern Populations

Premier Reference Source
Examining Biophilia and Societal Indifference to Environmental Protection

Premier Reference Source
Using Narratives and Storytelling to Promote Cultural Diversity on College Campuses

www.ingramcontent.com/pod-product-compliance
Lightning Source LLC
LaVergne TN
LVHW061550290926
849162LV00004B/961